Biology

for Cambridge International AS & A Level

WORKBOOK

Mary Jones & Matthew Parkin

CAMBRIDGE
UNIVERSITY PRESS

University Printing House, Cambridge CB2 8BS, United Kingdom

One Liberty Plaza, 20th Floor, New York, NY 10006, USA

477 Williamstown Road, Port Melbourne, VIC 3207, Australia

314–321, 3rd Floor, Plot 3, Splendor Forum, Jasola District Centre, New Delhi – 110025, India

79 Anson Road, #06–04/06, Singapore 079906

Cambridge University Press is part of the University of Cambridge.

It furthers the University's mission by disseminating knowledge in the pursuit of education, learning and research at the highest international levels of excellence.

www.cambridge.org
Information on this title: www.cambridge.org/9781108859424

© Cambridge University Press 2020

This publication is in copyright. Subject to statutory exception and to the provisions of relevant collective licensing agreements, no reproduction of any part may take place without the written permission of Cambridge University Press.

First edition 2016

Second edition 2020

20 19 18 17 16 15 14 13 12 11 10 9 8 7 6 5 4 3 2 1

Printed in Malaysia by Vivar Printing

A catalogue record for this publication is available from the British Library

ISBN 978-1-108-85942-4 Paperback with Digital Access (2 Years)

Additional resources for this publication at www.cambridge.org/9781108859424

Cambridge University Press has no responsibility for the persistence or accuracy of URLs for external or third-party Internet websites referred to in this publication, and does not guarantee that any content on such websites is, or will remain, accurate or appropriate. Information regarding prices, travel timetables, and other factual information given in this work is correct at the time of first printing but Cambridge University Press does not guarantee the accuracy of such information thereafter.

NOTICE TO TEACHERS IN THE UK
It is illegal to reproduce any part of this work in material form (including photocopying and electronic storage) except under the following circumstances:
(i) where you are abiding by a licence granted to your school or institution by the Copyright Licensing Agency;
(ii) where no such licence exists, or where you wish to exceed the terms of a licence, and you have gained the written permission of Cambridge University Press;
(iii) where you are allowed to reproduce without permission under the provisions of Chapter 3 of the Copyright, Designs and Patents Act 1988, which covers, for example, the reproduction of short passages within certain types of educational anthology and reproduction for the purposes of setting examination questions.

NOTICE TO TEACHERS
Cambridge International copyright material in this publication is reproduced under licence and remains the intellectual property of Cambridge Assessment International Education.

Exam-style questions and sample answers in this title were written by the authors. In examinations, the way marks are awarded may be different. References to assessment and/or assessment preparation are the publisher's interpretation of the syllabus requirements and may not fully reflect the approach of Cambridge Assessment International Education.

Cambridge International recommends that teachers consider using a range of teaching and learning resources in preparing learners for assessment, based on their own professional judgement of their students' needs.

DEDICATED TEACHER AWARDS

Teachers play an important part in shaping futures. Our Dedicated Teacher Awards recognise the hard work that teachers put in every day.

Thank you to everyone who nominated this year, we have been inspired and moved by all of your stories. Well done to all our nominees for your dedication to learning and for inspiring the next generation of thinkers, leaders and innovators.

Congratulations to our incredible winner and finalists

WINNER

Ahmed Saya
Cordoba School for A-Level,
Pakistan

Sharon Kong Foong
Sunway College,
Malaysia

Abhinandan Bhattacharya
JBCN International School Oshiwara,
India

Anthony Chelliah
Gateway College,
Sri Lanka

Candice Green
St Augustine's College,
Australia

Jimrey Buntas Dapin
University of San Jose-Recoletos,
Philippines

For more information about our dedicated teachers and their stories, go to
dedicatedteacher.cambridge.org

CAMBRIDGE UNIVERSITY PRESS

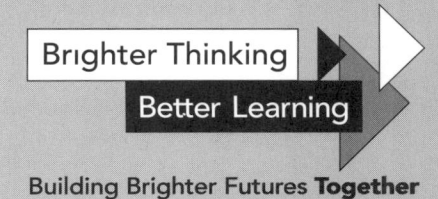

Brighter Thinking
Better Learning

Building Brighter Futures Together

Contents

How to use this series	vi
How to use this book	viii
Introduction	ix
1 Cell structure	1
2 Biological molecules	17
3 Enzymes	35
4 Cell membranes and transport	45
5 The mitotic cell cycle	56
6 Nucleic acids and protein synthesis	67
7 Transport in plants	77
8 Transport in mammals	93
9 Gas exchange	108
10 Infectious disease	122
11 Immunity	142
12 Energy and respiration	153
13 Photosynthesis	176
14 Homeostasis	194
15 Control and coordination	218
16 Inheritance	234
17 Selection and evolution	250
18 Classification, biodiversity and conservation	261
19 Genetic technology	275
Glossary	286
Skills Grids	290
Acknowledgements	297

How to use this series

This suite of resources supports students and teachers following the Cambridge International AS & A Level Biology syllabus (9700). All of the books in the series work together to help students develop the necessary knowledge and scientific skills required for this subject. With clear language and style, they are designed for international learners.

The coursebook provides comprehensive support for the full Cambridge International AS & A Level Biology syllabus (9700). It clearly explains facts, concepts and practical techniques, and uses real-world examples of scientific principles. Two chapters provide full guidance to help students develop investigative skills. Questions within each chapter help them to develop their understanding, while exam-style questions provide essential practice.

The workbook contains over 100 exercises and exam-style questions, carefully constructed to help learners develop the skills that they need as they progress through their Biology course. The exercises also help students develop understanding of the meaning of various command words used in questions, and provide practice in responding appropriately to these.

How to use this series

This write-in book provides students with a wealth of hands-on practical work, giving them full guidance and support that will help them to develop all of the essential investigative skills. These skills include planning investigations, selecting and handling apparatus, creating hypotheses, recording and displaying results, and analysing and evaluating data.

The teacher's resource supports and enhances the materials in the coursebook, workbook and practical workbook. It includes answers for every question and exercise in these three books. It also includes detailed lesson ideas, teaching notes for each topic area including a suggested teaching plan, ideas for active learning and formative assessment, links to resources, ideas for lesson starters and plenaries, differentiation, lists of common misconceptions and ideas for homework activities. The practical teacher's guide, included with this resource, includes detailed support for preparing and carrying out all of the investigations in the practical workbook, including tips for getting things to work well, and a set of sample results that can be used if students cannot do the experiment or fail to collect results.

CAMBRIDGE INTERNATIONAL AS & A LEVEL BIOLOGY: WORKBOOK

> How to use this book

Throughout this book, you will notice lots of different features that will help your learning. These are explained below.

CHAPTER OUTLINE

This appears at the start of every chapter. It lists the topics that are used as contexts for developing the skills covered in the chapter.

TIPS

The information in these boxes will help you complete the exercises, and give you support in areas that you might find difficult.

Exercises

Appearing throughout this book, these help you to practise skills that are important for studying AS & A Level Biology. Answers to these questions can be found in the digital Workbook.

EXAM-STYLE QUESTIONS

Questions at the end of each chapter are more demanding exam-style questions, some of which may require use of knowledge from previous chapters. Answers to these questions can be found in the digital Workbook.

KEY WORDS

Key vocabulary is highlighted in the text when it is first introduced. Definitions are then given in the margin, which explain the meanings of these words and phrases.

You will also find definitions of these words in the Glossary at the back of this book.

COMMAND WORDS

Command words that appear in the syllabus and might be used in exams are highlighted in the exam-style questions when they are first introduced. In the margin, you will find the Cambridge International definition.

You will also find these definitions in the Glossary at the back of this book with some further explanation on the meaning of these words.*

---

*The information in this section is taken from the Cambridge International syllabus (9700) for examination from 2022. You should always refer to the appropriate syllabus document for the year of your examination to confirm the details and for more information. The syllabus document is available on the Cambridge International website at www.cambridgeinternational.org.

> Introduction

This Workbook has been written to help you develop the skills you need for your AS & A Level Biology course. To succeed in this course, you need to have an excellent factual knowledge of all the topics in the syllabus and you also need to be able to think like a scientist. As you work through the book, chapter by chapter, you will develop the relevant scientific skills needed and gain the confidence to use them yourself. The exercises in this Workbook provide opportunities for you to practise the following skills:

- showing understanding of the scientific phenomena and theories that you are studying
- solving numerical and other problems
- thinking critically about experimental techniques and data
- making predictions and using scientific reasons to support your predictions
- planning experiments and investigations that will produce valid conclusions
- analysing data to draw conclusions
- selecting and using statistical tests to reach appropriate conclusions.

The exercises in each chapter will help you develop these skills by applying them to real life, interesting situations from each area of the syllabus. The chapters are arranged in the same order as the chapters in your Coursebook. Each exercise has an introduction that states the skills you will be using. At the end of each chapter a set of exam-style questions are provided to further support the skills you have practised in that chapter.

We hope that this book will help you succeed in the AS & A Level Biology course, give you the necessary scientific skills to help you with your future studies, and inspire you to have a love of biological science.

This Workbook is designed to support the Coursebook, with specially selected topics where students would benefit from further opportunities to apply skills, such as application, analysis and evaluation in addition to developing knowledge and understanding. (The Workbook does not cover all topics in the Cambridge International AS & A Level Biology syllabus (9700)).

Chapter 1
Cell structure

CHAPTER OUTLINE

The questions in this chapter cover the following topics:

- the structure of animal, plant and bacterial cells, and of viruses
- the use of light microscopes and electron microscopes to study cells
- drawing and measuring cell structures
- the variety of cell structures and their functions
- the organisation of cells into tissues and organs.

Exercise 1.1 Units for measuring small objects

Cells are small, and the organelles that they contain are sometimes very, very small. In this exercise, you will practise converting between the different units that we use for measuring very small objects. You will also make sure that you are able to write numbers in **standard form**.

KEY WORDS

standard form: a way of writing a number as a value that is always between 1 and 10, and using the power of ten to show how large or small the number is.

digit: a single whole number, e.g. 2

TIP

Standard form is a way of writing down large or small numbers simply.

The rules are:

- Write down the **digits** as a number between 1 and 10
- Then write $\times 10^{\text{power of the number}}$.

1 mm = 1000 µm = 10^3 µm

So 1 µm = 1/1000 mm = 10^{-3} mm

1. The units that we use when measuring cells are millimetres (mm), micrometres (µm) and nanometres (nm).

    Copy and complete:

    a   1 µm = 1000 nm = $10^{....}$ nm
    b   1 nm = ... µm = $10^{....}$ µm
    c   1 nm = ... mm = $10^{....}$ mm

TIP

Here are some examples of writing small numbers in standard form:

$0.678 = 6.78 \times 10^{-1}$

$0.012 = 1.2 \times 10^{-2}$

$0.0057 = 5.7 \times 10^{-3}$

> TIP
>
> To work out the correct power, imagine moving the decimal point to the right or left, until you get a number between 1 and 10. Count how many moves you have to make, and that is the power of ten you should write.
>
> For example, if your number is 4297, you would move the decimal point as shown.
>
> 4.297
>
> So we write this number as 4.297×10^3.
>
> Here are some examples of writing large numbers in standard form:
>
> $6000 = 6 \times 10^3$
>
> $6248 = 6.248 \times 10^3$
>
> $82\,910 = 8.291 \times 10^4$
>
> $547.5 = 5.475 \times 10^2$

2   Write these numbers in standard form:
   a   5000
   b   63
   c   63 000
   d   63 497
   e   8521.89

3   Write these numbers in standard form:
   a   0.1257
   b   0.0006
   c   0.0104

4   A cell measures 0.094 mm in diameter.
   a   Convert this to micrometres.
   b   Express this value in standard form.

5   A cell organelle is 12 nm long.
   Express this value in µm, in standard form.

6   A mitochondrion is 1.28×10^2 µm long.
   Express this value in nm.

7   A chloroplast is 2.7×10^3 nm in diameter.
   Express this value in µm.

Exercise 1.2 Magnification calculations

This exercise will help you to gain confidence in doing magnification calculations, as well as providing further practice in using different units and converting numbers to standard form. You will also need to think about selecting a suitable number of significant figures to give in your answers. In general, you should use the same number of significant figures as there are in the value with the smallest number of significant figures that you used in your calculations.

$$\text{magnification} = \frac{\text{image size}}{\text{actual size}}$$

TIP

Remember – magnification has no units.

1 A light micrograph of a plant cell shows the cell to be 5.63 cm long. The real size of the cell is 73 μm.

Follow the steps to find the **magnification** of the micrograph.

Step 1 Convert 5.63 cm to μm.

Step 2 Substitute into the magnification equation: magnification = ...

Step 3 Calculate the magnification. Write the answer as ×................

2 An electron micrograph of a nucleus shows it to be 44 mm in diameter. The actual diameter of the nucleus is 6 μm. Calculate the magnification of the electron micrograph.

3 An electron micrograph of a **mitochondrion** shows its diameter as 28 mm. The magnification of the image is given as ×22 700. Follow the steps to find the actual diameter of the mitochondrion.

Step 1 Convert 28 mm to μm.

Step 2 Rearrange the magnification equation, and then substitute into it:

$$\text{actual size} = \frac{\text{image size}}{\text{magnification}} = \text{..................}$$

Step 3 Calculate the actual diameter of the mitochondrion. Remember to give your answer to the same number of significant figures as the value with the smallest number of significant figures that you used in your calculation.

4 An image of a chloroplast in an electron micrograph is 36 mm long. The magnification of the micrograph is ×1285. Calculate the actual length of the chloroplast.

KEY WORDS

magnification: the number of times greater that an image is than the actual object; magnification = image size ÷ actual (real) size of the object

mitochondrion: the organelle in eukaryotes in which aerobic respiration takes place

5 The micrograph shows a group of *Legionella* bacteria. The image has been magnified ×980.

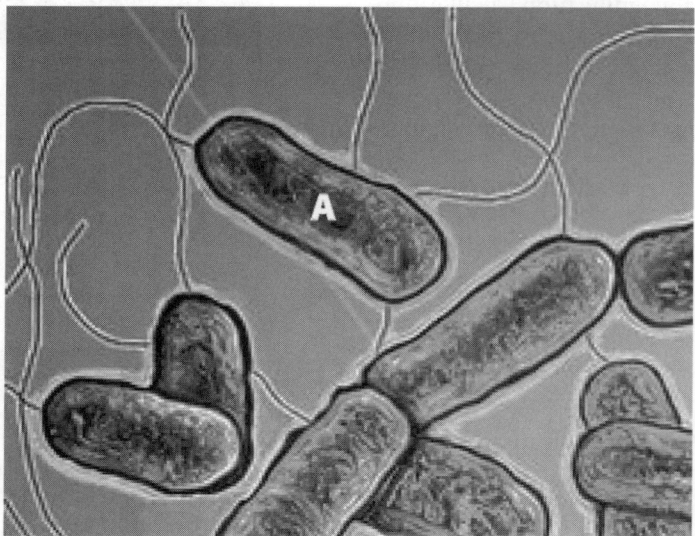

Figure 1.1: Micrograph of *Legionella* bacteria.

a   Measure the maximum length of bacterium **A**.
b   Calculate the actual length of this bacterium. Show all the steps in your working.

6 The micrograph shows some plant cells containing starch grains. There is a scale bar beneath the image.

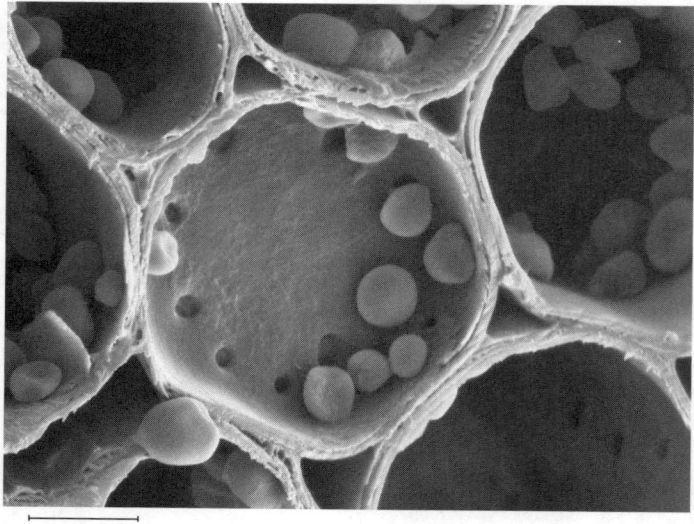

20 µm

Figure 1.2: Micrograph of plant cells with starch grains.

1 Cell structure

   a   Measure the length of the scale bar in mm.
   b   Convert this measurement to μm.
   c   Use this image size of the scale bar, and the actual size that we are told it represents, to calculate the magnification of the image.
   d   Measure the maximum diameter of the central cell in the micrograph.
   e   Use the value of the magnification you have calculated to find the actual diameter of this cell.

7 The micrograph shows a cell from the pancreas of a mammal. Several mitochondria are visible.

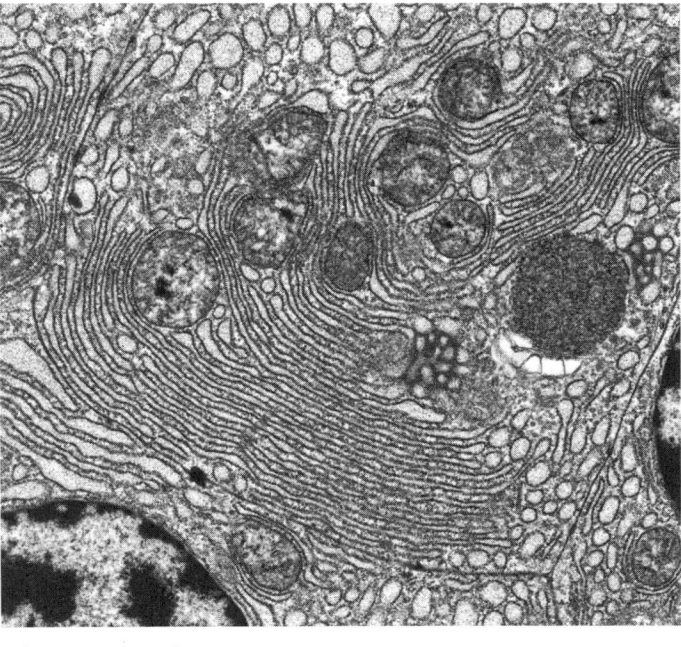

2 μm

Figure 1.3: Micrograph of a cell from the pancreas.

Use the scale bar to calculate the actual diameter of the largest mitochondrion.

Exercise 1.3 Drawing from light micrographs

Being able to draw good diagrams from micrographs or from what you can see when using a microscope is nothing to do with being good at art. Your task as a biologist is to make a clear, simple representation of what you can see. Use a sharp, medium-hard (HB) pencil and have a good eraser to hand. Each line should be clean and not have breaks in it – unless there really are breaks that you want to represent.

1 This drawing was made from the electron micrograph of the plant cells shown in Figure 1.2 (see question 6 in Exercise 1.2).

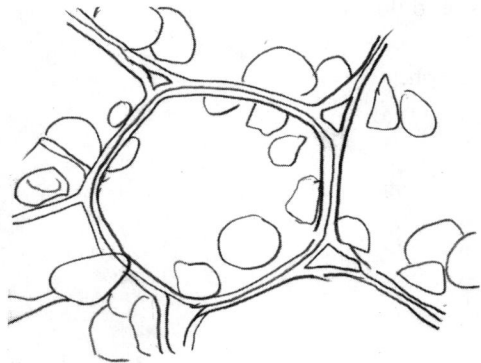

Figure 1.4: Drawing of plant cells.

a   Use these criteria to assess the quality of the drawing. Copy Table 1.1, and put a tick in one box in each row. You could also add a brief comment explaining why you made each decision.

Feature	Done very well	Done fairly well	Poorly done
suitably large diagram – makes good use of space available but does not extend over any text			
clean, clear, continuous lines			
overall shape and proportions look approximately correct			
correct number of starch grains shown, each carefully drawn the right shape and size			
relative sizes of starch grains and cell size correctly shown			
no shading has been used			
good and correct detail of cell walls shown			

Table 1.1: Assessment grid for drawing.

b   Now make your own drawing of the cells shown in Figure 1.2. Take care to meet all of the criteria fully.

2   The micrograph in Figure 1.5 shows a lymphocyte, a type of white blood cell found in mammals.

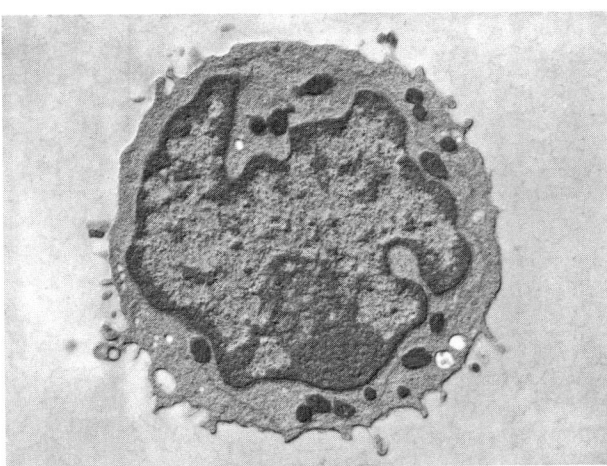

Figure 1.5: Micrograph of a lymphocyte.

a   Make a drawing of the lymphocyte.

b   Construct a list of criteria for your drawing, using the criteria from question **1a** as a guide.

c   Assess the standard of your drawing against your criteria. Alternatively, or as well, you could exchange your drawing with a partner, and assess each other's drawings.

d   The magnification of the micrograph of the lymphocyte is ×4750.

Calculate the actual diameter of the lymphocyte. Give your answer in μm, using standard form.

e   Use your answer to **d** to calculate the magnification of your drawing.

Exercise 1.4 Electron microscopes and optical (light) microscopes

The micrographs in this chapter have been made using different kinds of microscope. In this exercise, you will practise identifying features that distinguish images taken with different types of microscope, and then summarise the differences between what we can see using optical (light) microscopes and electron microscopes.

1   Copy and complete Table 1.2. In the 'type of microscope' column, choose from **optical microscope**, **transmission electron microscope** or **scanning electron microscope**.

Micrograph	Type of microscope used to produce the micrograph	Reason for your decision
Figure 1.2		
Figure 1.3		
Figure 1.5		

Table 1.2: Results table.

2   Copy and complete Table 1.3, to compare what can be seen in typical animal cells and plant cells using optical microscopes and electron microscopes. Put a tick or a cross in each box.

Organelle	Visible in plant cells		Visible in animal cells	
	Visible using optical microscope	Visible using electron microscope	Visible using optical microscope	Visible using electron microscope
nucleus				
mitochondrion				
membranes within mitochondrion				
Golgi body				
ribosomes				
endoplasmic reticulum				
chloroplast				
internal structure of chloroplast				
centriole				

Table 1.3: What can be seen in typical animal cells and plant cells using optical microscopes and electron microscopes.

> **KEY WORDS**
>
> **optical microscope:** a microscope that uses light to view a specimen; maximum resolution down to 200 nm
>
> **transmission electron microscope:** a microscope that uses a beam of electrons to view a very thin section; maximum resolution down to 0.5 nm
>
> **scanning electron microscope:** an electron microscope that provides a three-dimensional view of the surface of a specimen; maximum resolution down to 2 nm

Exercise 1.5 Using an eyepiece graticule and stage micrometer

An **eyepiece graticule**, calibrated using a stage micrometer, enables you to work out the actual size of objects you can see using a microscope. This exercise provides practice in this technique, and also involves decisions about how many **significant figures** to give in your answers.

Figure 1.6 shows a group of palisade cells, as they look through a light microscope with an eyepiece graticule fitted inside the eyepiece. The highest power objective lens of the microscope is being used.

TIP

When you are doing this using your own microscope, you will need to swivel the eyepiece and/or move the slide, so that your eyepiece graticule scale lies neatly over the thing you want to measure.

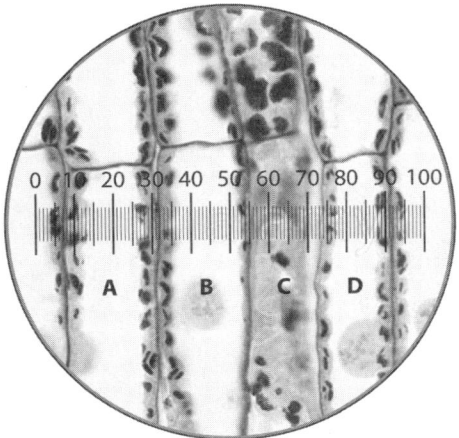

Figure 1.6: Micrograph of palisade cells seen using an eyepiece graticule. The small divisions on the graticule scale can be referred to as 'graticule units'.

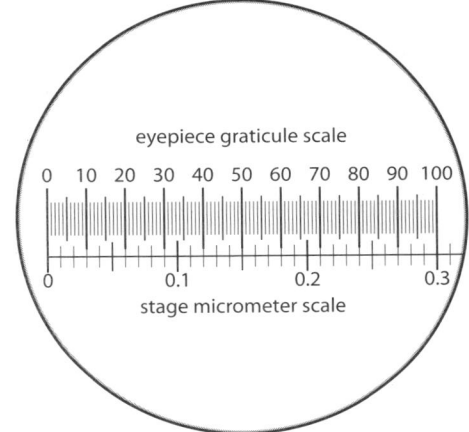

Figure 1.7: Stage micrometer seen using an eyepiece graticule.

KEY WORDS

significant figures: the digits that carry meaningful information about the size of the number

eyepiece graticule: small scale that is placed in a microscope eyepiece

stage micrometer: very small, accurately drawn scale of known dimensions, engraved on a microscope slide

1. Measure the total width of the four palisade cells A, B, C and D in graticule units.

   In order to find out the true size represented by one eyepiece graticule unit, we now need to calibrate the eyepiece graticule using a **stage micrometer**. This is a slide that is accurately marked off in small divisions of 0.01 mm. Figure 1.7 shows what is seen when the slide with the palisade cells is replaced on the microscope stage by a stage micrometer.

2. a Look for a good alignment of marks on the two scales, as far apart as possible. The 0s of both scales match up, and there is another good match at 80 small divisions on the eyepiece graticule.

   How many small divisions on the micrometer equal 80 small divisions on the eyepiece graticule?

   b Remember that one small division on the micrometer is 0.01 mm. Use your answer to **a** to calculate how many micrometres (μm) are represented by one small division on the eyepiece graticule.

TIP

It is essential to use the same objective lens – the one that you used when you measured the palisade cells in eyepiece graticule units. Again, you may need to swivel the eyepiece, and move the slide on the stage, to get them lined up against one another.

c Use your answer to **b** to find the total width of the four palisade cells in the micrograph.

d Now calculate the mean width of a palisade cell.

3 Explain why it is not possible to see both the palisade cells and the stage micrometer scale at the same time.

4 Figure 1.8 shows a light micrograph of some villi in the small intestine, seen using an eyepiece graticule.

Figure 1.9 shows the same eyepiece graticule, using the same objective lens, but this time with a stage micrometer on the microscope stage.

Use the two images to calculate the length of the villus that can be seen beneath the eyepiece graticule. Show each step in your working.

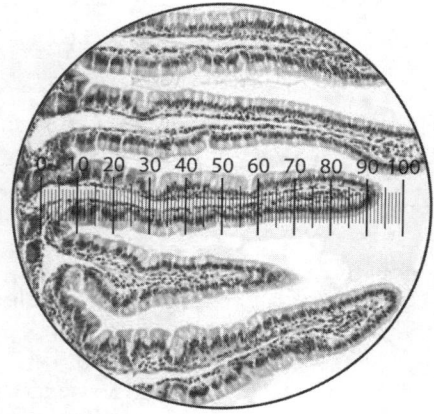

Figure 1.8: Light micrograph of villi seen using an eyepiece graticule.

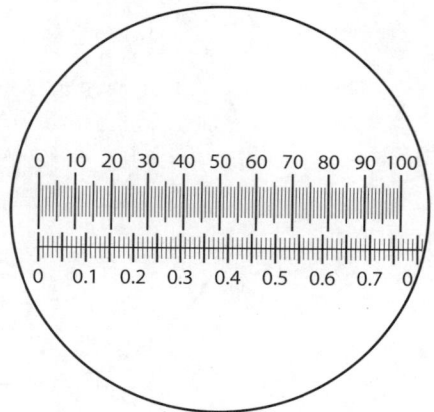

Figure 1.9: Stage micrometer seen using the eyepiece graticule.

Exercise 1.6 Membranes in different types of cell

Most cells in multicellular organisms become specialised for a particular set of functions. In this exercise, you will consider data relating to the membranes in two different types of cell, and use your biological knowledge to suggest explanations for patterns that you can pick out in these data.

All cells are surrounded by a cell surface membrane, and also contain many other membranes within them. Researchers estimated the total quantity of membranes in 20 liver cells and 20 exocrine pancreas cells, and then calculated the percentage of these membranes in all the different membrane-containing structures in the cells. Their results are shown in Table 1.4.

Source of membrane	Mean percentage of all membranes	
	Liver cells	Exocrine pancreas cells
cell surface membrane	1.8	4.7
mitochondrial membranes	39.4	22.3
nuclear membrane	0.5	0.7
rough endoplasmic reticulum	33.4	61.9
smooth endoplasmic reticulum	16.3	0.1
Golgi body	7.9	10.3
lysosomes	0.4	0
other small vesicles	0.3	0

Table 1.4: Percentage of membranes in all the different membrane-containing structures in the cells.

1 Explain why we cannot use these results to draw the conclusion that the mean quantity of cell surface membrane in liver cells is less than that in exocrine pancreas cells.

2 Which of the sources of membranes listed in the table are made up of two membranes (an envelope)?

3 Using the data in the table, state the organelle that contains the greatest mean percentage of membrane in:

   a liver cells
   b pancreas cells.

4 Liver cells have a wide variety of functions in metabolism, including synthesising proteins, breaking down toxins, synthesising cholesterol and producing bile. Exocrine pancreas cells have a single main role, which is the production and secretion of digestive enzymes.

   Use this information to suggest explanations for the differences between the percentages for mitochondria and rough endoplasmic reticulum in the liver cells and the pancreas cells.

Exercise 1.7 Command words

Command words are the instructional words in a question that tell you what you need to do. It is a good idea to identify the command word in each question part, and make sure that you understand what it means so that you can then write a strong answer.

1 Below are some of the Cambridge International definitions for command words which may be used in exams. The italic text provides some additional explanation on the meaning of these terms. You have already met several of them in this chapter.

assess   calculate   comment   compare   contrast   define
describe   discuss   explain   give   identify   outline
predict   sketch   state   suggest

Match the correct command word in the list above to the description given:

a express in clear terms – *that is, give a short, precise answer*

b give the precise meaning – *that is, provide a short but complete description of what the term means*

c state the points of a topic / give characteristics and main features – *for example, use words to say clearly what is shown by a graph, or give a step-by-step account of something*

d produce an answer from a given source or recall/memory – *for example, using information provided in the question, or from knowledge you have learnt during your course*

e make a simple drawing showing the key features

f set out the main points – *that is, give a brief account, picking out the most important points and omitting detail*

g set out purposes or reasons / make the relationships between things evident / provide why and/or how and support with relevant evidence – *note that you will often need to use your knowledge of biology to say why or how something happens*

h write about issue(s) or topic(s) in a structured way; *it is often a good idea to state points on both sides of an argument, for example, reasons for and against a particular viewpoint, or how a set of results could be interpreted to support or reject a hypothesis*

i make an informed judgement

j apply knowledge and understanding to situations where there is a range of valid responses, in order to make proposals – *that is, use information provided, and your biological knowledge, to put forward possible answers; there is often more than one possible correct answer*

k work out from given facts, figures or information – *it is usually a good idea to show all of the steps in your working*

l give an informed opinion – *you will often need to use your biological knowledge and understanding to make a range of statements about the topic*

m identify/comment on similarities and/or differences – *you can often use a table for this; if writing sentences, then use comparative words*

n identify/comment on differences – *you can often use a table for this; if writing sentences, then use comparative words*

o name/select/recognise – *for example, labelling a structure on a diagram or micrograph*

p suggest what may happen based on available information.

> **TIP**
>
> The information in this section is taken from the Cambridge International syllabus for examination from 2022. You should always refer to the appropriate syllabus document for the year of your examination to confirm the details and for more information. The syllabus document is available on the Cambridge International website at www.cambridgeinternational.org.

1 Cell structure

EXAM-STYLE QUESTIONS

1 This question relies on your recall of facts and concepts. You could answer part **b** either in words, or by using a labelled diagram. Note that the command word for **b** is 'outline'.

Table 1.5 lists some features of prokaryotic and eukaryotic cells.

Feature	Prokaryotic cell	Eukaryotic cell
cell surface membrane		
nucleus		
ribosomes		
mitochondria		
chloroplasts		

Table 1.5

a Copy and complete Table 1.5. If a feature can be present in the cell, write a tick in the box. If it cannot be present, write a cross. You should write either a tick or a cross in each box. [5]

b Viruses are not usually considered to be living organisms and are not made of cells. **Outline** the key features of the structure of a virus. [2]

[Total: 7]

COMMAND WORD

Outline: set out the main points.

2 This question asks you to identify structures within an animal cell. You should find this relatively straightforward, although you may have to think carefully about part **b**.

The diagram is a drawing of a cell from the body of a mammal.

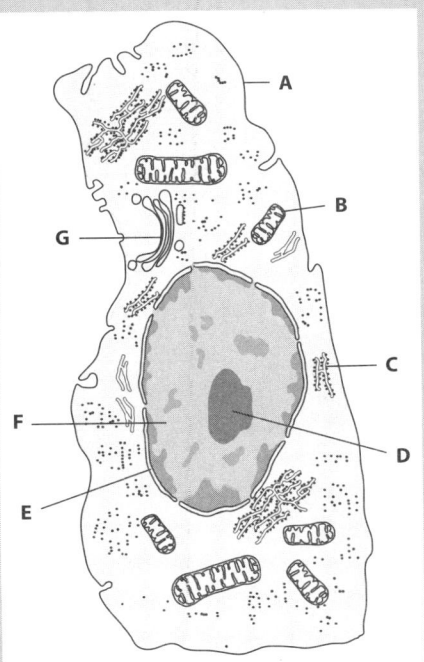

Figure 1.10

CONTINUED

a **State** the type of microscope that would be used to allow this amount of detail to be seen in the cell. [1]

b List the letters of the structures in the drawing which are made up of, or are surrounded by, phospholipid membranes. [3]

c **Describe** the functions of:
   i   structure **B** [2]
   ii structure **E** [2]
   iii structure **G**. [2]

[Total: 10]

3 The diagram shows a method for separating the different components of cells. This technique is called **ultracentrifugation**.

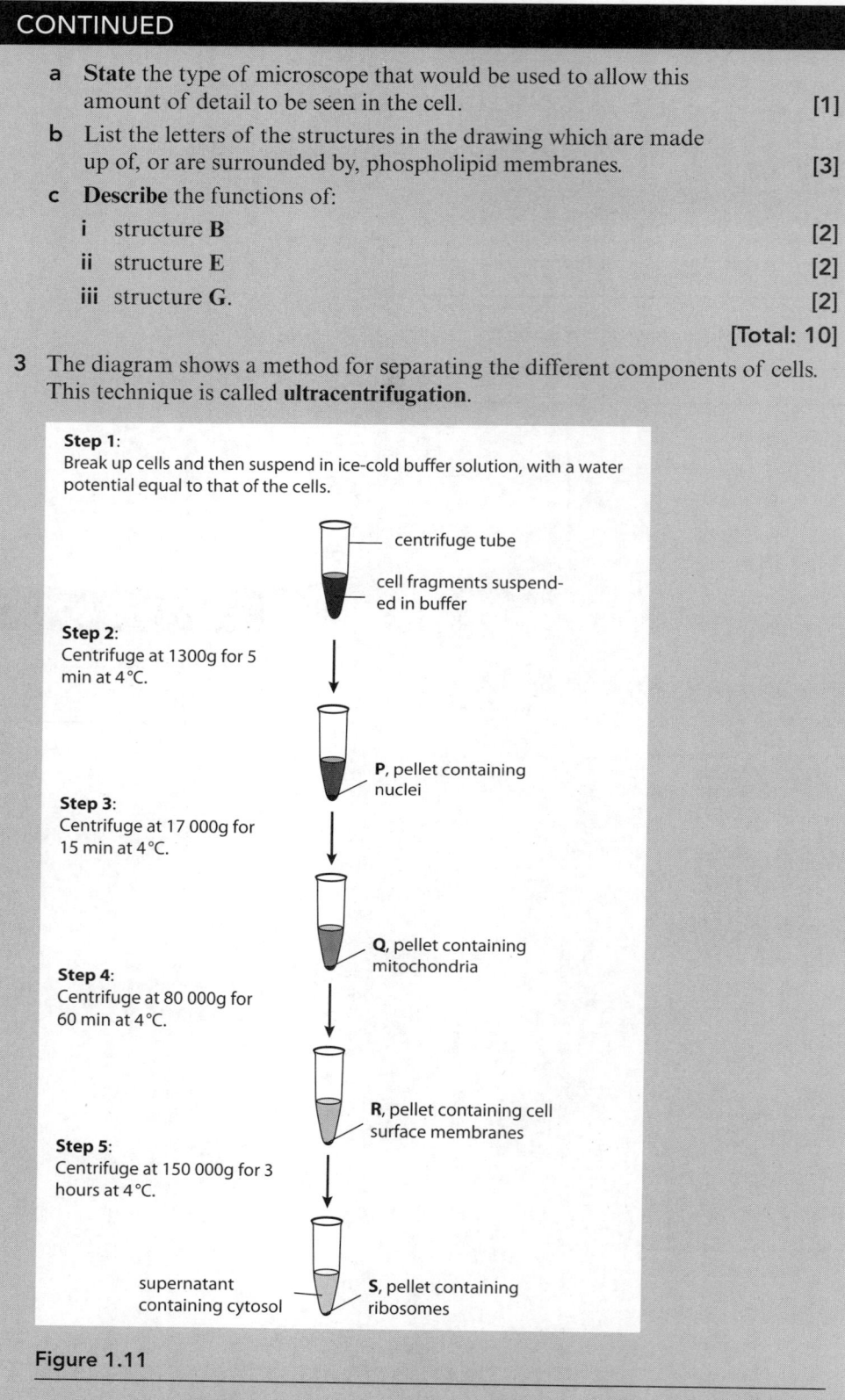

Figure 1.11

COMMAND WORDS

State: express in clear terms.

Describe: state the points of a topic / give characteristics and main features.

KEY WORD

ultracentrifugation: spinning a suspension at very high speed, so that more dense components settle to the bottom and can be separated

CONTINUED

   a **Suggest** why the solution in which the broken-up cells were suspended:

      i   was ice-cold [1]

      ii contained a buffer [2]

      iii had the same water potential as the cells. [2]

   b Suggest why ribosomes do not collect in the pellet until the final stage of the ultracentrifugation. [1]

   c Give the letter of the component or components in which you would expect to find:

      i   DNA [1]

      ii phospholipids. [1]

   d If this process were carried out using plant cells, which other cell organelles might you expect to find in the pellet containing mitochondria? Explain your answer. [2]

   [Total: 10]

COMMAND WORDS

Suggest: apply knowledge and understanding to situations where there are a range of valid responses, in order to make proposals / put forward considerations.

Identify: name/select/recognise.

Calculate: work out from given facts, figures or information.

Some questions will contain something new, such as an unfamiliar micrograph. But a combination of your own knowledge and the information in the question should help you work out suitable answers. Look carefully at the mark allocations, which guide you in how detailed an answer you need to give.

4 Figure 1.12 shows a micrograph of parts of two cells from the small intestine of a mammal. The structures along the surfaces of the two cells are microvilli.

   a State the type of microscope that was used to obtain this micrograph. Give a reason for your answer. [2]

   b **Identify** organelle A. [1]

   c The magnification of the micrograph is ×12 500.

      i **Calculate** the length of the microvillus between points **X** and **Y**. Show your working. [3]

      ii Microvilli greatly increase the surface areas of the cells. Suggest why the cells lining the small intestine have microvilli. [2]

   [Total: 8]

CAMBRIDGE INTERNATIONAL AS & A LEVEL BIOLOGY: WORKBOOK

CONTINUED

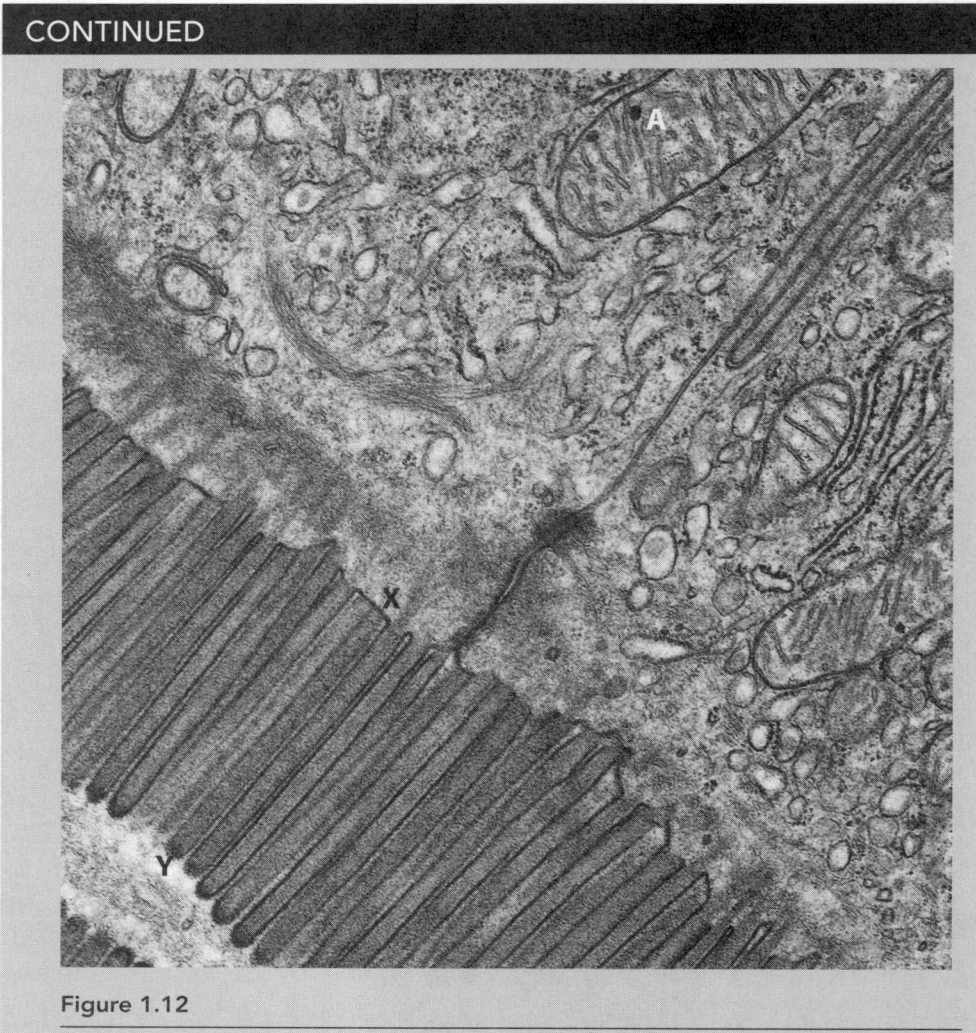

Figure 1.12

Chapter 2
Biological molecules

CHAPTER OUTLINE

The questions in this chapter cover the following topics:

- how large biological molecules are made from smaller molecules
- the structure and function of carbohydrates, lipids and proteins
- biochemical tests to identify carbohydrates, lipids and proteins
- some key properties of water that make life possible.

Exercise 2.1 Using summary tables

Note taking is an important skill, and you need to be able to extract the relevant information from complex writing. Students have many different approaches to making notes, but in every case, there is little value in copying down a whole passage of information into your own notes. One way to take notes is to create tables that summarise a topic. These tables can be useful for revision, and writing them out again can help move information into your long-term memory.

In this exercise, you will:

- read a complex passage about carbohydrate structure and identify the key points
- develop your note-taking skills by creating summary tables.

1   Read the passage below about carbohydrates carefully and then copy and complete Table 1.1 to make a summary of the structures and functions of the different **polysaccharides**.

Polysaccharides are **polymers** of **monosaccharides** and are made by joining the monosaccharides with **glycosidic bonds**. Glucose is the monomer for many polysaccharides. It is a highly soluble and reactive molecule and would affect the osmotic potential of cells and interfere with biochemical reactions. In plants, glucose is stored as starch, which is a mixture of two substances – amylose and amylopectin. In animals, glucose is stored as glycogen. Amylose is made of α-glucose molecules joined by α (1,4) glycosidic bonds, resulting in a very compact, helical molecule. Amylopectin, also made of α-glucose molecules, has many α (1,4) glycosidic bonds and has side branches joined by α (1,6) glycosidic bonds. Glycogen is made of α-glucose joined with both α (1,4) glycosidic and α (1,6) glycosidic bonds to make an even more branched molecule. Large numbers of branches enable rapid hydrolysis into glucose molecules. Cellulose is a polymer of β-glucose. In β-glucose, the –OH group on carbon number 1 points upwards, so that in cellulose molecules every other glucose molecule is inverted 180°. Cellulose molecules do not have branches and are long straight chains. About 60–70 cellulose molecules bind to each other using **hydrogen bonds** to form microfibrils. Several microfibrils

> **KEY WORDS**
>
> **polysaccharide:** a polymer whose monomer subunits are monosaccharides (single sugar units with the general formula $(CH_2O)n$) joined together by glycosidic bonds; a C–O–C link formed by a condensation reaction
>
> **hydrogen bonds:** a relatively weak bond formed by the attraction between a group with a small positive charge on a hydrogen atom and another group carrying a small negative charge, e.g. between two $-O^{\delta+}H^{\delta+}$ groups

bind together to form fibres, and these fibres make up the cell wall. Cellulose fibres have very high tensile strength, preventing cells from bursting, but allow free passage of water and water-soluble molecules.

Polysaccharide	Name of monomer	Bonds joining monomers	Description	Function
amylose	α-glucose	α (1,4)	• long • helical • compact	• cellular storage of glucose (energy) • does not affect osmotic potential • found in plants
amylopectin				
glycogen				
cellulose				

Table 2.1: Summary results table.

Exercise 2.2 Calculating the concentration of solutions and making dilutions

Getting the units right

It is essential to be able to accurately produce different concentrations of solutions. You can be asked to make up concentrations as either percentages (%) or as the number of **moles** per unit volume (mol dm^{-3}). In this exercise, you will:

- develop your understanding of different units
- develop your ability to calculate different concentrations of solutions as percentages and number of moles per unit volume.

> **KEY WORD**
>
> **mole** (plural **moles**): the amount of an element or substance that has a mass in grams numerically equal to the atomic or molecular weight of the substance

It is important to understand the units of volume and mass when making up solutions of different concentrations, and you may have to do some conversions.

The SI unit of mass is the kilogram (kg). In biology, you will often come across mass measured in kilograms (kg), grams (g), milligrams (mg), micrograms (μg) and sometimes even nanograms (ng).

- 1 kg is equivalent to 1000 g.
- 1 g is equivalent to 1000 mg.
- 1 mg is equivalent to 1000 μg.
- 1 μg is equivalent to 1000 ng.

In order to convert your value to the next smaller unit you need to multiply by 1000, and to convert to the next larger unit you need to divide by 1000.

For example, 0.005 kg is equivalent to: $0.005 \times 1000 = 5$ g and is also equivalent to: $0.005 \times 10^6 = 5000$ mg.

The SI unit of volume is the cubic metre (m^3). In biology, you will usually come across volumes as cubic decimetres (dm^3), cubic centimetres (cm^3) and cubic millimetres (mm^3).

- 1 cubic decimetre (dm^3) is equivalent to 1000 cubic centimetres (cm^3).
- 1 cubic centimetre (cm^3) is equivalent to 1000 cubic millimetres (mm^3).

In order to convert values given in cubic decimetres to cubic centimetres, you need to multiply by 1000 and to convert cubic centimetres into cubic decimetres you need to divide by 1000.

1 Carry out the following unit conversions:

- a 150 mg into g
- b 225 mg into kg
- c 0.005 kg into mg
- d 100 ng into g
- e 100 cm^3 into dm^3
- f 0.005 cm^3 into mm^3
- g 150 mm^3 into dm^3
- h 0.000005 dm^3 into mm^3.

Calculating concentrations as percentages of total volume

When the concentration of a solution is given as a percentage, for example, 1.0% sucrose solution, it means that 1 g of sucrose was dissolved in distilled water and made up to a total volume of 100 cm^3.

Percentage concentration may be calculated using the formula:

$$\text{percentage concentration mass of solute in grams} = \frac{\text{mass of solute in grams}}{\text{volume of solution in cm}^3} \times 100\%$$

2 Place the following statements about making up 100 cm³ of a 5% sucrose solution in the correct order:

  A Make up to 100 cm³ by adding distilled water.

  B Dissolve in a small amount of distilled water.

  C Place dissolved sucrose into a 100 cm³ volumetric flask.

  D Use a top pan balance to measure out 5 g of sucrose.

3 Calculate the sucrose concentrations in percentages of the following solutions:

  a 5 g of sucrose made up to a volume of 100 cm³
  b 10 g of sucrose made up to a volume of 500 cm³
  c 0.5 g sucrose made up to a volume of 500 cm³
  d 37.5 g sucrose made up to a volume of 150 cm³
  e 450 g sucrose made up to a volume of 1 dm³
  f 0.005 g sucrose made up to a volume of 100 mm³.

4 Calculate the mass of sucrose required to make the following solutions:

  a 100 cm³ of 0.5% sucrose solution
  b 250 cm³ of 12% sucrose solution
  c 500 cm³ of 25% sucrose solution
  d 750 cm³ of 50% sucrose solution
  e 500 mm³ of 10% sucrose solution
  f 1 dm³ of 45% sucrose solution.

Calculating concentrations as molarities

Concentrations of solutions are often given as molarities. This tells you how many moles of solute are dissolved in one cubic decimetre of water. A mole is the relative molecular mass of a substance in grams.

The molarity of a solution can be calculated using the following equation:

$$\text{molarity (mol dm}^{-3}) = \frac{\text{number of moles of solute}}{\text{total volume of solution in dm}^3}$$

To make a 1 mol dm⁻³ solution of a substance, you need to weigh out 1 mole of the substance, place it into a 1 dm³ volumetric flask and add distilled water up to a total volume of 1 dm³.

For example, if you wish to make a 1 mol dm⁻³ glucose solution:

Step 1 Find out the molecular formula of glucose ($C_6H_{12}O_6$).

Step 2 Determine the relative molecular mass of glucose.

  The relative atomic masses of the elements are: carbon 12, hydrogen 1, and oxygen 16, so the relative mass of glucose is:

  $(12 \times 6) + (1 \times 12) + (16 \times 6) = 72 + 12 + 96 = 180$.

Step 3 Weigh out 180 g of glucose.

Step 4 Dissolve the glucose in a small amount of distilled water.

Step 5 Place the glucose solution into a 1 dm³ volumetric flask and add distilled water to make a final volume of 1 dm³.

5   Calculate the masses of solutes required to make up the following solutions:
    a $1\,dm^3$ of $1\,mol\,dm^{-3}$ glucose solution
    b $1\,dm^3$ of $1\,mol\,dm^{-3}$ maltose solution
    c $1\,dm^3$ of $1\,mol\,dm^{-3}$ glycine ($C_2H_5O_2N$) solution
    d $100\,cm^3$ of $2\,mol\,dm^{-3}$ glucose solution
    e $100\,cm^3$ of $2\,mol\,dm^{-3}$ sucrose solution.

You should also be able work out what molarity of solution is made when you are given a mass of solution and the volume that it is made up to with distilled water.

$$\text{molarity (mol\,dm}^{-3}\text{)} = \frac{\text{number of moles of solute}}{\text{total volume of solution in dm}^3}$$

For example, 171 g of sucrose was dissolved by addition of distilled water and made up to $0.5\,dm^3$. To calculate the molarity of the solution, you must follow the steps below:

Step 1   Calculate the mass of 1 mole of sucrose ($C_{12}H_{22}O_{11}$):
$(12 \times 12) + (22 \times 1) + (11 \times 16) = 144 + 22 + 176 = 342\,g$

Step 2   Calculate how many moles are present in 171 g of sucrose:
$(171 \div 342) = 0.5$ moles

Step 3   Calculate the molarity by dividing the number of moles by the volume of water in dm^3:
$(0.5 \div 0.5) = 1\,mol\,dm^{-3}$

6   Now try to work out the molar concentrations of the following solutions:
    a   171 g sucrose dissolved in $1\,dm^3$ distilled water
    b   150 g glycine ($C_2H_5O_2N$) dissolved in $1\,dm^3$ distilled water
    c   4.5 g glucose dissolved in $100\,cm^3$ distilled water
    d   1.8 g glucose dissolved in $1\,dm^3$ distilled water
    e   342 mg sucrose dissolved in $1\,cm^3$ distilled water.

Exercise 2.3 Graph plotting and using calibration curves

You need to know the biochemical tests for starch, reducing sugars, non-reducing sugars, proteins and lipids. **Benedict's test** is a test for reducing sugars. Benedict's reagent is initially blue due to the presence of copper (II) sulphate. When heated in the presence of a reducing sugar, the copper (II) sulphate is reduced into insoluble red copper (I) oxide. Higher concentrations of reducing sugar will cause the solution to have less blue colour. The degree of blue colour can be estimated semi-quantitatively by comparing the colour of the solution with the colour of known standards. A more accurate method is to determine the degree of blue colour can be quantified by using a **colorimeter**. This is a piece of equipment that gives us a definite value of how blue the solution is. In this exercise, you will:
- develop your understanding of Benedict's test
- develop your understanding of how to make dilutions
- develop your understanding of how to use a calibration graph.

> **TIP**
>
> Be careful to make sure that you convert units! Often dilute solutions are given as $mmol\,dm^{-3}$ where 1000 mmoles is equivalent to 1 mole.

> **KEY WORDS**
>
> **Benedict's test:** a test for the presence of reducing sugars; the unknown substance is heated with Benedict's reagent, and a change from a clear blue solution to the production of yellow, red or brown precipitate indicates the presence of reducing sugars such as glucose
>
> **colorimeter:** an instrument that measures the intensity of colour of a solution, by passing light of a particular wavelength through it and measuring how much light is transmitted

Making dilutions

To decide what concentration of reducing sugar is in a particular solution, we must produce a calibration curve graph using a range of known concentrations. These are created by diluting stocks of reducing sugar solutions of known concentrations.

1   Copy and complete Table 2.2 to show how to make a range of concentrations when given a 1% standard solution of sucrose. The first two are done for you.

Concentration of sucrose / %	Volume of 1% glucose solution / cm³	Volume of added water / cm³
0.9	9	1
0.8	8	2
0.7		
0.6		
0.5		
0.4		
0.3		
0.2		
0.1		

Table 2.2: How to make a range of concentrations when given a 1% standard solution of sucrose.

To make wider ranges of dilutions you will need to be able to perform serial dilutions. An example of this is shown in the right-hand diagram below.

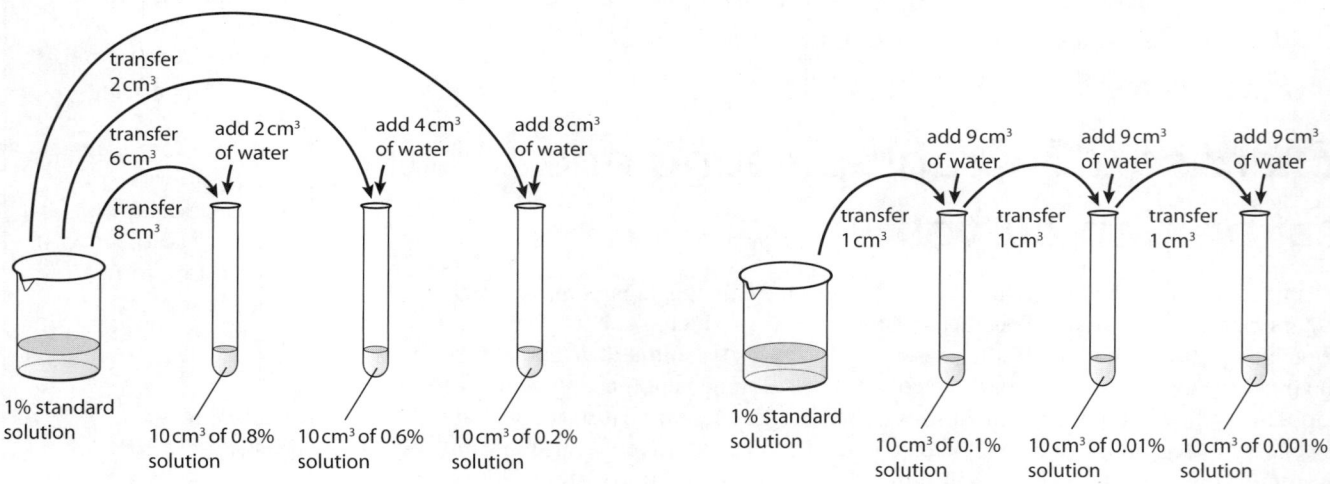

Figure 2.1: Producing a range of concentrations from standard solutions.

2   Describe how you could make up the following glucose concentrations when given a 1% standard solution (there may be several different ways for each):

   a   0.02%   b   0.003%   c   0.0005%.

2 Biological molecules

Plotting a calibration curve

A calibration curve is a graph with the known concentrations plotted against their percentage absorbance.

Table 2.3 shows the percentage of red light absorbed using a colorimeter from Benedict's tests performed on a variety of glucose concentrations.

Concentration of glucose / %	Absorption of light / %
1.0	5
0.8	35
0.6	45
0.4	64
0.2	85
0.1	92
0.0	96

Table 2.3: Percentage of red light absorbed using a colorimeter from Benedict's tests performed on a variety of glucose concentrations.

3 Plot a calibration curve of concentration of glucose against absorption of light.

   Follow these steps to plot your graph.

   Step 1 Label the *x*-axis with the independent variable, in this case, concentration of glucose / %.

   Step 2 Label the *y*-axis with the dependent variable, in this case, absorption of light / %.

   Step 3 Select appropriate linear scales on both axes. Make sure that you can fit all the plots along each axis but try to use at least half of each axis. You do not need to start the axes at 0 if it is not appropriate.

   Step 4 Using a sharp pencil, carefully plot the points in the form of small crosses.

   Step 5 Draw a smooth line or curve of best fit through the points. It is appropriate to use a line or curve of best fit when we can safely assume the intermediate points would lie along the line. If we cannot be sure of this, we should join the points with straight lines.

4 A mystery sample of glucose was analysed using a Benedict's test. The colorimeter gave an absorption reading of 74%.

   a Using a ruler, draw a straight line across from 74% to meet the curve then drop the line down to the *x*-axis. Record the concentration of glucose that would give an absorption of 74%.

   b Repeat this for an absorption of 53%.

5 When comparing different solutions of glucose using this method, it is important to control many factors in order to make a valid comparison.

   List **three** factors that need to be kept constant.

Exercise 2.4 Processing and analysing data

You will need to be able to apply your factual knowledge in order to interpret unfamiliar data. Do not be put off by the fact that you have not seen the information in that context before – look carefully at the information and use your knowledge. In this exercise, you will:

- develop your understanding of the different types of fatty acids
- develop your analytical skills when handling unfamiliar data.

1   Look carefully at the fatty acids in Figure 2.2 and answer the questions.

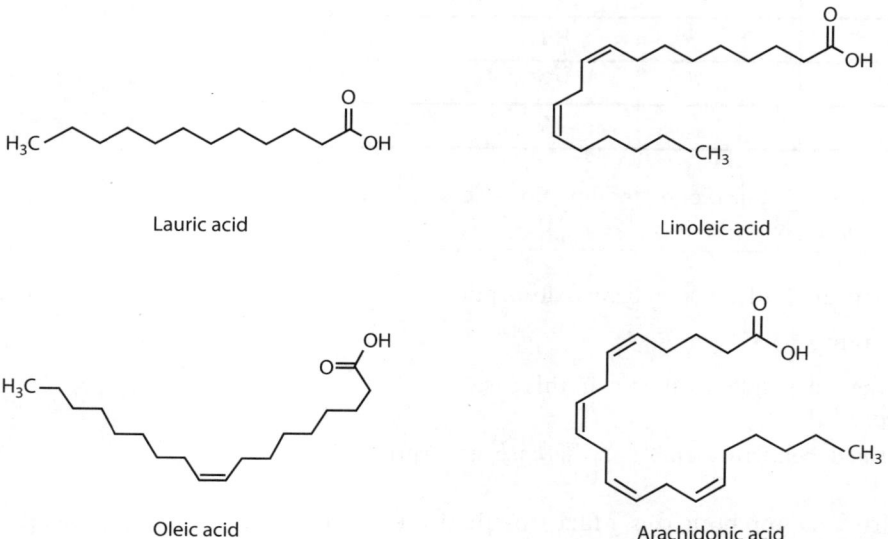

Figure 2.2: The structure of four different fatty acids.

a   Copy and complete Table 2.4 identifying which fatty acids are saturated, monounsaturated and polyunsaturated.

Fatty acid	Type of fatty acid	Melting point / °C
lauric acid		
oleic acid		
linoleic acid		
arachidonic acid		

Table 2.4: Fatty acid types and their melting points.

b The melting points of these fatty acids are: 45 °C, 13 °C, −49 °C, −11 °C. The presence of C=C bonds causes the melting point of fatty acids to be lowered. Match each fatty acid with its corresponding melting point and enter the data in the table.

c Suggest why the melting points of the different fatty acids are different.

Now look at Table 2.5 which shows the proportions of saturated, monounsaturated and polyunsaturated fatty acids found in lipids extracted from different organisms.

Organism	Total mass of lipid material tested / g	Saturated fatty acids / g	Monounsaturated fatty acids / g	Polyunsaturated fatty acids / g
sheep (animal)	100	40.8	43.8	9.6
cow (butter) (animal)	100	54.0	19.8	2.6
duck (animal)	50	16.7	24.5	6.8
mackerel (animal)	10	2.4	3.2	2.3
olive oil (plant)	200	28.0	139.4	22.4
corn oil (plant)	100	2.7	24.7	57.8
sunflower oil (plant)	100	11.9	20.2	63.0
hemp oil (plant)	75	7.5	10.0	50.0
coconut oil (plant)	150	127.8	9.9	2.5

Table 2.5: Proportions of saturated, monounsaturated and polyunsaturated fatty acids found in lipids extracted from different organisms.

When you compare data, you need to make sure that it is a fair comparison. The masses of fatty acids for the different species in the table were taken from different starting masses of total lipid. You will need to make each class of fatty acid proportional to the total mass of lipid, in other words grams (g) per 100 g of total lipid.

To convert the mass of each type of fat into mass per 100 grams, carry out the following steps.

Step 1 Calculate the mass of each type of fat per gram of total lipid:

$$\text{mass per gram} = \frac{\text{mass of fat}}{\text{total lipid mass}}$$

Step 2 Calculate the mass of each type of fat per 100 g:

$$\text{mass per 100 g of lipid} = \text{mass per gram} \times 100\,\text{g}$$

For example, in duck fat there are 16.7 g of saturated fat in 50 g of total lipid.

- In 1 g of total lipid there will be (16.7 ÷ 50) = 0.334 g saturated fat per gram of total lipid.
- If there is 0.334 g of saturated fat in 1 g of fatty material, there will be 0.334 g × 100 = 33.4 g of saturated fat in 100 g of total lipid.

2 a Copy and complete Table 2.6.

Organism	Saturated fatty acids / g per 100 g total lipid	Monounsaturated fatty acids / g per 100 g total lipid	Polyunsaturated fatty acids / g per 100 g total lipid
sheep (animal)	40.8	43.8	9.6
cow (butter) (animal)	54.0	19.8	2.6
duck (animal)			
mackerel (animal)			
olive oil (plant)			
corn oil (plant)	12.7	24.7	57.8
sunflower oil (plant)	11.9	20.2	63.0
hemp oil (plant)			
coconut oil (plant)			

Table 2.6: Results table.

Now that you have all the masses in the form of grams per 100 g of total lipid, you can start to evaluate the data.

b Identify any patterns that you can see in the data.

c Are there any results that do not seem to fit in with the patterns you have identified?

d Can you now suggest any explanations for any of the patterns? You will now need to use some of your knowledge about fatty acids here.

> **TIP**
>
> Think about the following points:
> - The higher the level of saturation of a fatty acid, the higher the energy content.
> - The more unsaturated the fatty acid, the lower the melting point.
> - **Homeothermic** animals maintain a constant internal body temperature.

> **KEY WORD**
>
> **homeothermic:** maintaining a constant body temperature

Exercise 2.5 Drawing molecular structures

Most biochemical reactions are best represented by drawing them out. Accuracy when drawing molecular structures is important, and it is easy to make mistakes.

You need to be able to understand the formation of glycosidic bonds in disaccharides and polysaccharides and be able to draw chemical structures.

1   a   Copy the diagram of two glucose molecules shown in Figure 2.3.

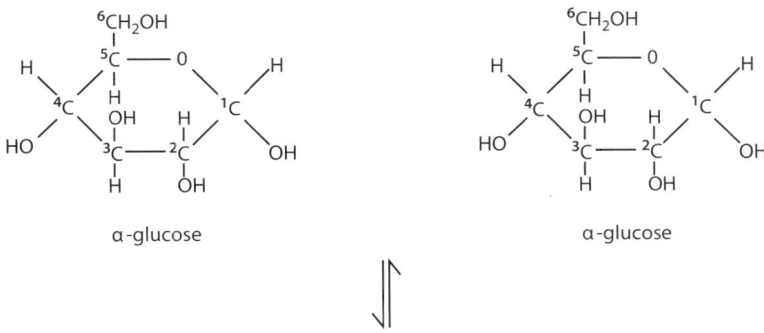

Figure 2.3: Two α-glucose molecules.

Step 1   Draw a ring around the hydrogen in the –OH group on carbon number 1 of the α-glucose on the left.

Step 2   Now draw a ring around the –OH group on carbon number 4 of the α-glucose on the right.

Step 3   Draw the reversible arrows as shown in the diagram.

Step 4   Draw out the disaccharide maltose and label it underneath the arrows. Also label the glycosidic bond.

Step 5   Show on the arrows where water (H₂O) is added or taken away.

Step 6   Next to the arrows, write the names of the reactions that would have taken place in each direction (**hydrolysis** and **condensation**).

b   Use your knowledge of dipeptide formation to repeat this procedure in order to show the formation of a **peptide bond** between the two amino acids shown in Figure 2.4.

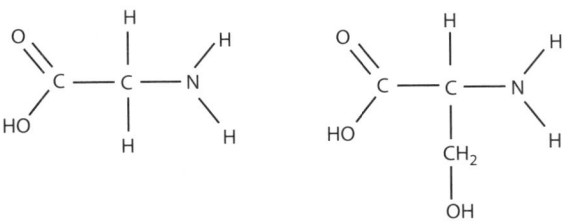

Figure 2.4: Molecular structure of the amino acids glycine (left) and serine (right).

KEY WORDS

hydrolysis: a reaction in which a complex molecule is broken down to simpler ones, involving the addition of water

condensation: a chemical reaction involving the joining together of two molecules by removal of a water molecule

peptide bond: a C–N link between two amino acid molecules, formed by a condensation reaction

2   Below is an example of an exam-style question and an example answer. Use the attached mark scheme to identify what is correct and what is incorrect.

   i   Complete the diagram below showing the formation of a **triglyceride**. [2]

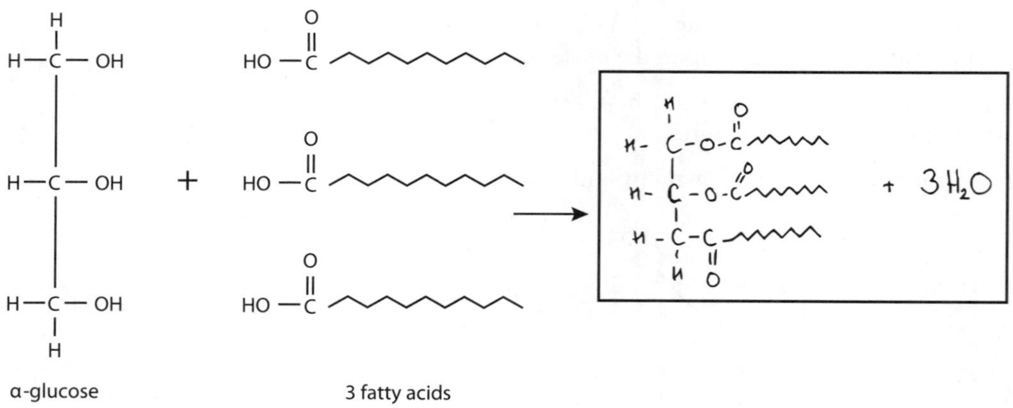

Figure 2.5: The formation of a triglyceride.

   ii   Name the type of reaction that would have taken place.
   condensation/hydrolysis [1]
   iii   When you have marked the example answer, write your own, corrected answer. [1]

Mark scheme	
i 3 water molecules on right of arrow	1 mark
correct bonds drawn on the triglyceride	1 mark
ii condensation	1 mark

Exercise 2.6 Planning experiments that generate reliable results

You need to be able to plan experiments that will enable you to generate valid conclusions. When planning an experiment, you need to be able to identify the independent, dependent and standardised variables. The **independent variable** is the variable that you will investigate and change. The **dependent variable** is the variable that you will measure. **Standardised variables** are other variables that would affect the dependent variable and need to be kept constant. To ensure that your experiment will produce a valid conclusion, you need to:

- include replicates – to improve reliability and make it easier to spot **anomalous** values
- have one variable changed at a time. All other variables must be controlled.

KEY WORDS

triglyceride: a lipid whose molecules are made up of a glycerol molecule and three fatty acids

independent variable: the variable (factor) that is deliberately changed in an experiment

dependent variable: in an experiment, the variable that changes as a result of changing the independent variable

standardised variable: variables (factors) that are kept constant in an experiment; only the independent variable should be changed

anomalous: a result or value that lies well outside the range of other results or values

In this exercise, you will:

- develop your experimental planning skills by evaluating an experimental plan and writing your own plan.

1   Read this experimental plan and then answer the questions.

An experiment to determine the temperature at which egg white protein (albumin) denatures.

Apparatus:

4 test tubes
test tube rack
measuring cylinder
4 eggs

Thermometer
glass marker pen
Bunsen burner
stop clock

Method:

The test tubes will be placed into the test tube rack. Into each test tube I will pour the clear liquid egg white from one egg. I will place a thermometer into the egg white of each tube. I will heat the tubes to different temperatures using the Bunsen burner for five minutes and see in which ones the albumin goes white. I will carry out the experiment at 20°C, 25°C, 50°C and 55°C. The lowest temperature at which the albumin goes white is the temperature at which it denatures.

   a   State the independent variable the experiment was testing.
   b   State the dependent variable the experiment was measuring.
   c   Will this experiment generate valid data? Explain your answer.
   d   Write an improved method that would produce more reliable, valid data. In your method you should think about:
       i   how you will change the independent variable, what range you will use and how you will monitor it
       ii   how you will measure the dependent variable more accurately and reliably
       iii what variables you will control and how you will control each one
       iv   how you will make the results more reliable.

Exercise 2.7 Extended writing skills

Any higher-level study of biology requires you to be able to write well-organised, factual essays. A good essay has factual detail in keeping with A level standard and is planned so that it flows and does not jump around from topic to topic. The number of words is a guide, although a shorter essay with more scientific detail and relevance to the question is better than a longer one with less relevance.

In this exercise, you will:

- develop your extended writing skills by planning a short essay about the biological role of water.

1   Plan an essay on the biological importance of water.

Start by planning it to include the following:
- water as a solvent
- water as a transport medium
- high specific heat capacity
- high latent heat of vaporisation
- density and freezing properties
- high surface tension and cohesion
- hydrogen bonding between water molecules
- water as a reagent.

For each of these properties you need to:
- explain how water achieves it (hydrogen bonding is very important)
- give biological examples.

Try to give examples of both plants and animals.

EXAM-STYLE QUESTIONS

1   a   Chemical bonds are found in all biological molecules. Complete Table 2.7 to match up different bonds with some biological molecules. [4]

	Hydrogen bond	Disulfide bond	Ionic bond	α (1,4) glycosidic bond
found in the tertiary structure of a protein				
found in amylose				
found in cellulose				
found in the secondary structure of a protein				

Table 2.7

CONTINUED

b Figure 2.6 shows two amino acids.

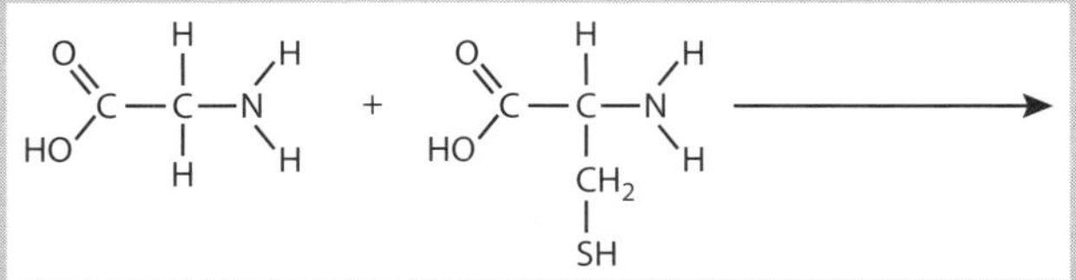

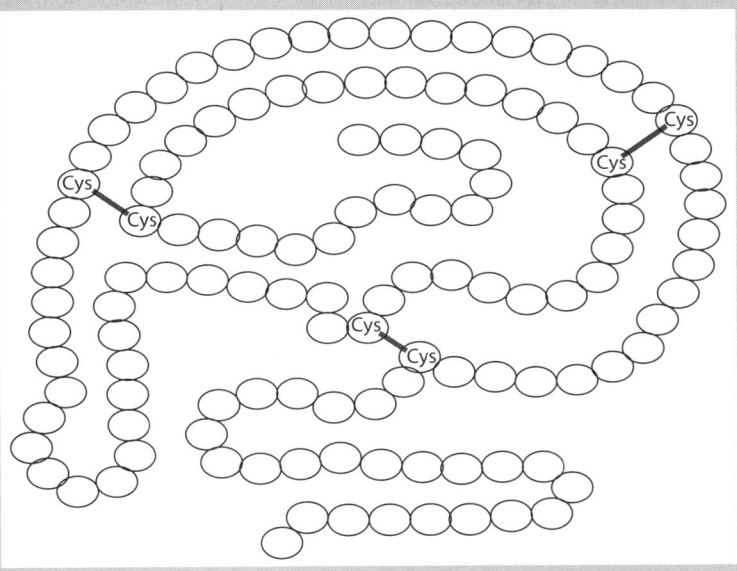

Figure 2.6

    i Complete the diagram to show the dipeptide formed by joining these two amino acids. [2]

    ii Name the type of reaction that occurs to join the two amino acids. [1]

[Total: 7]

2 Figure 2.7 shows a diagram of the enzyme RNase.

Figure 2.7

a State the level of protein structure shown in Figure 2.7. [1]

b Addition of a substance called β-mercaptoethanol causes the reduction of disulfide bonds.

Explain the effect this would have on RNase activity. [4]

[Total: 5]

COMMAND WORDS

State: express in clear terms.

Explain: set out purposes or reasons / make the relationships between things evident / provide why and/or how and support with relevant evidence.

CONTINUED

3 A student set up a series of reactions to test the activity of the enzymes amylase and sucrase. Biochemical tests were carried out on each of the reaction mixtures after incubating at 37 °C for three hours.

Table 2.8 shows the reactions and tests carried out.

Tube	Contents	Iodine test	Benedict's test	Biuret test
A	starch and amylase			
B	starch and sucrase			
C	sucrose and sucrase			
D	sucrose and amylase			

Table 2.8

    a Complete Table 2.8 by indicating which tests would give positive (+) and negative (−) results. [3]

    b Explain how you could carry out a biochemical test to show that a solution contained a mixture of both glucose and sucrose. [4]

[Total: 7]

4 Figure 2.8 shows the structure of the disaccharide sucrose.

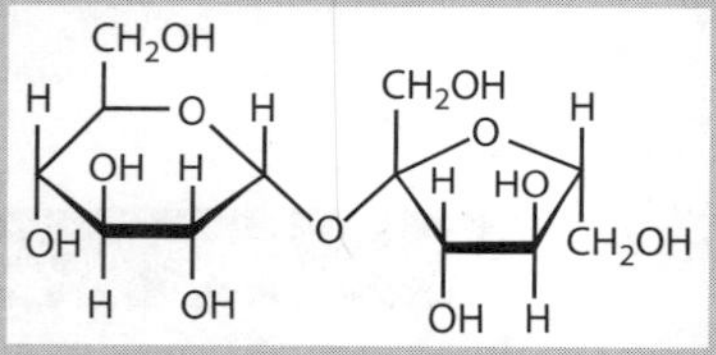

Figure 2.8

    a **i** Name the two monosaccharides produced by the breakdown of sucrose. [1]

       ii Draw the structure of the two monosaccharides produced by the breakdown of sucrose. [2]

       iii Name the type of reaction that is shown by the breakdown of sucrose. [1]

CONTINUED

b Figure 2.9 shows a light micrograph of potato tuber cells.

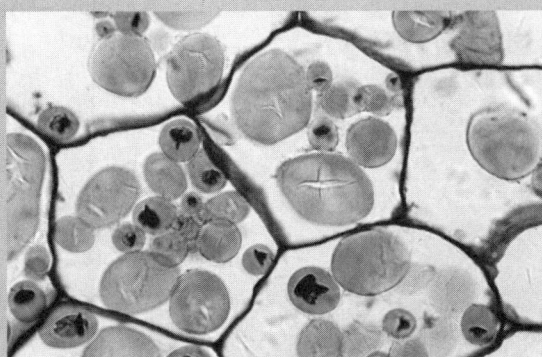

Figure 2.9

   i Add labels on the diagram to show the locations of cellulose and starch. [2]

   ii Explain how the structure of cellulose enables it to carry out its function. [4]

   iii Explain the advantages of storing starch rather than glucose in plant cells. [3]

[Total: 13]

5 a **Describe** how you could test a sample of sesame seeds for the presence of lipids. [3]

  b Figure 2.10 shows the general structure of a triglyceride.

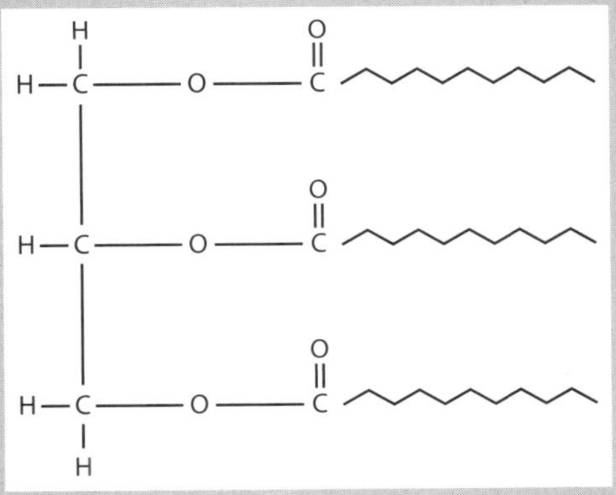

Figure 2.10

Complete the diagram to show the products of hydrolysis of this triglyceride. [2]

> **COMMAND WORD**
>
> **Describe:** state the points of a topic / give characteristics and main features.

CONTINUED

c Table 2.9 shows the molecular formulae and melting points of three fatty acids.

Fatty acid	Molecular formula	Melting point / °C
linoleic acid	$C_{18}H_{32}O_2$	−5
oleic acid	$C_{18}H_{34}O_2$	13
stearic acid	$C_{18}H_{36}O_2$	69

Table 2.9

Using the information in Table 2.9, **suggest** reasons for the different melting points of the fatty acids. [4]

[Total: 9]

6 a The p53 gene is a tumour suppressor gene. The gene codes for the p53 protein which binds to other cellular proteins to regulate cell division. Tumour cells often contain mutated p53 protein where an amino acid called arginine is replaced by a different amino acid called proline. Figure 2.11 shows the structures of arginine and proline.

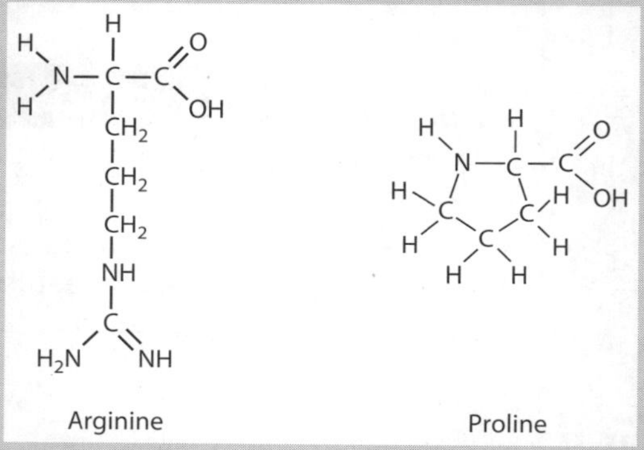

Figure 2.11

Arginine is a hydrophilic amino acid that has a net positive charge. Proline is a hydrophobic uncharged amino acid.

i Draw the R-group of arginine. [1]

ii Using the example above, explain how changing the primary sequence of a polypeptide could affect its function. [3]

b Collagen is a fibrous protein found in many animal tissues including bones and cartilage. Explain how the structure of fibrous proteins such as collagen gives them high tensile strength. [3]

[Total: 7]

COMMAND WORD

Suggest: apply knowledge and understanding to situations where there are a range of valid responses, in order to make proposals / put forward considerations.

> Chapter 3
Enzymes

CHAPTER OUTLINE

The questions in this chapter cover the following topics:
- how enzymes work
- the factors that affect enzyme activity, including inhibitors
- finding and using V_{max} and K_m
- how to carry out experiments with enzymes
- using immobilised enzymes.

Exercise 3.1 Answering questions about graphs

Questions sometimes provide a graph, and then ask various questions that relate to it. In this exercise, you will answer questions involving three different command words, all about the same graph. You then calculate a rate by drawing a **tangent** to one of the curves.

When you are asked to *describe* a graph, your task is to change the information shown by the line into words. You don't use your biological knowledge for this. Concentrate on the overall trend, then any obvious changes in gradient. Quote figures, referring to both the *x*- and *y*-axes, and remember to give units.

Explaining a graph, however, requires to you to say *why* the graph is the shape that it is.

When asked to *compare* two graphs, try to make statements that include words such as 'but', 'however', and comparative terms such as 'faster', 'higher', 'greater'. Quote comparative figures for particular values that look significant to you. For example, you could subtract one from the other, or calculate how many times greater one is than the other.

A student extracted catalase from 100 g of carrot and 100 g of apple. He added the two extracts to two tubes of hydrogen peroxide solution and measured the oxygen given off over a period of ten minutes.

Figure 3.1 shows the results the student obtained.

1. a Describe the curve for the catalase from carrot.
   b Explain the shape of this curve.
2. Compare the curves for catalase from carrot and catalase from apple.

KEY WORD

tangent: a straight line that just touches a curve at a particular point

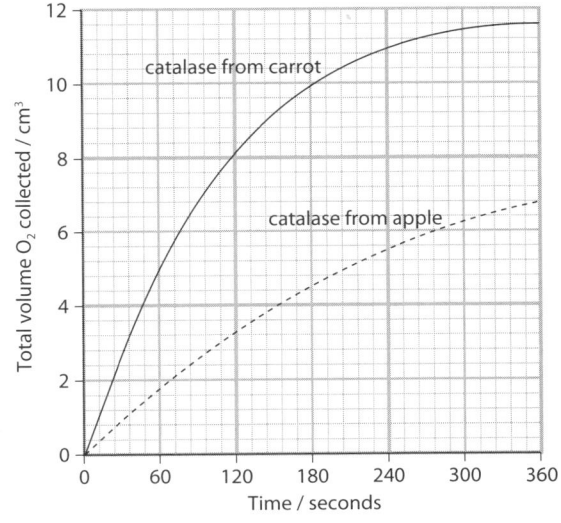

Figure 3.1: Graph showing the time course of the reaction for two sources of catalase.

We can calculate the rate at which oxygen is given off at a particular time. Here's how to do this for the apple catalase curve:

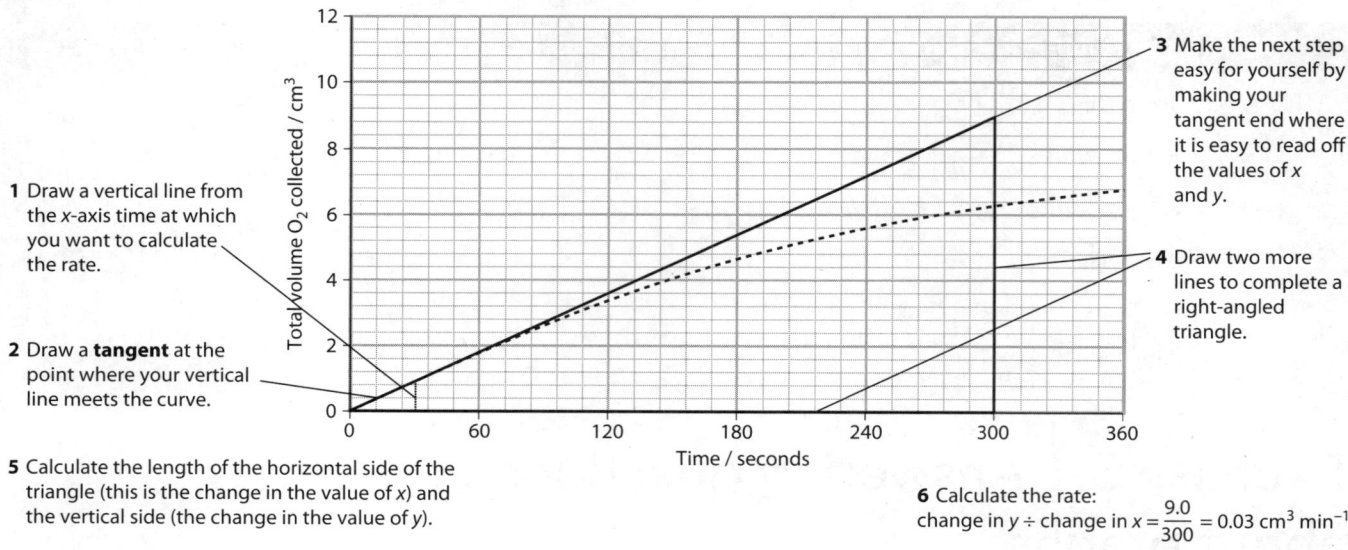

Figure 3.2: Calculating a gradient by drawing a tangent.

3   Calculate the rate at which oxygen is given off at 30 s for the carrot catalase curve.

Exercise 3.2 Effects of temperature and pH on enzyme activity

Like Exercise 3.1, this exercise involves interpreting graphs. Here, however, some of the data are rather unexpected, and you are asked to think about what might be causing these unexpected results.

Spondias is a South American tree whose fruit is used for making juices and ice cream. Like the cells of many fruits, the cells of *Spondias* contain an enzyme called polyphenol oxidase. The substrate for this enzyme is a group of colourless compounds, which the enzyme converts to brown quinones. Normally, the enzyme and its substrate are kept in separate cellular compartments, but when the fruit is damaged they meet and the damaged tissue turns brown.

Producers of fruit juice and other products from *Spondias* need to understand how this enzyme works, so that they can prevent browning in their products. A team of researchers added an extract from *Spondias* fruits to a solution of pyrocatechol, which is a substrate for polyphenol oxidase. They measured the degree of browning using a colorimeter.

The graphs in Figures 3.3 and 3.4 show how temperature and pH affected the rate of activity of polyphenol oxidase from *Spondias* fruits.

1   a   Describe the effect of temperature on the rate of activity of polyphenol oxidase from *Spondias* fruits.

b Explain the effects you have described in your answer to **a**.

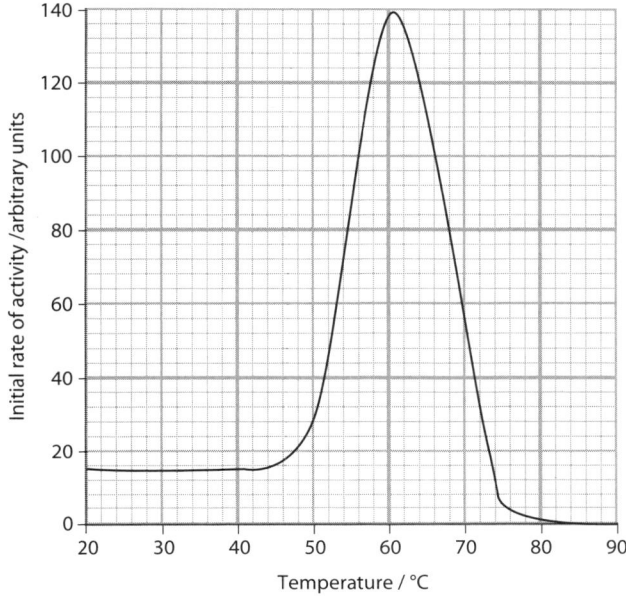

Figure 3.3: The effect of temperature on the rate of activity of polyphenol oxidase.

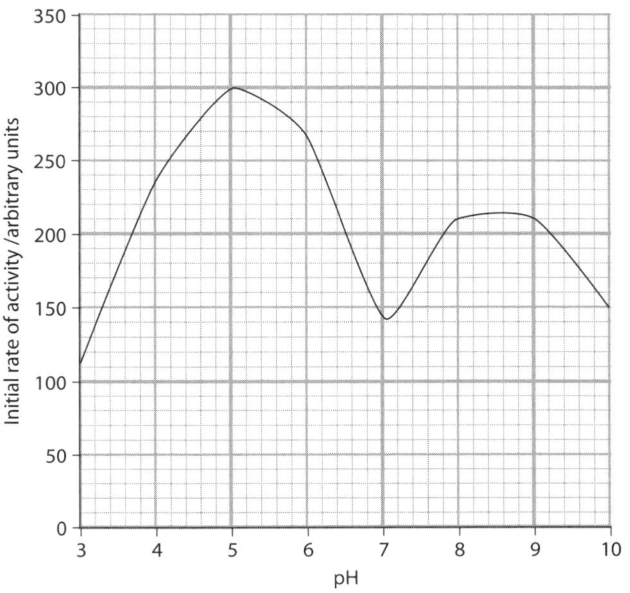

Figure 3.4: The effect of pH on the rate of activity of polyphenol oxidase.

2 a Describe how the effect of pH on the rate of activity of polyphenol oxidase from *Spondias* fruits differs from the usual effect of pH on the activity of an enzyme.

b The researchers concluded that the extract they had made from the *Spondias* fruits probably contained more than one type of polyphenol oxidase. What is the evidence for this conclusion?

3 The researchers recommended that the food manufacturers could reduce the browning of products made from *Spondias* fruits by adding lemon juice during their processing.

Use the data in the graphs to explain how this might work.

Exercise 3.3 Finding V_{max} and K_m

In this exercise, you will: plot a graph, and then use it to find V_{max} and the **Michaelis–Menten constant, K_m**.

Lipase is an enzyme that hydrolyses lipids to fatty acids and glycerol. A researcher extracted lipase from *Burkholderia cepacia*, a species of bacterium. The lipase was added to an emulsion of olive oil. The rate of activity of the enzyme was measured by finding how much fatty acid was released from the olive oil in one minute. The researcher investigated the effect of substrate concentration on the initial rate of reaction, V, of the lipase. The results are shown in Table 3.1.

KEY WORDS

V_{max}: the theoretical maximum rate of an enzyme-controlled reaction, obtained when all the active sites are occupied

Michaelis–Menten constant, K_m: the substrate concentration at which an enzyme works at half its maximum rate (½V_{max}), used as a measure of the efficiency of an enzyme; the lower the value of K_m, the more efficient the enzyme

Concentration of substrate / %	Initial rate of activity of lipase / μmol of fatty acid produced per minute
0	0
5	8
10	16
15	22
20	26
25	29
30	31
35	32
40	33
45	33
50	33

Table 3.1: The effect of substrate concentration on the initial rate of reaction, V, of lipase.

1 Which *two* of the variables below would the researcher keep constant in the experiments?

- the concentration of the lipase
- the concentration of the olive oil
- the temperature.

2 On a sheet of graph paper, construct a line graph to display these data. Draw a best fit line.

3 V_{max} is the initial rate of reaction at which substrate concentration is no longer limiting the reaction rate. Use your graph to find V_{max}.

4 The Michaelis–Menten constant, K_m, for the enzyme is the substrate concentration at which the rate of activity of the enzyme is ½ V_{max}.

Step 1 On the *y*-axis of your graph, find the value that is exactly half the value of V_{max}.

Step 2 Draw a horizontal line from this point on the *y*-axis, until it hits the curve.

Step 3 Draw a vertical line from this point on the curve, down to the *x*-axis.

Step 4 Read off the value of *x* at this point. This is K_m.

5 The researcher repeated this experiment using lipase extracted from a different species of bacterium. He found that K_m for this species was lower than that for *B. cepacia*.

What can you conclude about the relative affinity for their substrates of the lipases from the two species of bacteria?

> **TIP**
>
> Find V_{max} by carefully drawing a horizontal line from the highest position of the line on the graph, to meet the *y*-axis. Read off the value of V from the *y*-axis scale.

Exercise 3.4 Planning an investigation into the effect of inhibitor concentration on urease activity

Many experiments involve finding the effect of one variable (the independent variable) on another (the dependent variable). Any other variables that might affect your results should be kept the same – these are often known as standardised variables. It's often necessary to do some preliminary trials, to try to decide on the values of the independent and standardised variables that might work best. In this exercise, you will think about how you could do such an experiment.

Here is some information about urease, its substrate and an inhibitor that slows down its action.

- Urease is an enzyme that converts urea to ammonia:

$$CO(NH_2)_2 + H_2O \rightarrow 2NH_3 + CO_2$$

- Farmers often add fertilisers containing urea to the soil, which is a way of providing slow-release nitrogen compounds for their crops. These fertilisers sometimes also contain a urease inhibitor. This reduces the production of ammonia.

- You can obtain an extract containing urease by soaking soya beans in water overnight, then liquidising the mixture in a blender (or use a pestle and mortar). Filter the mixture, and keep the filtrate to use as your enzyme extract.

- Urea can be bought as a powder, which dissolves in water.

- The inhibitor can also be bought as a powder, which dissolves in water.

- Ammonia dissolves in water to form a solution with a pH above 7.

- The rate of production of ammonia can be followed by measuring the pH of the enzyme–substrate mixture, using an indicator solution or a pH meter.

Answer the questions below to explain how you would plan an experiment to investigate how the concentration of the inhibitor affects the rate of activity of the enzyme.

1   a   What is your independent variable?
    b   What is your dependent variable?
    c   List the important variables that you must keep the same.

2   What risks can you identify? What can you do to keep yourself and others safe?

3   You will need to do some preliminary work to find out a suitable concentration of substrate to use, so that you can obtain results in a reasonable length of time.

    a   Imagine that you have been provided with a bottle of urea powder. Describe how you could make up a 10% solution of urea in water.

    b   Imagine that you have made a solution containing urease from soya beans (as described above). You try adding 5 cm³ of this urease solution to 5 cm³ of the 10% urea solution, and measure the pH change. You find that the change in pH is very slow.

Complete the sentence to suggest how you could make the reaction happen more quickly:

You could use a .... concentrated solution of urease.

4 Imagine you have been provided with a 10% solution of the inhibitor.

   a Describe how you could use this solution to make up a range of solutions with different concentrations.

   b Suggest a suitable **range** and **interval** for the concentration of the inhibitor. Remember that the range is the spread between your lowest and highest value, and the interval is the difference between each value. You should have at least five values.

   c Decide if you will use **replicates** and if so, how many.

5 Describe how you will carry out your experiment.

6 Draw a results chart, with full headings, that you would be able to fill in as you collect your results.

   Remember:

   - The results chart should be drawn using a ruler. Each row and column should be separated from the next by a ruled line.
   - The independent variable should be in the first column of your table.
   - The replicate readings for the dependent variable should be in the next columns.
   - The mean values of the dependent variable should be in the last column of the table.

7 Sketch the curve that you would expect to obtain if the inhibitor is a non-competitive inhibitor.

> **KEY WORDS**
>
> **range:** the minimum value to maximum value in a set of data
>
> **interval:** the 'gap' between the values chosen within the range
>
> **replicates:** several experiments done in the same situation with the same apparatus and materials

Exercise 3.5 Calculating actual and percentage error

We can never be absolutely certain that a value that we measure is a true value. It is possible to estimate the actual error in a reading quantitatively. We can then use this to estimate the percentage error in a reading.

Whenever you make a measurement with an instrument such as a measuring cylinder, pipette or thermometer you cannot be certain that the reading is exactly correct. There is an uncertainty or error in your measurement.

In general, we say that the size of the error is equivalent to half the size of the smallest division on the scale you are reading from.

For example, imagine you have a thermometer marked off in °C. When you read the thermometer, you can read the scale to the nearest 0.5 °C. However, we must accept that there is an uncertainty in this reading of 0.5 °C. So, if you read a temperature of 17.5 °C, you can show this by writing it as 17.5 ± 0.5 °C. This is the error in our measurement.

1. A 1000 cm³ measuring cylinder is marked off in intervals of 10 cm³. What is the actual error in each reading of volume that you make using this measuring cylinder?

If you are measuring a *change* in a quantity, then you must take into account that there are uncertainties in the readings for both the starting value and the final value. For example, if you measure a temperature at the start of an experiment as 22.0 °C, and at the end as 27.5 °C, then you must add together the 0.5 °C uncertainties for each reading. The change in temperature is therefore given as an increase of 5.5 ± 1 °C.

2. A 100 cm³ measuring cylinder is marked off in intervals of 2 cm³. What is the size of the actual error if you measure a change in volume using this measuring cylinder?

When we have worked out the size of the error in the measurements, we can use this to calculate the **percentage error**:

$$\text{percentage error} = \frac{\text{size of the error in the reading}}{\text{the reading}} \times 100$$

So, if we measure a volume of 28 cm³ using a 50 cm³ measuring cylinder marked off in intervals of 1 cm³, the error in that measurement is ± 0.5 cm³, and the percentage error is:

$$\frac{0.5}{28} \times 100 = 1.8\%$$

3. A student measures out a volume of 10 cm³ of an enzyme solution, using a graduated pipette marked off in intervals of 0.1 cm³. Calculate the percentage error in her measurement of this volume.

4. A student measures a change in the length of a seedling over a period of three days. He measures the length with a ruler marked off in 1 mm intervals. The initial measurement is 2.15 cm, and the final measurement is 5.50 cm. What is the percentage error in his measurement of the change in length of the seedling?

5. A student measures a change in the mass of a potato strip after it has been immersed in a sucrose solution. He uses a top pan balance that gives readings to the nearest 0.01 g. He measures the initial mass of the strip as 1.63 g, and the final mass as 1.56 g. What is the percentage error in his measurement of the change in mass?

KEY WORDS

percentage error: the actual error in a measurement calculated as a percentage of the measured value

EXAM-STYLE QUESTIONS

Take care when answering these questions to focus on the command words for each part. For example, the command word 'suggest' means that you should use your biological knowledge and skills to work out a suitable answer.

1 Amylase digests starch to the reducing sugar maltose. Figure 3.5 shows the activity of a sample of amylase produced by a fungus, at different pH values and at three different temperatures.

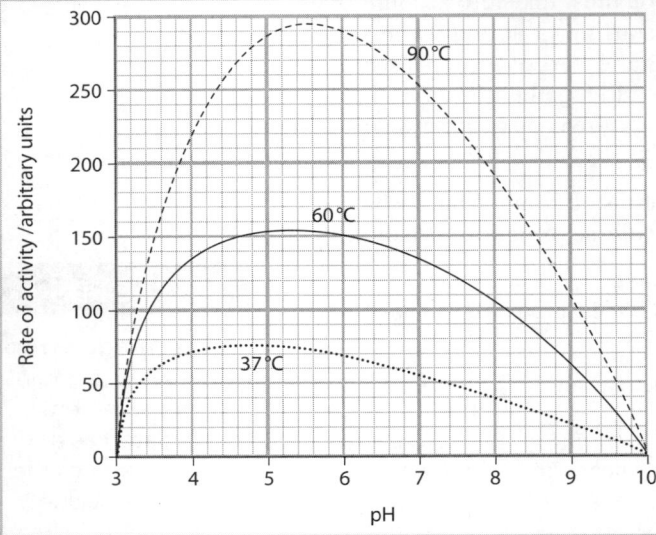

Figure 3.5

   a **Suggest** how the researchers could measure the rate of activity of the amylase sample. [3]
   b i With reference to the graph, **describe** the effect of pH on the activity of amylase at 60 °C. [2]
   ii **Explain** the reasons for the effects that you have described in **i**. [3]
   c i Describe the effect of temperature on the activity of this sample of amylase. [2]
   ii Suggest how the optimum temperature of this sample of amylase could be determined. [3]

   [Total: 13]

2 Enzymes have a huge number of uses, and this question is about a medical application of knowledge that we have about an enzyme made by a bacterium.

Stomach ulcers are caused by a bacterium called *Helicobacter pylori*. This bacterium produces the enzyme urease, which converts urea to ammonia and carbon dioxide.

COMMAND WORDS

Suggest: apply knowledge and understanding to situations where there is a range of valid responses, in order to make proposals / put forward considerations.

Describe: state the points of a topic / give characteristics and main features.

Explain: set out purposes or reasons / make the relationships between things evident / provide why and/or how and support with relevant evidence.

CONTINUED

To diagnose the presence of *H. pylori* in a person's stomach, the urea breath test is used. The person swallows a tablet of urea labelled with ^{13}C. The presence of ^{13}C in their breath indicates infection with *H. pylori*.

a **Explain** why the urea is labelled with ^{13}C. [1]

b **Outline** how carbon dioxide produced in the stomach is eventually exhaled in the breath. [3]

c Figure 3.6 shows the results of urea breath tests carried out in two people, **A** and **B**.

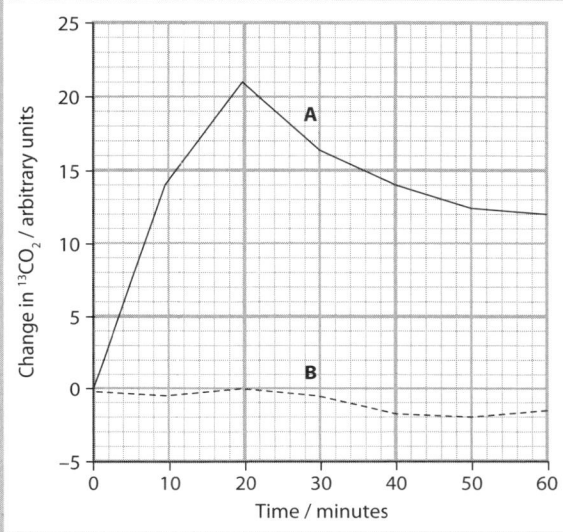

Figure 3.6

Compare the results in Figure 3.6 for person **A** and person **B**. [2]

d Suggest why the urea breath test is a good way of diagnosing the presence of *H. pylori* in a person's stomach. [1]

[Total: 7]

> **COMMAND WORDS**
>
> **Outline:** set out the main points.
>
> **Compare:** identify/comment on similarities and/or differences.

3 Immobilised enzymes are very widely used in industry. Here you are asked to interpret data about the activity of free and immobilised enzymes, and then to use your own knowledge to answer the final part of the question.

Pectinase is an enzyme that breaks down pectin, a component of plant tissues that helps to hold adjacent cell walls together. Pectinase is used in the food industry to increase the quantity of juice that can be extracted from fruit, and also to clarify fruit juice.

Pectinase obtained from a fungus, *Aspergillus niger*, was immobilised onto an alginate support. This enzyme is known to be damaged by temperatures above 40 °C.

CONTINUED

An experiment was carried out to compare the effect of exposure to a high temperature on the activity of free and immobilised pectinase. The enzymes were kept at a temperature of 50 °C for 30 minutes. The activity of the enzymes was then tested, and was recorded as a percentage of their activity before heating. The results are shown in Figure 3.7.

Figure 3.7

a Compare the effect of exposure to a temperature of 50 °C on the free and immobilised samples of pectinase. [3]

b Suggest *two* variables that should have been standardised when comparing the activity of the free and immobilised pectinase after exposure to 50 °C. [2]

c Explain how these results indicate that using immobilised pectinase could be advantageous for commercial manufacturers of fruit juice. [2]

d Use your knowledge of immobilised enzymes to list *two* other advantages of using immobilised pectinase rather than free pectinase. [2]

[Total: 9]

> Chapter 4
Cell membranes and transport

CHAPTER OUTLINE

The questions in this chapter cover the following topics:
- surface area:volume ratios
- the structure of cell membranes
- how substances move through membranes
- finding the water potential of plant tissues through experiment.

Exercise 4.1 Investigating the effect of surface area-to-volume ratio on the time taken for diffusion

A simple way of investigating how surface area-to-volume ratio affects the time taken for diffusion is to use blocks of agar with different surface areas and volumes. You need to be able to calculate surface areas, volumes and surface area:volume ratios. This exercise also gives you practice in constructing results charts and graphs, and identifying systematic and random errors.

1. A student made up a quantity of agar jelly by dissolving agar powder in slightly acidic water. She put a couple of drops of universal indicator solution into the water, which made the jelly red.

   The student used a ruler to measure the jelly, and cut cubes from it using a sharp scalpel. She cut cubes with sides of 4.0 cm, 3.0 cm, 2.0 cm, 1.0 cm and 0.5 cm.

   a. Calculate the surface area of each cube. Show your working.
   b. Calculate the volume of each cube. Show your working.
   c. Calculate the surface area:volume ratio of each cube. Show your working.
   d. Plot a graph of surface area:volume ratio (y-axis) against length of side of the cube (x-axis).

   Join the points with a smooth curve.

2. The student made three sets of cubes. She then placed one cube into a glass dish, and poured dilute sodium hydroxide solution around it. As the sodium hydroxide diffused into the agar cube, the colour changed to green and then blue.

The student measured the time for the cube to turn completely blue. Although the timer measured to 0.01 s, the student decided to measure the time to the nearest quarter of a minute. She repeated this with each cube.

These are the results she wrote down.

4 cm cube: 12.25 minutes, 12.5 minutes, 12 minutes
3 cm cube: 8.5 minutes, 9.25 minutes, 8.5 minutes
2 cm cube: 6 minutes, 5.75 minutes, 6 minutes
1 cm cube: 3.25 minutes, 3.5 minutes, 3 minutes
0.5 cm cube: 1 minute, 0.75 minutes, 1.25 minutes

a Do you agree with the student's decision to measure the time to the nearest quarter of a minute? Explain your answer.

b Construct a results chart, and record the student's results.

Remember:

- Draw the chart with a ruler.
- Head each column and/or row fully, including units.
- Make sure that each entry for a particular set of readings is given to the same number of decimal places, which is determined by the method of measuring the value.
- When calculating a mean value, the number of significant figures should be the same as (or one more than) the value with the least number of significant figures used in your calculation.

3 a Plot a graph of mean time taken for the colour to change (*y*-axis) against the surface area : volume ratio of the cube (*x*-axis).

b Use your graph to write a brief conclusion for the student's experiment. Your conclusion should summarise how the dependent variable is affected by the independent variable in the student's experiment.

c The student stated that the rate of diffusion – that is the distance travelled by the dye in a certain amount of time – was probably the same for all of the cubes. Do you agree with this statement?

4 The student was asked to identify significant sources of error in her experiment, and to state whether each one was random or systematic. She was told:

- **Systematic errors** are generally caused by lack of accuracy or precision in measuring instruments, and by the limitations of reading the scale. They tend to be always the same size, and act in the same direction, on all of the readings and results.
- **Random errors** are often caused by difficulties in keeping standardised variables constant. They can also be caused by difficulties in measuring the dependent variable. They may be of different sizes, and act in different directions, at different times or stages of the experiment.

These are the sources of error that the student listed. For each one:

- State whether you think this was likely to be a significant source of error or not.
- If you think it was a significant source of error, state whether it was a random or a systematic error.

> **KEY WORDS**
>
> **systematic errors:** uncertainties in experimental results that are caused by lack of precision or accuracy in a measuring instrument; they always act in the same direction
>
> **random errors:** uncertainties in experimental results that are caused by variations in variables that should be standardised, or by difficulties in measuring the independent or dependent variables; they can act in different directions for different parts of the experiment

a It was difficult to cut the agar blocks to the exact size.

b Temperature changes might have made diffusion happen faster or slower.

c One side of the block was resting on the base of the dish, so was not exposed to the sodium hydroxide solution.

d It was very difficult to decide exactly when the whole block had changed colour.

Exercise 4.2 Planning an experiment to determine how concentration affects diffusion rate

This exercise involves planning an experiment. You may need to look back at Chapter 2 to help you with some of the questions. You may be able to carry out your experiment after your teacher has checked your plan.

You are going to plan an investigation to find out how the concentration of a glucose solution affects the rate of diffusion.

The diagram shows the apparatus you can use. Visking (dialysis) tubing allows both water molecules and glucose molecules to move through it.

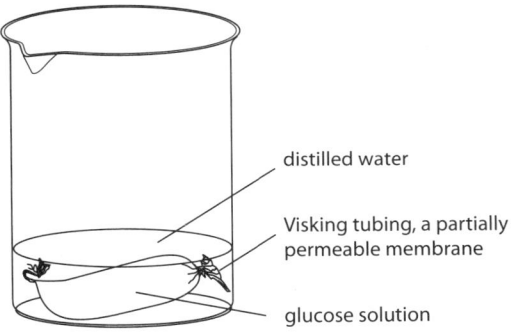

Figure 4.1: Apparatus for diffusion experiment.

1 Suggest a testable **null hypothesis** for this experiment.

2 Identify the independent and dependent variables.

3 You will be given a 10% solution of glucose.

   a Suggest the values of concentrations of glucose solution that you will use.

   b Describe how you will make up these concentrations of glucose solutions.

4 Which of the following variables should be standardised? For each variable that you decide should be standardised, explain how you would do this.

   a temperature of the solution and the water

   b light intensity

   c pH

   d time for which the apparatus is left.

> **KEY WORDS**
>
> **null hypothesis:** a hypothesis that assumes there is no relationship between two variables, or that there is no significant difference between two samples

5 You can measure your dependent variable by finding the concentration of glucose in the water outside the Visking tubing.

    a Explain how you can use the Benedict's test and a colorimeter to do this.

    b Suggest how you could estimate the concentration of glucose if you do not have access to a colorimeter.

6 Construct a results chart, with full headings, in which you could write your results. Remember to include spaces for replicates.

7 Draw and label the axes of the graph that you would plot. (You will not be able to draw a scale for the *y*-axis, because you cannot predict the values of the results.)

Exercise 4.3 Determining the water potential of a plant tissue

In this exercise, you will practise constructing good results tables, and will also think back to earlier work on calculating magnification.

A student carried out an experiment, using flower petals, to estimate the water potential of the cells in the epidermis of the petals.

He made up a range of sucrose solutions of different concentrations between 0 and 1 mol dm^{-3}. He cut small pieces of flower petal epidermis and immersed three pieces in each solution, and also in distilled water. He left all of the pieces for 20 minutes.

He then mounted each piece of tissue on a slide. He recorded the number of cells he could see that were **plasmolysed** and non-plasmolysed. He moved the slide a little and continued to count until he had recorded results for 100 cells for each tissue sample.

Figure 4.2 shows a light micrograph of a piece of the petal epidermis that was immersed in 1 mol dm^{-3} sucrose solution.

> **KEY WORD**
>
> **plasmolysed:** a cell where the cytoplasm and vacuole have shrunk so much that the cell membrane has pulled away from the cell wall

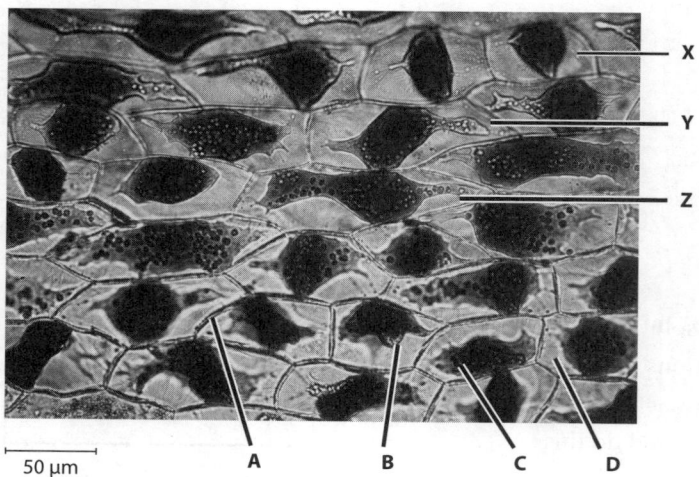

Figure 4.2: Plant epidermis cells after immersion in 1 mol dm^{-3} sucrose solution.

4 Cell membranes and transport

1 Use the scale bar to calculate the mean length of cells **X**, **Y** and **Z**. Show all of your working.

2 a Name the parts labelled **A**, **B** and **C** on the micrograph.

b What is present in the part labelled **D**? Explain your answer.

c State the percentage of cells in this sample that are plasmolysed.

d Explain why these cells are plasmolysed.

3 Table 4.1 is the table the student completed.

Number of plasmolysed cells	Number of non-plasmolysed cells	Concentration of solution
0	100	0
0	100	0
0	100	0
14	86	0.2
19	81	0.2
12	88	0.2
29	71	0.4
63	37	0.4
31	69	0.4
45	55	0.6
43	57	0.6
46	54	0.6
82	18	0.8
89	11	0.8
87	13	0.8
100	0	1.0
100	0	1.0
100	0	1.0

Table 4.1: Student results table.

a Which set of results could be omitted from the results chart, without losing any useful information?

b Identify the result that appears to be anomalous.

c Identify the mistakes that the student made in designing his results chart.

d Draw a fully correct results chart to show these results. (Remember that the anomalous result should be disregarded – it should not be included when calculating a mean.)

4 Plot a graph to display these results.

5   Figure 4.3 shows the relationship between the concentration of a sucrose solution and its water potential.

When a cell is just starting to become plasmolysed (known as incipient plasmolysis), the water potentials of the solutions outside and inside the cell surface membrane are equal. We can assume that, on average, each cell is at incipient plasmolysis when 50% of the cells in the tissue are plasmolysed.

Use the graph that you drew in **4**, and the graph below, to estimate the water potential of the cytoplasm of the epidermal cells of the flower petals.

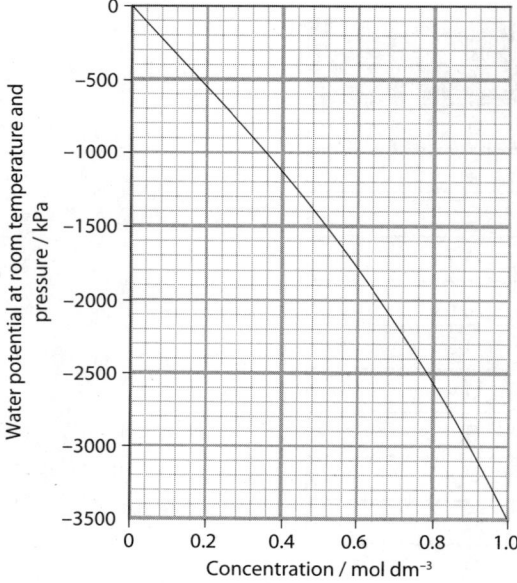

Figure 4.3: Relationship between concentration and water potential of a sucrose solution.

6   The student wanted to obtain a more precise value for the water potential of these cells. Which one of the following would be the best way to do this?

   A   Repeat the experiment using glucose instead of sucrose.
   B   Repeat the experiment using a wider range of sucrose concentrations.
   C   Repeat the experiment at a higher temperature, to speed up the rate of osmosis.
   D   Repeat the experiment using smaller intervals of sucrose concentration between the range 0.5 and 0.8 mol dm^{-3}.

Exercise 4.4 Comparing lipid composition of membranes

In Chapter 2, Exercise 2.4, you practised comparing data about the types of lipids found in different organisms. In this exercise, you will compare the types of lipids found in the cell membranes of normal red blood cells and those infected by the malarial parasite. First, though, you will check your understanding of the structure of a cell membrane and the roles of its various components.

4 Cell membranes and transport

1 Figure 4.4 represents a phospholipid molecule.

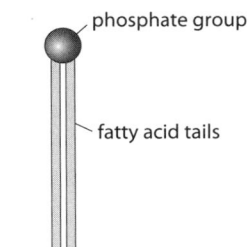

Figure 4.4: A phospholipid molecule.

  a Using this symbol for a phospholipid molecule, draw a diagram to show a phospholipid bilayer, as found in a cell membrane. Explain why the molecules arrange themselves like this when in a watery environment.
  b Explain why ions such as Na^+ and Cl^- are unable to move freely through a phospholipid bilayer.
  c Return to the diagram you drew for **a**. Add a channel protein to your diagram.
  d Explain how channel proteins can help ions such as Na^+ and Cl^- to move through the membrane.

When you are asked to make a comparison, it is a good idea to use comparative words, such as 'less' or 'more'. When you are asked to compare numbers, it may be useful to calculate a difference (e.g. by subtracting one number from another) or to calculate how many times greater one number is than another. In some cases, it might be appropriate or useful to calculate the percentage difference between two numbers. Note that in this question the numbers in Table 4.2 are already expressed as percentages.

2 Malaria is an infectious disease caused by a single-celled parasite called *Plasmodium*. This parasite enters red blood cells and replicates there. After infection, the composition of the lipids in the red blood cell membrane changes. Some of the differences between the uninfected and infected red blood cell membranes are shown in Table 4.2.

Molecule	Type of substance	Percentage in uninfected red blood cell membrane	Percentage in infected red blood cell membrane
phosphatidylcholine	phospholipid	31.70	38.70
sphingomyelin	phospholipid	28.00	14.60
palmitic acid	polyunsaturated fatty acid	21.88	31.21
oleic acid	monounsaturated fatty acid	14.64	24.60
arachidonic acid	polyunsaturated fatty acid	17.36	7.85

Table 4.2: Differences in composition of uninfected and infected red blood cell membranes.

  a Explain the difference between:
   - a phospholipid and a fatty acid
   - a polyunsaturated fatty acid and a monounsaturated fatty acid.
  b Compare the phospholipid composition of the uninfected and infected red blood cell membrane.
  c Compare the fatty acid composition of the uninfected and infected red blood cell membrane.

3 Red blood cells do not contain endoplasmic reticulum. Suggest how the changes in the composition of the infected red blood cell membrane are brought about.

EXAM-STYLE QUESTIONS

1 This question asks you to use your knowledge of the structure and function of cell membranes to interpret and explain data.

Figure 4.5 is a representation of the fluid mosaic model of membrane structure. The membrane shown is a cell surface (plasma) membrane.

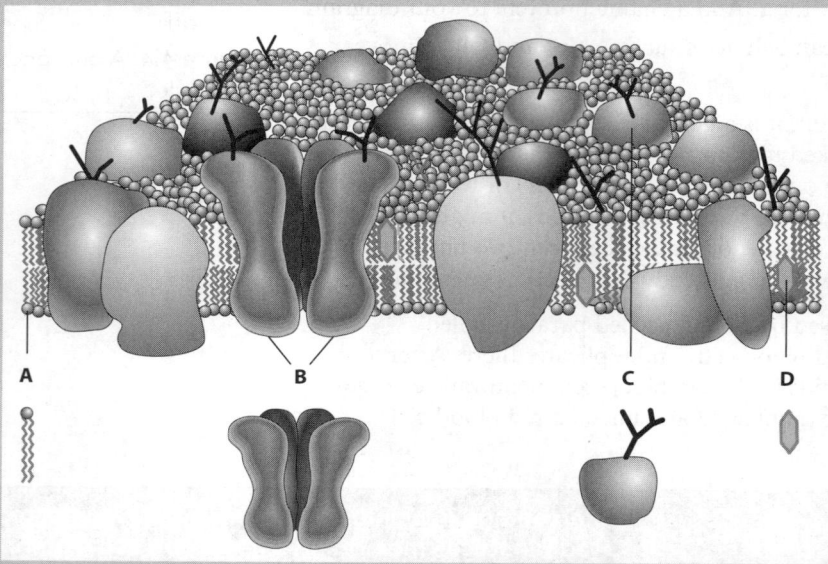

Figure 4.5

Table 4.3 shows the concentrations of several substances outside the cell (extracellular) and inside the cell (intracellular).

Substance	Extracellular concentration / mmol dm^{-3}	Intracellular concentration / mmol dm^{-3}
sodium ions, Na$^+$	150	15
potassium ions, K$^+$	5	150
amino acids	2	8
glucose	5.6	1.0
ATP	0	4

Table 4.3

a Name the types of molecule labelled **A**, **B**, **C** and **D** on Figure 4.5. [2]
b **Outline** how molecule **C** is involved in cell signalling. [3]
c **Explain** why sodium ions cannot diffuse through the part of the membrane made up of molecule **A**. [2]

COMMAND WORDS

Outline: set out the main points.

Explain: set out purposes or reasons / make the relationships between things evident / provide why and/or how and support with relevant evidence.

CONTINUED

d Sodium ions and potassium ions move through the membrane by active transport.
   i **Give** the letter of the part of the membrane through which these ions could move. [1]
   ii **Describe** the evidence in Table 4.3 that supports the statement that these ions move through the membrane by active transport. [2]

e Glucose is able to move through gated channels in the membrane by facilitated diffusion.
   i Explain the meanings of the terms gated channel and facilitated diffusion. [2]
   ii **Suggest** two reasons why, despite glucose being able to move through the membrane by facilitated diffusion, the concentration of glucose is not the same on the two sides of the membrane. [2]

f Explain why no ATP is found outside the cell. [2]

[Total: 16]

COMMAND WORDS

Give: produce an answer from a given source or recall/memory.

Describe: state the points of a topic / give characteristics and main features.

Suggest: apply knowledge and understanding to situations where there is a range of valid responses, in order to make proposals.

In Chapter 3, Exam-style question 2, you learnt how the presence of *Helicobacter pylori* in the stomach can be detected by testing for urease activity. For Question 2, you will need to combine knowledge about enzymes with understanding of how substances can pass through membranes.

2 The presence of the bacterium *Helicobacter pylori* can lead to the development of stomach ulcers. This bacterium produces the enzyme urease, which breaks down urea to carbon dioxide and ammonia. Sodium acetate is also a substrate for this enzyme. A patient suspected to have *H. pylori* in their stomach can be asked to swallow a tablet containing sodium acetate labelled with ^{13}C. Their breath is then tested for $^{13}CO_2$.

The effect of pH on the activity of urease from *H. pylori* was measured in intact bacteria and in a homogenate made from the bacteria. A homogenate is produced by liquidising the bacteria, so that the enzyme is released from the cells. Figure 4.6 shows the results.

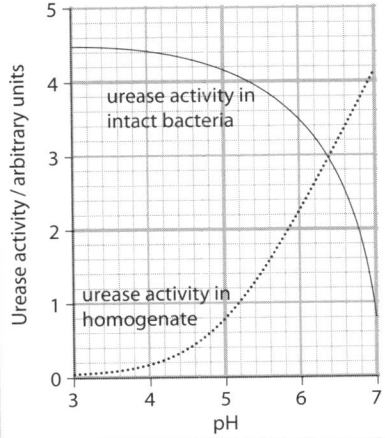

Figure 4.6

CONTINUED

a **Compare** the effect of pH on urease in intact bacteria and urease in a homogenate. [3]

b It has been suggested that urea enters bacterial cells by facilitated diffusion, and that the urea channel in the bacterial cell membrane is activated by low pH.

Explain how this could account for the differences that you have described in your answer to **a**. [3]

c Before taking the urease breath test, patients are often asked to drink apple juice, which contains a variety of organic acids. Suggest why this is done. [2]

[Total: 8]

COMMAND WORD

Compare: identify/comment on similarities and/or differences.

3 You should be getting confident now in describing graphs. In this question, as well as describing a graph, you also need to suggest explanations for a set of results and to consider whether these results support a particular hypothesis.

In birds, the main nitrogenous excretory product is uric acid, which forms salts called urates. Figure 4.7 shows an ion of urate.

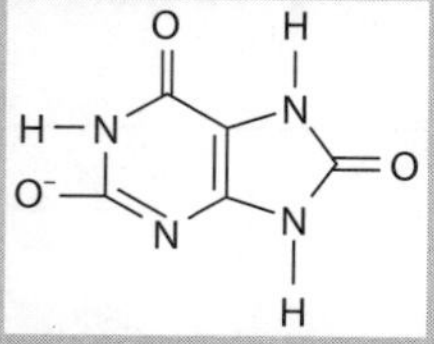

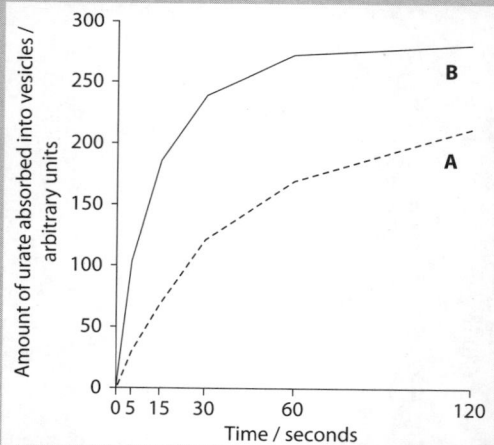

Figure 4.7

Urate is taken up by cells in the kidneys of birds.

Membrane-bound vesicles were isolated from turkey kidneys. One set of vesicles, set **A**, contained urate. The other set, **B**, did not. Both sets of vesicles were placed in a solution containing urate. The amount of urate taken up into the vesicles was measured over the next two minutes. The results are shown in Figure 4.8.

Figure 4.8

CONTINUED

a With reference to Figure 4.8, describe the effect of concentration gradient on the uptake of urate. [3]

b With reference to the procedure that was used for this experiment, explain why the uptake of urate cannot be by active transport. [3]

c **Assess** whether these results support the hypothesis that urate is taken into the vesicles by facilitated diffusion. Explain your answer. [2]

d It was found that urate was taken up more rapidly when there were excess potassium ions, K^+, inside the vesicles. Suggest reasons for this. [2]

[Total: 10]

> **COMMAND WORD**
>
> **Assess:** make an informed judgement.

4 You will often be provided with information in a question that requires you to work through it logically in order to arrive at an answer. In this question, you will use what you have learnt about the roles of cell membranes to explain events. You might find it useful to sketch diagrams or flow charts first, before then writing your answers.

a The cell surface membranes of the cells lining the walls of the small intestine contain ion channels through which chloride ions, Cl^-, can be actively transported out of the cells. When the channels are open, chloride ions accumulate outside the cells, causing water and sodium ions, Na^+, to also move out of the cells.

  i Explain what is meant by the terms ion channel and active transport. [2]

  ii Using the term water potential, explain how a build-up of chloride ions outside a cell causes water to move out of it. [3]

b The bacterium *Vibrio cholerae* causes cholera. When this bacterium gets into the small intestine, it produces a toxin that binds to receptors on the cell surface membranes of the cells in the intestinal lining. The cells then take up the toxin by endocytosis.

The uptake of the toxin sets off a cell signalling mechanism that results in the Cl^- channels remaining permanently open. The cells therefore lose a great deal of water, resulting in diarrhoea and dehydration.

  i Describe how a cell could take up the toxin by endocytosis. [2]

  ii Explain what is meant by a *cell signalling mechanism*. [3]

c The cell surface membranes also contain glucose–sodium co-transporter proteins. These use active transport to move glucose molecules and sodium ions into the cells from the lumen of the intestine. People with cholera are often treated by giving them drinks containing glucose and sodium ions. This results in the cells reabsorbing water from the small intestine, reducing diarrhoea and speeding recovery.

Suggest how a drink containing glucose and sodium ions can cause the cells in the intestine to reabsorb water. [3]

[Total: 13]

> Chapter 5
The mitotic cell cycle

> **CHAPTER OUTLINE**
>
> The questions in this chapter cover the following topics:
> - the structure of chromosomes
> - the stages of the cell cycle, including the interpretation of diagrams and micrographs
> - the roles of mitosis
> - telomeres and stem cells.

Exercise 5.1 Interpreting a micrograph showing mitosis

You need to be able to recognise the stages of mitosis from micrographs and microscope slides. This exercise also gives you practice in calculating real dimensions when given the magnification of an image. Look back at Chapter 1 if you need reminding how to do this.

Figure 5.1 is a photomicrograph of a stained root tip squash of broad bean, *Vicia faba*.

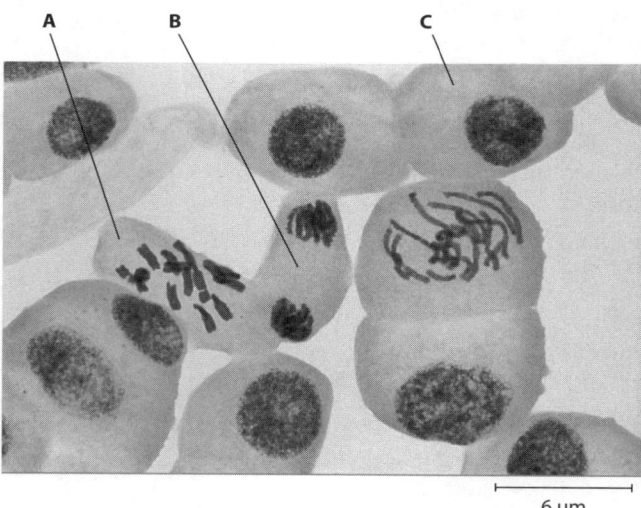

6 µm

Figure 5.1: Light micrograph of root tip squash of *Vicia faba*.

1   Count the number of chromosomes in cell **A**, and use your result to state the diploid number of chromosomes in *Vicia faba*.

> **TIP**
>
> Remember that a diploid number must be an even number.

56

2 Give the letter of a cell that is:

   a in interphase

   b in prophase

   c in telophase.

3 Measure the maximum width of cell **C** on the micrograph. Use the scale bar to calculate the actual width of this cell. Show all of your working.

Exercise 5.2 Answering multiple-choice questions on the cell cycle

In this exercise, you are encouraged to think about how you can maximise your chance of getting a multiple-choice question correct. The example question is about what happens during anaphase.

Ex. What happens during anaphase of mitosis?

   A Chromatids line up on the equator of the spindle.

   B Chromosomes/Chromatids are pulled to opposite ends of the cell.

   C Spindle fibres pull chromatids towards the poles of the spindle.

   D Nuclear membranes begin to reform.

Option A is not correct, because it is whole chromosomes (each made of two chromatids joined together) that line up on the equator. Option B is not correct, because each chromosome is pulled apart into its two chromatids during anaphase, and it is these individual chromatids that are pulled to opposite ends of the cell. D is not correct, because nuclear membranes do not reform until late telophase.

The correct answer is C. However, when this question was set as part of a test, many students answered B. This is probably because they were in a hurry, and gave the first answer that seemed to be correct, without taking the time to read all the options.

You may find the following approach helps you when answering multiple-choice questions:

Step 1 Read the question very carefully (before looking at the options) to make sure you understand exactly what it is asking you. Many incorrect answers are a result of skim-reading the question too quickly.

Step 2 Next, read all of the options carefully, or study diagrams in detail.

Step 3 Decide if any of the options are definitely not correct.

Step 4 Now concentrate on the remaining options, and consider each one carefully. Think about the exact wording of each one (chromosomes or chromatids?) – or exactly what is shown in the diagrams if the options are illustrations – and make your decision.

Step 5 Never leave a multiple-choice question unanswered. If you cannot decide between two options, just write down one of them.

Now apply this technique to the following multiple-choice question.

1 Figure 5.2 shows the chromosomes in a cell in prophase of mitosis.

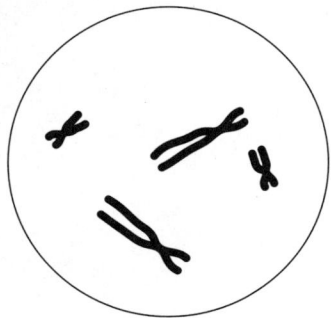

Figure 5.2: Chromosomes during prophase.

Now look at Figure 5.3. Which is the correct diagram that shows the cells that would be produced after cytokinesis from the cell in Figure 5.2?

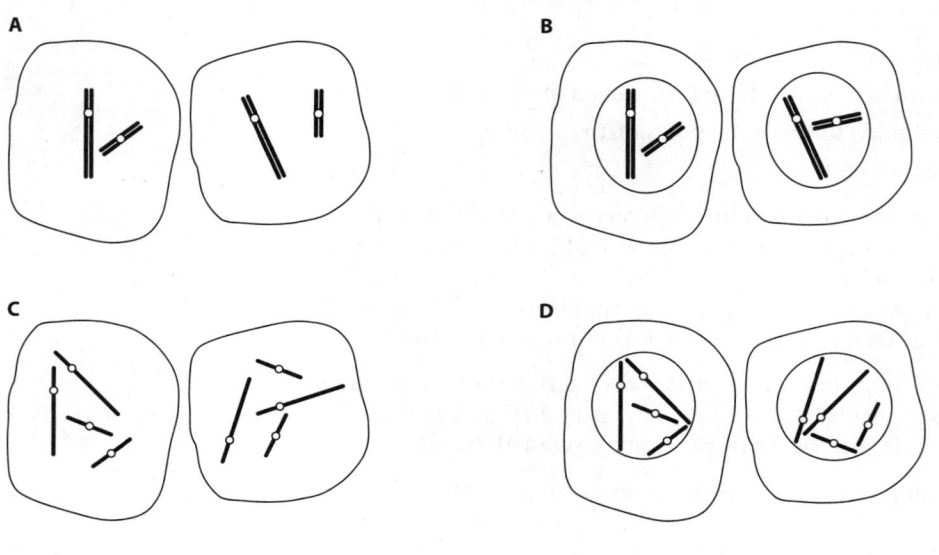

Figure 5.3: Multiple-choice options.

TIP

In deciding which answer is correct, ask yourself: what are the differences between the diagrams?

Some show nuclear membranes and some do not – which is correct?

Some show chromosomes made up of two chromatids and some show separate chromatids – which is correct?

2 Writing your own multiple-choice questions can be a good way to help you to think clearly about the possible answers.

Here are two sets of options for multiple-choice questions. Decide which will be the correct answer, and then write the question. Try out your questions on someone else in the class.

Option set 1

　A centriole
　B centromere
　C chromatid
　D chromosome

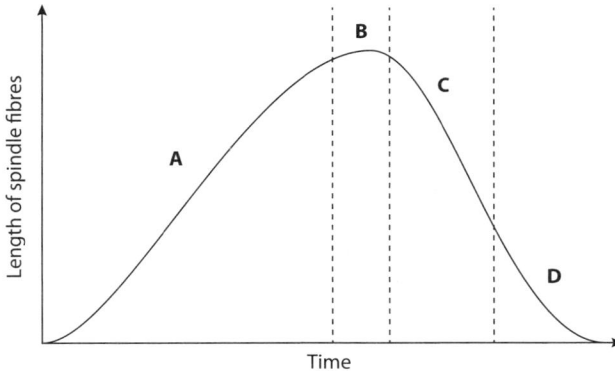

Figure 5.4: Option set 2.

Exercise 5.3 Tally counts and testing a hypothesis

KEY WORDS

tally count: recording numbers by making a mark for each item counted, using a diagonal mark for each fifth item

Tally counts are a good way of counting a lot of objects. You simply draw a tally mark for each object counted, and put a cross through a group of four marks to represent the fifth. Later you can convert the tally marks to numbers. Before you answer question **1b**, make sure that you know which stages of the cell cycle are part of mitosis and which are not.

1   A student made a stained root tip squash of a bean plant, *Vicia faba*, following this procedure:

- She cut off the tip of a bean root, and placed it in acidified acetic orcein stain in a watch glass. This stain colours DNA red.

- She warmed the watch glass gently for five minutes to allow the acid to separate the cells in the root tip.

- She put the stained root tip onto a clean microscope slide and cut it in half, discarding the part nearest the tip – this part does not contain dividing cells.

- She added two drops of acetic orcein to the remaining root tip, and gently broke it apart and spread the root tissues out using a mounted needle.

- She covered the root tip with a coverslip, wrapped a piece of filter paper around it and very gently tapped it with the blunt end of a pencil until the tissue had spread out into a very thin layer.

- Finally, to intensify the stain, she moved the root tip squash preparation quickly through a Bunsen flame a few times, holding the slide with her fingers to make sure she did not let it get too hot.

- She observed her preparation using a light microscope. She counted the number of cells in each stage of the cell cycle in several different fields of view and recorded her results in a tally chart (Table 5.1).

Stage of cell cycle	Tally count
interphase	ℳ ℳ ℳ ℳ ℳ ℳ ℳ ℳ ℳ ℳ ℳ ℳ III
prophase	ℳ ℳ ℳ III
metaphase	ℳ ℳ I
anaphase	ℳ ℳ
telophase	ℳ
cytokinesis	III

Table 5.1: Tally chart.

a How many cells did the student count in total?
b Calculate the percentage of cells that were in a stage of mitosis.
c Display the student's results as a **bar chart**.
d The student wanted to test this hypothesis:

   The cell cycle of Vicia faba follows a diurnal (daily) rhythm, in which mitosis is more likely to take place at some times of day than at others.

   Outline how the student could test this hypothesis. In your plan, you should refer to:

   - the independent variable: what it is, how the student can vary it, and what values she should use
   - the dependent variable: what it is, how the student can measure and record it
   - variables that should be standardised (kept the same) and how this can be done
   - if replicates should be done
   - how the results will be recorded, processed and displayed
   - how the results can be used to draw a conclusion.

KEY WORDS

bar chart: a type of graph drawn where the *x*-axis variable is discontinuous; bars do not touch

TIP

Remember:

In a bar chart, the bars should not touch.

Draw each bar very accurately, using a ruler and a sharp HB (medium hard) pencil.

Do not shade the bars.

Exercise 5.4 Factors affecting telomere length

Investigating how a particular lifestyle factor affects humans is very difficult, because we cannot usually do controlled experiments. Instead, researchers measure particular variables in a large sample of people, and look for correlations between them. This means that we are rarely able to prove that one variable actually causes a change in another. In this exercise, you will practise describing and explaining a graph, and also discuss the strength of the evidence to support a conclusion made by researchers.

5 The mitotic cell cycle

A chromosome is a very long molecule of DNA, and DNA molecules are made up of nucleotides linked together into long chains. Each nucleotide contains one of four bases: A, C, T or G.

Telomeres are protective caps of short, repeating DNA sequences on the ends of chromosomes. Each time the chromosomes replicate just before cell division, the telomeres get a little shorter. Eventually, they become so short that the cells can no longer divide. It is thought that loss of telomeres is a significant contribution to the effects of ageing, because tissues are not able to repair and renew themselves effectively.

1 A study was carried out to find the effects of age, obesity and smoking on the length of telomeres in humans.

   Figure 5.5 shows the results of a study that measured the mean telomere length in the cells of 1122 women between the ages of 18 and 76. The length of the telomeres was measured in kilobases (kb). (The line was drawn after doing a calculation called linear regression, which uses the individual data points to predict the overall relationship between the two variables.)

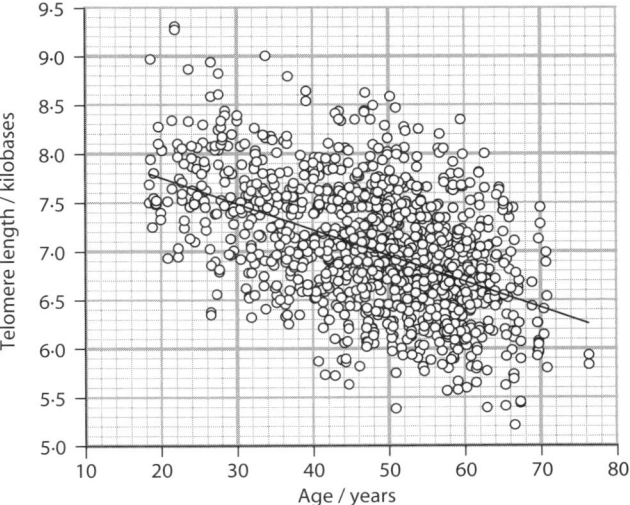

Figure 5.5: Results of a study into age and telomere length.

   a Describe the pattern shown in the graph.
   b Explain the reasons for the patterns that you have described in **a**.

> ### TIP
>
> Remember, when answering **a**:
>
> - If you are asked to 'describe', you should not try to give any reasons or explanations. You are simply translating what you can see in the graph into words.
> - Start by describing the general trend – what happens to the *y*-axis variable as the *x*-axis variable increases?
> - Make a comment about how strong this trend is – in this case, you should comment on the wide variation around the line that has been drawn.
> - Quote figures e.g., the range of telomere lengths at a particular age, or the telomere length indicated by the line at age 18 and age 76.
> - Use the figures to do a calculation that provides information about the changes with age e.g., the mean change in telomere length as a person ages by 10 years.
> - For all the figures that you quote, remember to state the units.

2   Figure 5.6a shows the relationship between BMI and mean telomere length. BMI stands for body mass index, which is a measure of a person's mass in relation to their height. In general, a person with a BMI between 18.5 and 25.0 kg m^{-2} is considered to be of normal weight, while BMIs above or below that range indicate being overweight or underweight.

Figure 5.6b shows the relationship between smoking and mean telomere length.

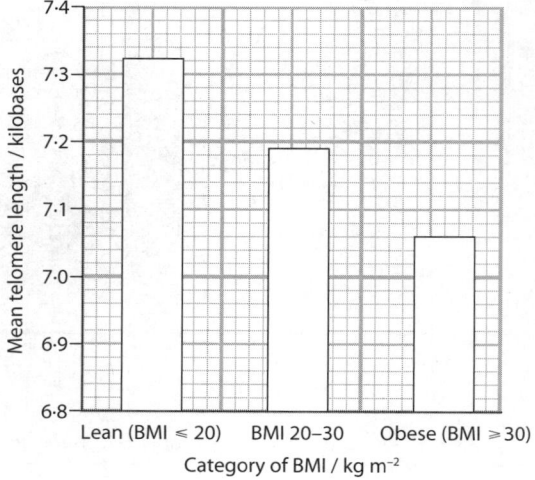

Figure 5.6a: Results of a study into BMI and telomere length.

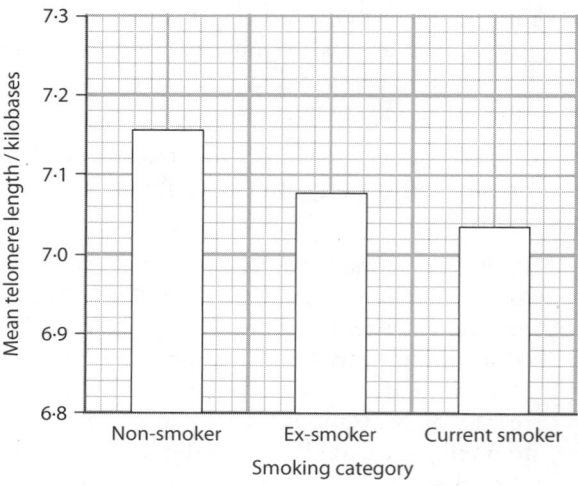

Figure 5.6b: Results of a study into smoking and telomere length.

The researchers suggested that their results show that obesity and smoking accelerate ageing. Discuss the extent to which the data in Figures 5.6a and 5.6b support this idea.

> ### TIP
>
> Remember:
>
> - When you are asked to 'discuss', you should try to make points on both sides of the argument. Here, you should state evidence that supports the researchers' suggestion, and also state the shortcomings in this evidence.
>
> - To answer this question, you will need to pull together information from Figures 5.6a and b, and also use some of the information in the paragraph about telomeres and ageing at the beginning of the question.
>
> - Note the points you want to make – you could have two columns, one for 'supports the suggestion' and one for 'does not support / more evidence needed' – and sort out your ideas into a sensible sequence before you begin to write your answer.

5 The mitotic cell cycle

EXAM-STYLE QUESTIONS

1 This question tests your understanding of chromosome structure. Each part requires only a short answer.

   a A chromosome is a structure made up of a long molecule of DNA, associated with proteins. When a cell is not dividing, the chromosomes are not visible with an optical (light) microscope.

   i Name the type of proteins that are associated with DNA to form a chromosome. [1]

   ii **Explain** why chromosomes are not visible with an optical microscope when a cell is not dividing. [2]

   b The diagram shows the structure of a chromosome, just before the cell enters mitosis.

   i Name the parts of the chromosome labelled **A** and **B**. [2]

   ii Make a copy of Figure 5.7, and label the parts of the chromosome where telomeres would be found. [1]

   iii **Outline** the function of telomeres. [2]

   [Total: 8]

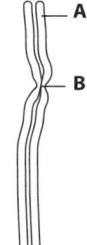

Figure 5.7: A chromosome.

COMMAND WORDS

Explain: set out purposes or reasons / make the relationships between things evident / provide why and/or how and support with relevant evidence.

Outline: set out the main points.

2 This question tests your knowledge of the roles of mitosis, and your ability to interpret diagrams of cells in different stages of the mitotic cell cycle.

   a Outline the functions of mitosis in a multicellular organism such as a human. [3]

   b Figure 5.8 is a pie chart representing the mitotic cell cycle. The diagrams in Figure 5.9 show cells in various stages of the cell cycle.

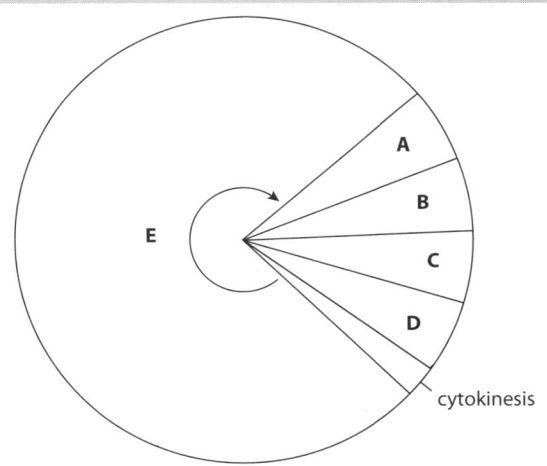

Figure 5.8

CONTINUED

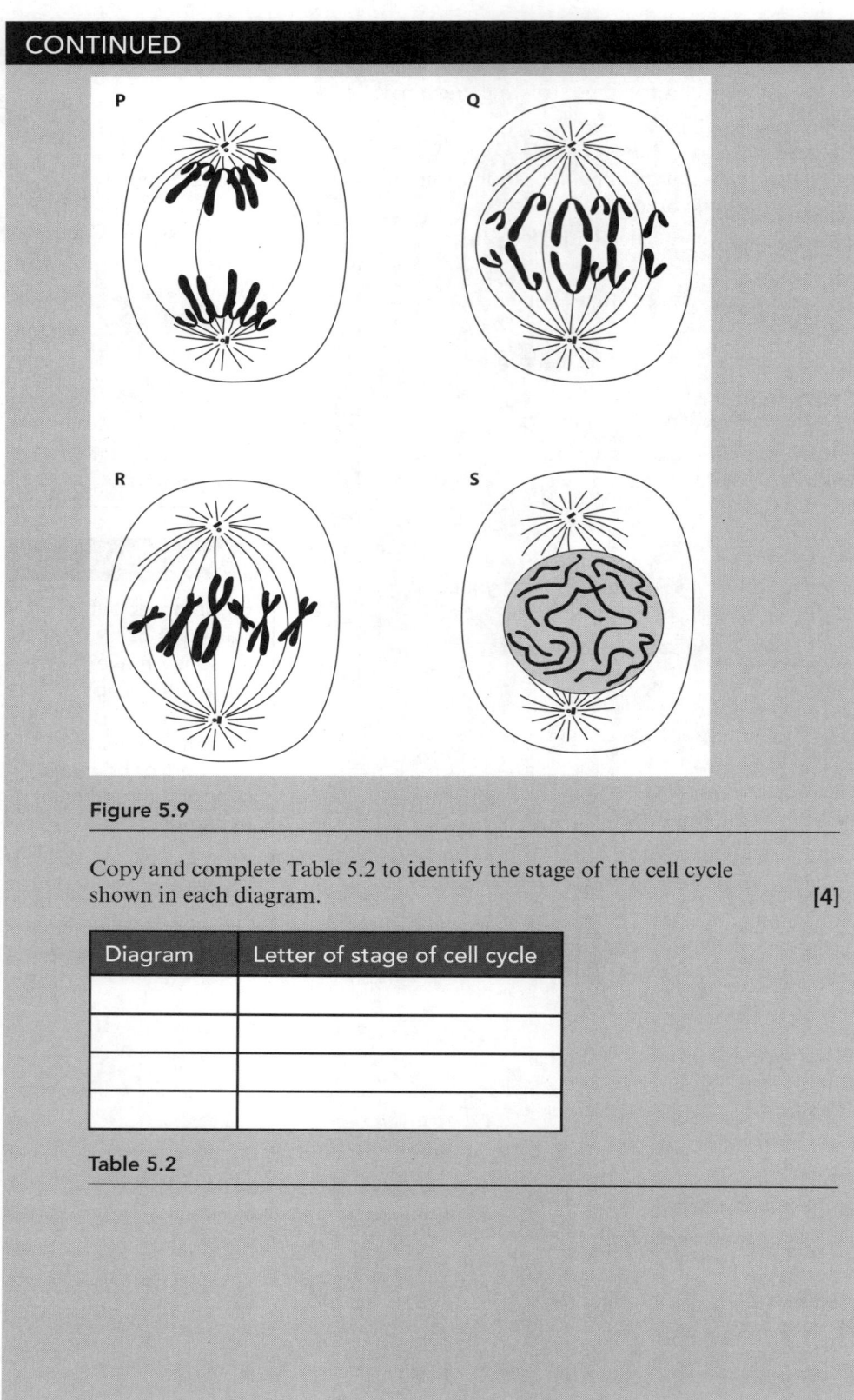

Figure 5.9

Copy and complete Table 5.2 to identify the stage of the cell cycle shown in each diagram. [4]

Diagram	Letter of stage of cell cycle

Table 5.2

CONTINUED

c During a normal cell cycle, a cell will not continue with mitosis if the spindle fibres are not attached correctly to the centromeres. Several proteins, known as spindle checkpoint proteins, are involved in this control process. In the cells in cancer tissue, these spindle checkpoint proteins may not work properly.

Suggest how this can explain the observation that many cells in cancer tissue contain *more* or *fewer* chromosomes than normal cells. [3]

[Total: 10]

3 In this question, you first practise your ability to describe a graph, and to compare two sets of data. Then you are asked to apply your knowledge of stem cells and telomeres in an unfamiliar context – a challenging question which is a good test of how well you understand these two topics.

The human heart is made of cardiac muscle tissue. This tissue contains a small proportion of stem cells. These cardiac stem cells are thought to be able to produce new cells to replace damaged cardiac muscle tissue.

a Explain what is meant by the term *stem cell*. [2]

b Chronic (long-term) heart failure occurs when the heart is not able to pump blood effectively. This may be caused by damaged cardiac muscle tissue.

Data were collected about the telomere length in cardiac stem cells in the hearts of people who had chronic heart failure and people with healthy hearts. The results are shown in Figure 5.10.

COMMAND WORD

Suggest: apply knowledge and understanding to situations where there is a range of valid responses, in order to make proposals / put forward considerations.

CONTINUED

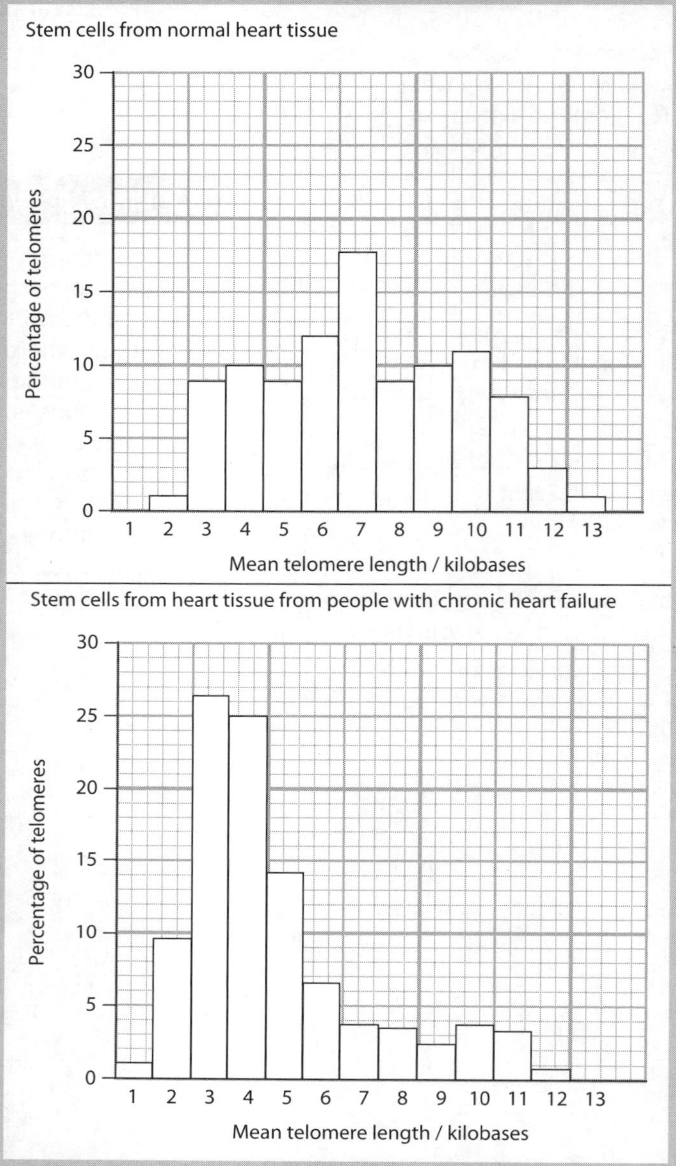

Figure 5.10

i Describe the distribution of telomere lengths in stem cells from healthy human hearts. [2]

ii **Compare** the telomere lengths in stem cells from people with chronic heart failure and people with healthy hearts. [3]

iii Suggest explanations for the differences you have described in your answer to ii. [4]

[Total: 11]

> **COMMAND WORD**
>
> **Compare:** identify/comment on similarities and/or differences.

> Chapter 6
Nucleic acids and protein synthesis

CHAPTER OUTLINE

The questions in this chapter cover the following topics:
- the structure and replication of DNA
- the structure and roles of RNA
- how proteins are synthesised in cells
- mutations.

Exercise 6.1 Answering multiple-choice questions about DNA

In Chapter 5, we looked at an approach for answering multiple-choice questions. Here are some more examples for you to try.

TIP

Some questions are more complex than others, so it is a good idea to answer the straightforward ones first, and then go back and tackle the more complex ones that need more time.

Set your phone or a timer to time one minute. Press 'start' and then sit motionless and silent until the minute is over. That's how long you should try to spend on a simple multiple-choice question. You can allow more time for more complex questions, or you can work out how much time on average you have to answer each question, based on the total time available.

Look at Question 1. Don't rush – the aim at this point is to get the right answer, however long it takes. Read the question, then follow the steps to help you to arrive at an answer.

1   Table 6.1 shows the percentages of each type of base in the DNA of cells from different organisms.

Organism	Percentage of base			
	adenine	cytosine	guanine	thymine
human	30.9	19.8	19.9	29.4
grasshopper	29.3	20.7	20.5	29.3
yeast	31.3	17.1	18.7	32.9
sea urchin	32.8	17.3	17.7	32.1
wheat	27.3	22.8	22.7	27.1

Table 6.1: Percentages of each type of base in the DNA of cells from different organisms.

What can be concluded from the data shown in Table 6.1?

A   Plants have less adenine and thymine than animals have.

B   Small differences in the base sequences in DNA have large effects on the characteristics of an organism.

C   There is approximately the same quantity of DNA in the cells of all organisms.

D   The total percentage of purines is approximately the same as the total percentage of pyrimidines.

Step 1   First, look carefully at the table of data. Make sure you understand what it is showing.

Step 2   Now look at the question. Note that it is asking what can be 'concluded from these data', so you need to think about what the data are showing, not anything else that you might happen to know.

Step 3   Now for the options. Start with option **A**. It is about plants – which entries in the table are plants? There is only one, so we really can't conclude anything about plants. Discard that option.

Step 4   Now look at option **B**. You know that statement is true, but can we conclude it from the data?

Step 5   Now option **C**. Look at the headings of the table. Do the data tell us anything about the actual quantity of DNA in the cells in the organisms?

Step 6   Now look at option **D**. You'll need to remember which bases are purines and which are pyrimidines. (In case you've forgotten, A and G are purines, and C and T are pyrimidines.) Test it out on a couple of rows in the table.

So, which of the statements listed are correct?

For Questions 2 to 6 that follow, you might like to underline any words in the information or question that seem to be particularly important. There are also some suggestions in the margin that might help you.

2   Which of the following statements describes the replication of DNA?

A   It copies information from DNA to mRNA.

B   It is semi-conservative.

C   It produces tRNA.

D   It takes place at a ribosome.

> **TIP**
>
> What are the important words to concentrate on in this question?

3 A sample of DNA is analysed to determine the relative proportions of its components. What would you expect in the results?

   A equal proportions of phosphates, pentose sugars, adenine, guanine, cytosine and thymine
   B equal proportions of purines and pyrimidines, and half the number of sugars and phosphates
   C twice as many bases as phosphate and pentose sugars
   D twice as many phosphates and bases as pentose sugars.

TIP

Try making a rough sketch of a small part of a DNA molecule before working through each option in turn.

4 This list shows the events that take place during DNA replication.

   1 Free bases pair up with exposed bases on each DNA strand.
   2 The hydrogen bonds between the bases are broken.
   3 DNA polymerase links adjacent pentose and phosphate groups.
   4 Bonds form between complementary bases.

   In what order do these events take place?

   A 1, 4, 3, 2
   B 2, 1, 4, 3
   C 3, 2, 1, 4
   D 4, 1, 2, 3

TIP

Sort the four events into the correct order yourself, before looking at the options.

5 Figure 6.1 shows two bases, P and Q, linked by hydrogen bonds in a DNA molecule.

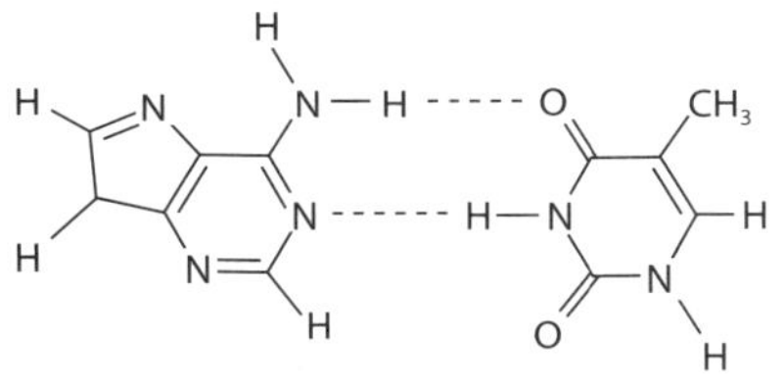

Figure 6.1: Two bases linked by hydrogen bonds.

Table 6.2 shows four possible answers for the correct identities of P and Q shown in Figure 6.1. Which is the correct combination?

	P	Q
A	adenine	thymine
B	cytosine	guanine
C	guanine	thymine
D	thymine	adenine

Table 6.2: Possible identities of P and Q.

TIP

You are not expected to know the molecular structures of the four bases, so you probably can't work the answer out from that. But you should know that purines have a double ring structure, so you can tell which of the two bases is a purine. You should also know that adenine and guanine are purines. And lastly, you should know which bases pair together.

6   Nitrogen exists as two different isotopes, ^{14}N and ^{15}N. Bacteria were grown in a medium containing ^{14}N, for many generations. They were then transferred to a medium containing only ^{15}N.

Which curve in Figure 6.2 shows the percentage of the bacterial DNA strands that contain ^{14}N in the next five generations?

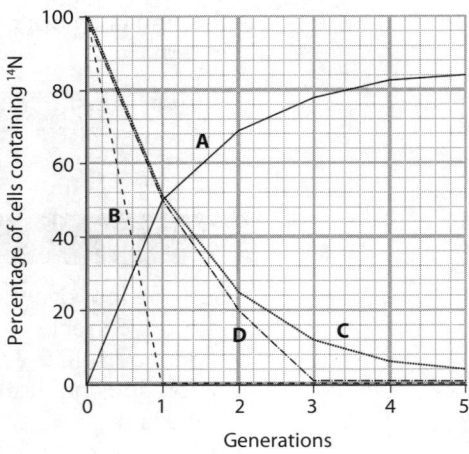

TIP

Try drawing a diagram of the original DNA – two strands, both all ^{14}N. Then draw the two molecules that will be produced when this DNA replicates in ^{15}N – remembering that it replicates semi-conservatively. Then draw the four molecules that will be produced when each of these two replicate in ^{15}N. That should help you to see which curve is correct.

Figure 6.2: Graph showing ^{14}N content in several generations of bacteria.

Exercise 6.2 Writing good answers to questions

Read the sample question 1 which is about normal and sickle cell haemoglobin. Then read three example answers. Review these answers and then answer the questions.

1   The sequences of DNA coding for the first seven amino acids in the gene for normal haemoglobin and the gene for sickle cell haemoglobin are as follows:

DNA for normal haemoglobin   CAC GTG GAC TGA GGA CTC CTC
DNA for sickle cell haemoglobin   CAC GTG GAC TGA GGA CAC CTC

Table 6.3 shows some of the DNA triplets that code for six different amino acids.

Use the information to describe the similarities and differences between the structure of a normal haemoglobin molecule and a sickle cell haemoglobin molecule. [3]

DNA triplet	Amino acid
CTC, CTT	glutamic acid
GTA, GTG	histidine
AAC, GAC	leucine
GGA, GGC	proline
TGA, TGG	threonine
CAA, CAC	valine

Table 6.3: DNA triplets.

Example answer X

The two kinds of haemoglobin have the same triplet code for their first 5 amino acids and the seventh amino acid. However, they differ in their sixth amino acid code CTC : CAC. This is the result of a substitution mutation. This shows that while the molecules remain the same, their base sequences can vary.

Example answer Y

The sickle cell haemoglobin has CAC instead of CTC, so it has a valine instead of a glutamic acid. This makes the haemoglobin molecules a different shape, so when the person is short of oxygen the haemoglobin molecules stick together and make the red blood cells into a sickle shape. This is very painful and the person is said to be having a sickle cell crisis.

Example answer Z

Both kinds of haemoglobin have a sequence of five amino acids that are identical, which is valine, histidine, valine, threonine and proline. The sixth amino acid is different; it is glutamic acid in the normal haemoglobin and valine in the sickle cell haemoglobin. This change in the primary structure will affect the way that the molecule folds, and will therefore affect its tertiary structure.

a   Example answer X does not answer the question that was asked. Explain what it has done incorrectly.

b   Example answer Y gives information that is correct, but that is not required by the question. Which information is this?

c   Example answer Z is the best answer, but there is one mistake. Correct the mistake.

Exercise 6.3 Making links across topic areas

Questions often require you to think about several different areas of the syllabus, moving from one topic to another. In this exercise, you will think about protein structure, the cell cycle and the involvement of DNA in protein synthesis.

1   With reference to the structure of haemoglobin, explain what is meant by each of the following terms:

   a   globular protein
   b   quaternary structure.

2   The gene for the α polypeptide of haemoglobin is found on chromosome 16, and the gene for the β polypeptide is found on chromosome 11.

   How many copies of the α polypeptide gene would be present in a cell at the following stages of the cell cycle:

   a   early interphase
   b   prophase?

3 The gene for the β polypeptide occurs in several different forms (alleles) with slightly different base sequences. This results in different amino acid sequences in the β polypeptide that is synthesised. One of these alleles, Hb^S, differs from the normal allele, Hb^A, in only one base, but the protein that is produced behaves so differently that it causes a genetic disease called sickle cell anaemia. This disease can be fatal if medical treatment is not available.

Outline how a difference in a single base in DNA can result in a form of a protein that behaves very differently from the normal protein.

4 The allele Hb^S is especially common in populations of people living in areas where malaria is present.

An investigation was carried out in a rural part of Kenya, where malaria is common, to find out whether children who had one copy of this allele and one copy of the normal allele (Hb^AHb^S) had a better chance of survival than those who had two copies of the normal allele (Hb^AHb^A) or two copies of the sickle cell allele (Hb^SHb^S).

The results are shown in Figure 6.3.

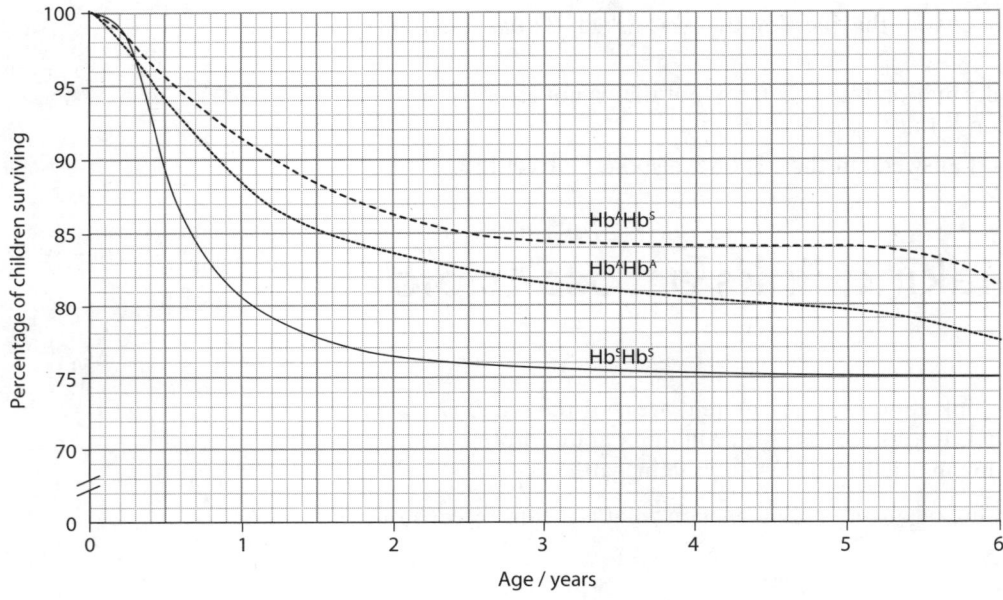

Figure 6.3: Survival curves for children with different genotypes for the haemoglobin β polypeptide gene.

a Describe the survival pattern of the children with two copies of the normal allele, Hb^AHb^A.

b Compare the survival pattern of the children with two copies of the sickle cell allele, Hb^SHb^S, with that of the children with two copies of the normal allele.

c These results support the hypothesis that having one copy of the sickle cell allele increases the chances of survival in an area where malaria is present. Suggest further investigations that could be carried out, or more information that could be obtained, to attempt to strengthen the evidence for this hypothesis.

> **TIP**
>
> Think carefully about the three different command words in the three parts of question 4 – describe, compare, suggest. Make sure you are trying to respond to these appropriately in your answers.

EXAM-STYLE QUESTIONS

1 Tyrosinase is an enzyme that catalyses reactions involved in the production of melanin, a brown pigment in skin that gives protection from the damaging effects of sunlight.

Figure 6.4 outlines how tyrosinase is produced in human cells.

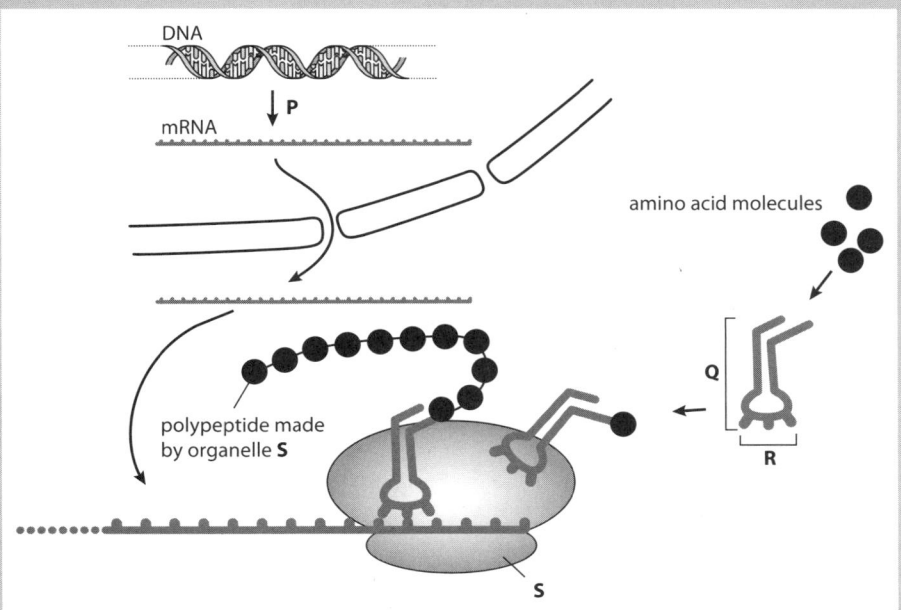

Figure 6.4

a i Name process **P**. [1]
  ii Name molecule **Q**. [1]
  iii Name organelle **S**. [1]
  iv **Outline** the importance of structure **R** in the synthesis of tyrosinase. [3]

b In some people, a mutation results in an altered base sequence of the gene for tyrosinase, *TYR*. In one such mutation, a nucleotide containing the base cytosine is deleted from the DNA molecule that makes up this gene.

Explain why a person with this mutation is unable to make melanin. [4]

[Total: 10]

> **TIP**
> The introduction to question 1 will make you think about enzymes, but in fact the answers you are asked to write are about protein synthesis and mutations. Only part **b** requires any information about enzymes, and even then it is quite a minor part of the question.

> **COMMAND WORDS**
> **Outline:** set out the main points.
>
> **Explain:** set out purposes or reasons / make the relationships between things evident / provide why and/or how and support with relevant evidence.

CONTINUED

> **TIP**
>
> Here is another question that crosses between several different areas of the syllabus. You may remember something about insulin from your previous studies, but don't worry if you do not – all the information that you need to know about it is given in the question.

2 Insulin is a protein synthesised in β cells in the pancreas. Insulin is a hormone that brings about actions in liver cells that reduce blood glucose concentrations. A lack of functioning insulin results in diabetes.

This is the base sequence of part of the mRNA coding for the formation of insulin extracted from β cells:

GGUAUCGUGCAAUGUUGCACUUCCAUU

a i Write the nine base triplets in the DNA from which this mRNA was transcribed. [2]

   ii Write the anticodons of the tRNA that will form temporary bonds with the mRNA during translation at a ribosome. [2]

b Insulin does not enter liver cells, but brings about its effects by binding with a receptor on the liver cell membrane. This causes phosphate groups to bind with certain proteins within the cell which, in turn, leads to a series of events that increase the number of glucose transporter proteins in the cell membrane.

   i With reference to the structure of the cell surface membrane, **suggest** why insulin molecules are not able to enter liver cells. [2]

   ii **State** the type of molecule in a cell membrane that could act as a receptor for insulin. [1]

   iii Explain how an increase in the number of glucose transporter proteins in the liver cell membrane can result in a decrease in blood glucose concentration. [2]

c In some people, a mutation in one triplet of the insulin gene causes leucine to be incorporated into the protein instead of valine. This insulin is unable to bind with its receptor.

   i Explain what is meant by the term *mutation*. [2]

   ii Explain how this mutation could result in a form of insulin that is unable to bind with its receptor. [3]

   iii People with a chromosome containing this mutation do not normally show any symptoms of diabetes. Suggest why this is so. [2]

[Total: 16]

> **COMMAND WORDS**
>
> **Suggest:** apply knowledge and understanding to situations where there is a range of valid responses, in order to make proposals / put forward considerations.
>
> **State:** express in clear terms.

CONTINUED

3 a i State how the structure of ribosomes differs between prokaryotic and eukaryotic cells. [1]

 ii State how the distribution of ribosomes differs between prokaryotic and eukaryotic cell. [2]

b Figure 6.5 shows a ribosome, on which the synthesis of a protein is taking place.

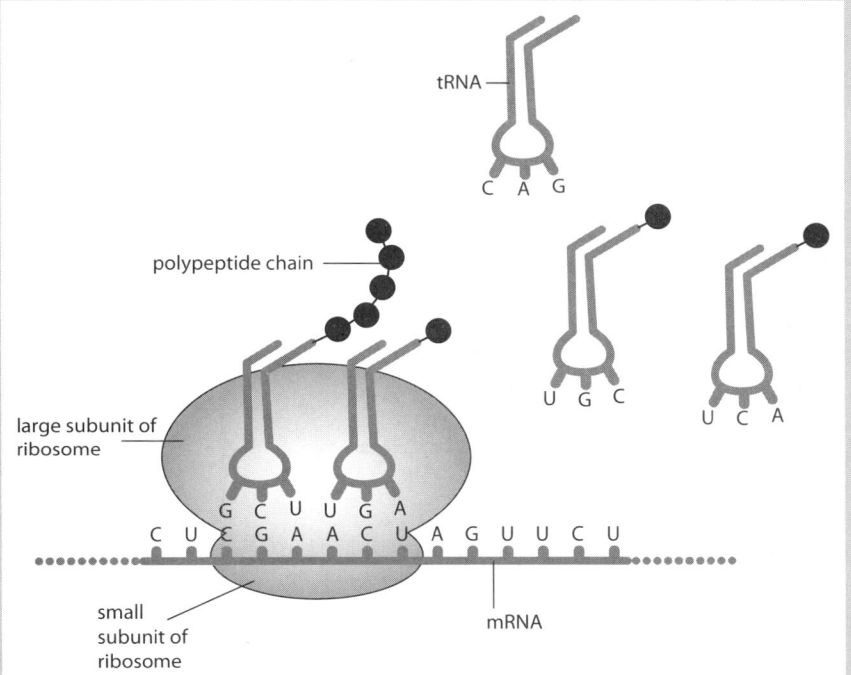

Figure 6.5

Outline the roles of the ribosome in the synthesis of a protein. [4]

TIP

This question contains one 4-mark section and one 5-mark section. These can be difficult to deal with. Common errors are to include material that goes beyond what the question is asking, or saying the same thing more than once rather than moving on to something new in your answer. You should make sure that you give at least four or five (preferably five or six) different, strong, relevant points in your answer.

CONTINUED

c An investigation was carried out to track the movement of amino acids through different organelles in a cell. Pancreatic cells were provided with a small quantity of amino acids containing some tritium atoms, ^3H, instead of normal hydrogen atoms. Tritium is radioactive, and molecules containing it can be detected, so the amino acids containing tritium are said to be 'labelled'. After a short time, the pancreatic cells were provided with a large quantity of unlabelled amino acids.

Samples of the cells were extracted at particular times, and the levels of radioactivity in the rough endoplasmic reticulum, Golgi apparatus and secretory vesicles were determined. The results are shown in Figure 6.6.

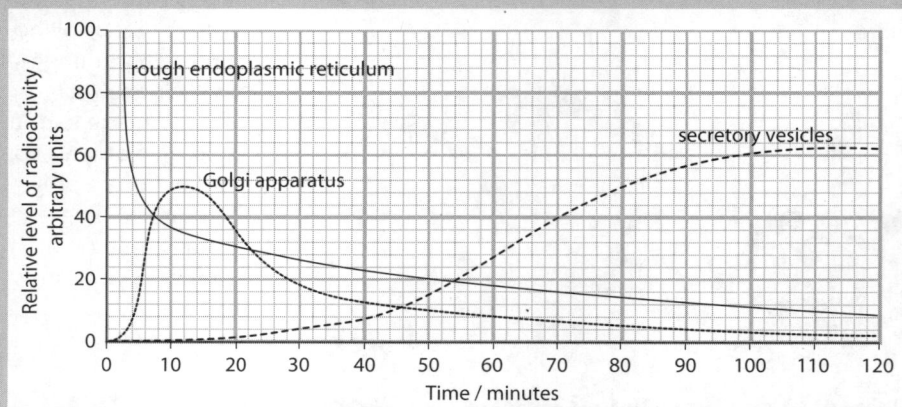

Figure 6.6

    i Suggest why pancreatic cells were chosen for this experiment. [1]

    ii Explain why the pancreatic cells were provided with a large quantity of unlabelled amino acids, after being given a small quantity of labelled amino acids. [2]

    iii Use the information in Figure 6.6 to state the sequence in which the labelled amino acids moved through the three organelles. [1]

    iv Explain the pattern shown by these results. [5]

[Total: 16]

> Chapter 7
Transport in plants

CHAPTER OUTLINE

The questions in this chapter cover the following topics:
- drawing, labelling and describing the structure of stems, roots and leaves
- the structure of the two transport tissues in plants: xylem and phloem
- how the structures of xylem and phloem are related to their functions
- how water and mineral salts move through plants
- the investigation of water loss from leaves
- the adaptations of the leaves of xerophytic plants
- how organic molecules such as sucrose and amino acids move through plants.

Exercise 7.1 Drawing, labelling and describing plant structures

In Chapter 1 you explored how to make diagrams of individual cells and groups of cells and how to carry out magnification calculations. You also need to be able to draw plan diagrams of tissue groups using lower magnifications. Plan diagrams aim to show tissues rather than individual cells. In this exercise, you will:

- develop your drawing skills by drawing plan diagrams of plant tissues
- develop your understanding of plant transport tissues.

Figure 7.1 shows a transverse section through part of the stem of *Helianthus*, the sunflower.

1   The image has a magnification of 21. Calculate the actual width of the entire specimen as shown along A–B on the image.

   Before drawing a tissue map (plan diagram), take note of the following points.

   - Take a very sharp HB pencil, ruler, eraser and piece of plain paper.
   - Do not shade in regions and make sure that you use firm, solid lines.
   - Draw a large diagram, at least 10 cm in width.
   - Print labels neatly.
   - Use a ruler to draw label lines accurately. Do not use arrows.
   - Do not be frightened to take your time. Being familiar with how to draw a plan diagram will help ensure you can do this quickly and not waste time. You could try practising this under timed conditions.

CAMBRIDGE INTERNATIONAL AS & A LEVEL BIOLOGY: WORKBOOK

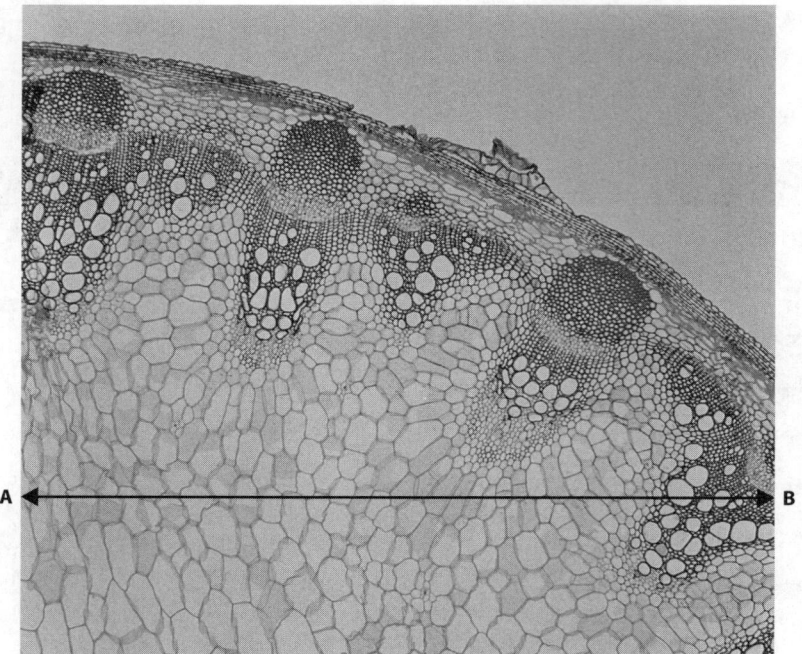

Figure 7.1: Transverse section through *Helianthus* stem. Magnification ×21.

> **KEY WORDS**
>
> **sclerenchyma fibre:** a supportive plant tissue that consists of thick walled, usually lignified cells; sclerenchyma cells are either fibres or sclereids

2 Now draw a plan diagram of Figure 7.1 using the following steps.

Step 1 Draw out the outline of the stem showing the epidermis as two lines. Ensure that the outline is a minimum of 5 cm wide.

Step 2 Draw out the vascular bundles clearly showing where the xylem, cambium and phloem tissues are. Do *not* draw individual cells.

Step 3 Draw out an outline of where the cortex tissue is found. Again do not draw individual cells.

Step 4 Add straight label lines and labels to show one example of xylem, phloem, cambium, cortex, **sclerenchyma fibres** and epidermis.

Step 5 Write a legend for your diagram: 'TS stem *Helianthus*' and state the magnification (×25).

Step 6 Now add a scale bar. You already know the actual width of the whole specimen. Determine the length of line that would represent 1 mm and draw it underneath your diagram. Compare your diagram with the answer.

3 Figure 7.2 shows a higher magnification of one of the vascular bundles from the *Helianthus* stem.

  a Calculate the length of the vascular bundle labelled A–B in Figure 7.2.

  b Draw a tissue map of Figure 7.2 (not showing individual cells), clearly labelling the sclerenchyma fibres, phloem, cambium, xylem, cortex and epidermis. Add a legend, magnification and scale bar that shows a 1 mm line.

  Check your diagram with the answers and be critical!

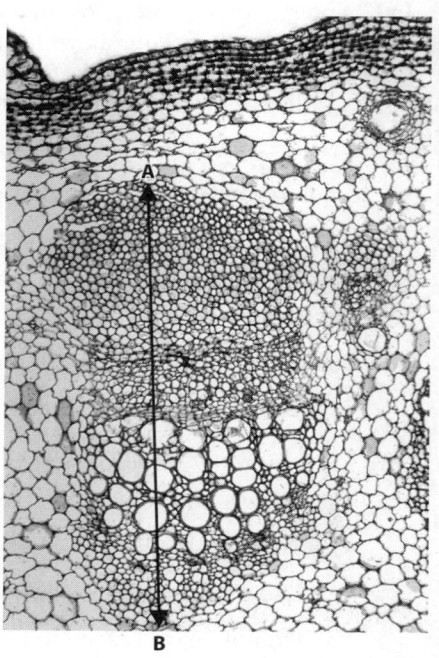

Figure 7.2: Vascular bundle from stem of *Helianthus*. Magnification ×21.

4 a As well as drawing plan diagrams of tissues, you need to able to recognise particular cell types. Figure 7.3 shows a tissue plan diagram of a cross section of the centre of a buttercup (*Ranunculus*) root. Copy the photograph and add labels for: xylem, endodermis, cortex and phloem tissues. Include a scale bar representing 0.1 mm.

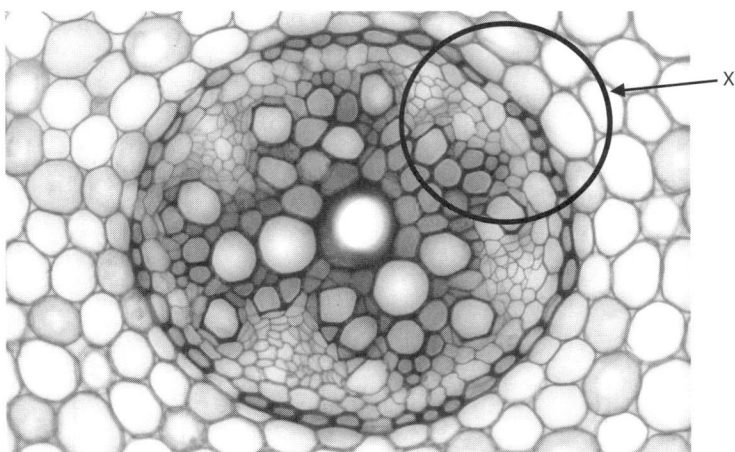

Figure 7.3: Photograph of a transverse section through *Ranunculus* root showing the stele. Magnification ×195.

b Now draw a diagram showing individual cells for the area labelled X. Ensure that your diagram is about 4 cm in width. Identify and label the following cell types: xylem, phloem, endodermis, passage cell, pericycle and cortex.

5 Mark the response to an exam-style question below.

The light micrograph below shows a section of lower epidermis from a kidney bean plant.

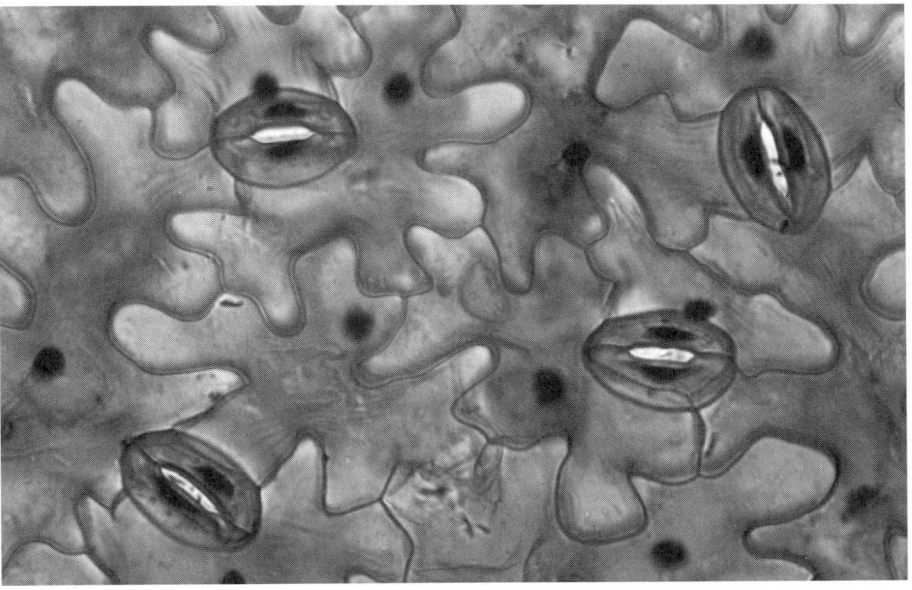

Figure 7.4: Lower epidermis from leaf of kidney bean. Magnification ×233.

> **KEY WORDS**
>
> **guard cell:** a sausage-shaped epidermal cell found with another, in a pair bounding a stoma and controlling its opening or closure
>
> **stoma (plural stomata):** a pore in the epidermis of a leaf, bounded by two **guard cells** and needed for efficient gas exchange

Make a copy of the diagram including labels to show: **guard cells**, **stomata**, epidermis cells, cell walls and cell nuclei.

Figure 7.5: Example drawing.

> **Mark scheme**
>
> - Diagram minimum of 4 cm width + sharp pencil + no shading ;
> - All cells correctly and clearly labelled using straight lines ;
> - Guard cells in proportion to epidermis cells and stomata are open ;
> - Epidermal cell walls smoothly drawn with accurate curves ; all 19 nuclei drawn ;

Exercise 7.2 Comparing two specimens

It is important to be able to look at biological specimens and compare them. **Xerophytes** have adaptations that enable them to survive in areas of low water. Figure 7.6 shows drawings of sections through the leaves of an oleander plant and a *Camellia*. In this exercise, you will:

- compare the leaf structures of two plants
- develop your understanding of xerophytic plants.

KEY WORD

xerophyte: plants adapted to survive in conditions where water is in short supply

7 Transport in plants

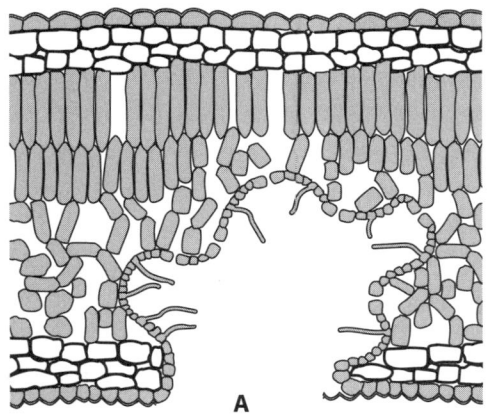 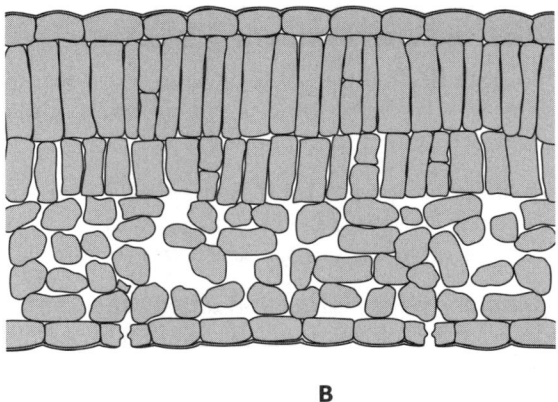

Figure 7.6: Drawing of transverse sections through leaves from **A** oleander and **B** *Camellia*.

1 Copy out Table 7.1 and complete it by comparing the structures in each leaf.

Structure	Oleander	*Camellia*
upper epidermis		
palisade mesophyll		
spongy mesophyll		
lower epidermis		

Table 7.1: Comparison of leaf structures in oleander and *Camellia*.

2 Suggest which plant is adapted to live in areas of very low water and explain how the leaf is designed to minimise water loss.

Exercise 7.3 Identifying plant cell functions

There are many different types of plant cell with different structures and functions. You need to be able to recognise these and understand how they are adapted to their particular functions. In this exercise, you will:

- develop your understanding of plant cell structure and functions.

> CAMBRIDGE INTERNATIONAL AS & A LEVEL BIOLOGY: WORKBOOK

> **KEY WORDS**
>
> **lignin (lignified):** a complex organic compound that binds to cellulose fibres and hardens and strengthens the cell walls of plants (lignification); chief noncarbohydrate constituent of wood
>
> **sieve element** or **sieve tube element:** a cell found in phloem tissue, with non-thickened cellulose walls, very little cytoplasm, no nucleus and end walls perforated to form sieve plates, through which sap containing sucrose is transported
>
> **companion cell:** a cell with an unthickened cellulose wall and dense cytoplasm that is found in close association with a phloem sieve element to which it is directly linked via many plasmodesmata; the companion cell and the sieve element form a functional unit
>
> **collenchyma:** a modified form of parenchyma in which the corners of the cells have extra cellulose thickening, providing support, as in the midrib of leaves and at the corners of square stems; in three dimensions the tissue occurs in strands (as in celery petioles)
>
> **parenchyma:** a basic plant tissue typically used as packing tissue between more specialised structures; it is metabolically active and may have a variety of functions such as food storage, support and transport via symplast and apoplast pathways

1   Copy and complete Table 7.2 to match up the cell types with the statements given for functions, locations and structures. Some statements may be used several times.

The first cell type is done for you.

Cell type	Structure	Location	Function
xylem vessels	• dead, hollow cells • **lignified** cell walls which are impermeable to water • cells joined end on end with no cross walls	• associated with vascular bundles	• transport of water from root to stem and leaf • support of the stem
sieve tube element			
companion cell			
sclerenchyma			
collenchyma			
parenchyma			

Table 7.2: Function, location and structures of different plant cell types.

Structures
- dead, hollow cells
- cell walls contain extra cellulose at the corners of cells
- living cells
- cells separated by sieve plates
- cells joined end on end with no cross walls
- **lignified** cell walls which are impermeable to water
- have cellulose cell walls
- no nucleus or ribosomes and only a thin layer of cytoplasm
- all typical cell organelles present

Locations
- associated with vascular bundles
- attached to sieve tube elements by **plasmodesmata**
- leaf mesophyll tissues
- found in corners of some stems
- found in leaf midrib
- root and stem cortex tissue

Functions
- transport of water from root to stem and leaf
- support of the stem
- starch storage
- support of leaf
- transport of sucrose and amino acids
- carries out metabolic processes for attached sieve tube element
- photosynthesis.

Exercise 7.4 Measuring transpiration rates

Use of a **potometer** to investigate **transpiration** rates is a standard method that you should be aware of. You may encounter a range of different potometers, and they can measure water loss by weight and volume. Regardless of their design, the purpose of all potometers is to measure the loss of water from plants. They are often used to investigate the effect of environmental factors on the rate of transpiration. In this exercise, you will:

- develop your understanding of how to ensure that experiments produce valid results
- develop your understanding of the factors that affect transpiration rates
- develop your mathematical skills by carrying out a series of calculations.

> **KEY WORDS**
>
> **plasmodesma** (plural: **plasmodesmata**): a pore-like structure found in plant cell walls; plasmodesmata of neighbouring plant cells line up to form tube-like pores through the cell walls, allowing the controlled passage of materials from one cell to the other; the pores contain ER and are lined with the cell surface membrane
>
> **potometer**: a piece of apparatus used to measure the rate of uptake of water by a plant
>
> **transpiration**: the loss of water vapour from a plant to its environment by diffusion down a water potential gradient; most transpiration takes place through the stomata in the leaves

1 Plant transpiration rates can be measured using a potometer as shown in Figure 7.7.

A student was interested in comparing the effect of increasing wind speed on the rate of transpiration of two different plant species, laurel and rosemary. He set up the potometer as shown and investigated the distance that an air bubble moved along the capillary tube over a ten-minute period at each wind speed. This was repeated for both plants.

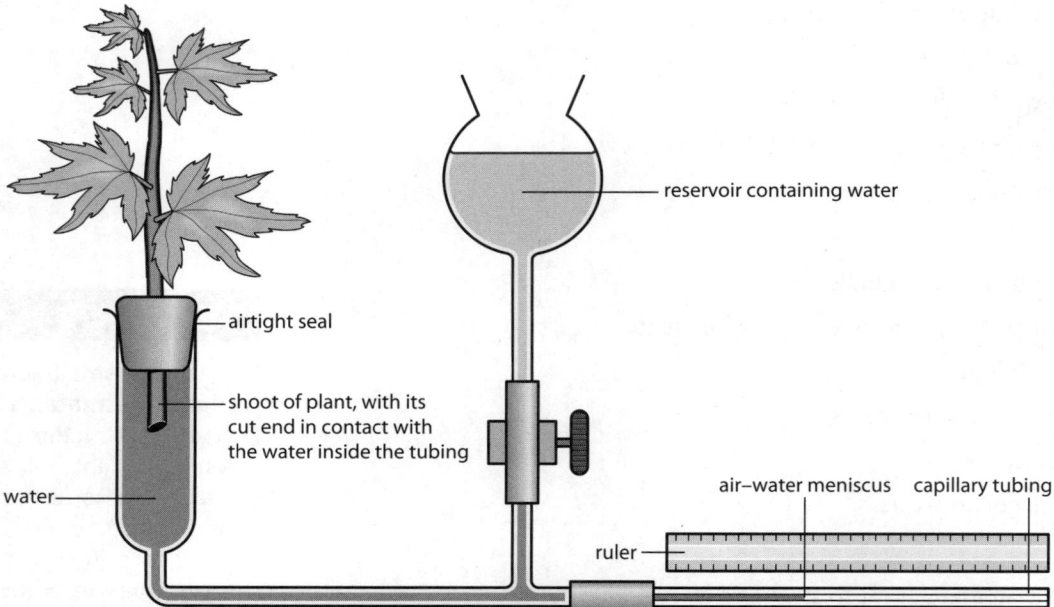

Figure 7.7: A potometer.

a Suggest and describe the precautions the student would need to take in order to ensure that there were no air bubbles in the potometer.
b Suggest how he could alter wind speed.
c In this experiment there are two independent variables; what are they?
d What is the dependent variable?
e Transpiration is affected by many variables. List **all** the variables that need to be standardised.
f What assumption do we make about the link between water movement in the capillary tube and transpiration?

In order to compare the transpiration rate of the two plants, the student wanted to calculate the rate of water loss in mm^3 per square centimetre of leaf. The table below shows the distance moved by the air bubble over a ten-minute period for the rosemary plant. The total surface area of the leaves was determined to be $52\,cm^2$.

g Copy the results table, and use the following steps to complete the student's calculations:

Step 1 The volume of water lost is actually a cylinder of water where the length of the cylinder is the distance, d, moved by the air bubble, and the radius, r, is the radius of the capillary tubing, which in these experiments was 0.75 mm.

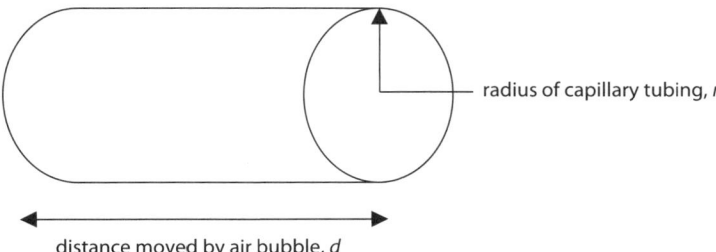

Figure 7.8: Calculating the volume of a cylinder of water in the potometer.

So, the volume for each wind speed is $\pi r^2 d$.

Step 2 Now, calculate the water volume lost per minute. The experiment was run for ten minutes, so divide the volume lost by ten. Fill in your answer in the table.

Step 3 To calculate the rate of water loss per square centimetre of leaf, divide your last answer by the surface area of the leaves (52 cm²). Fill in your answer in Table 7.3.

Wind speed / arbitrary units	Distance moved by water / mm	Volume of water lost / mm³	Rate of water loss / mm³ min⁻¹	Rate of water loss per cm² leaf area / mm³ min⁻¹ cm⁻²
0	8	14.13	1.41	0.03
10	15			
20	24			
30	27			
40	31			
50	33			

Table 7.3: Results table for rosemary.

h   The student repeated the experiment with the laurel plant. The time taken for the air bubble to move was five minutes, the total leaf surface area was 160 cm² and the radius of the capillary was 0.75 mm. Copy and complete Table 7.4.

Wind speed / arbitrary units	Distance moved by water / mm	Volume of water lost / mm³	Rate of water loss / mm³ min⁻¹	Rate of water loss per cm² leaf area / mm³ min⁻¹ cm⁻²
0	22			
10	36			
20	46			
30	57			
40	71			
50	85			

Table 7.4: Results table for laurel.

i   Plot a graph to compare the rate of water loss per square centimetre of the two plants at different wind speeds.

j   Why is it appropriate to draw lines of best fit for these series of data?

k   Another student criticised the data by pointing out that the time taken for the distance moved by the air bubble was different for the two different plants. Is the criticism valid?

> **TIP**
>
> Ensure that you do the following:
>
> - Use continuous axes with wind speed on the *x*-axis and rate of water loss per square centimetre on the *y*-axis.
>
> - Plot both sets of data using the same axes. Draw appropriate curves or lines of best fit and label the lines for the two different plants with a key.

EXAM-STYLE QUESTIONS

1 Figure 7.9 shows a light micrograph of stomata from a *Tradescantia* plant.

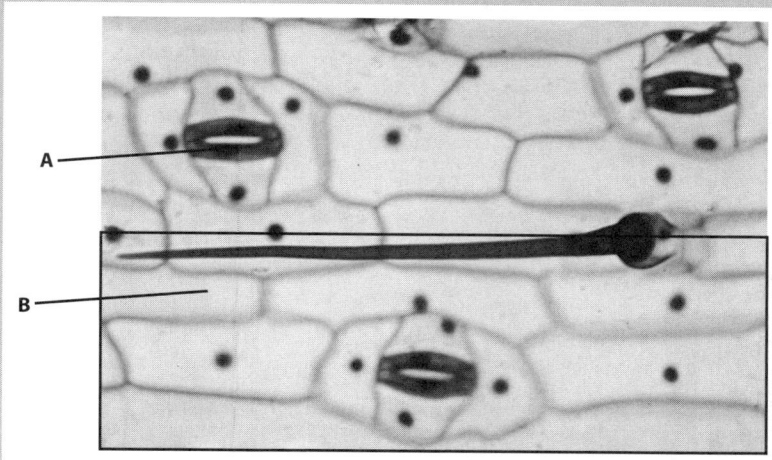

Figure 7.9

a The magnification of the light micrograph is ×400. Calculate the length in μm of the stoma labelled **A** on the diagram. [2]

b Draw a copy of the area on the diagram labelled **B**. Add labels to clearly show the stomata, guard cells and lower epidermis cells. [4]

[Total: 6]

2 Figure 7.10 shows part of a cross section through a root of the buttercup, *Ranunculus*.

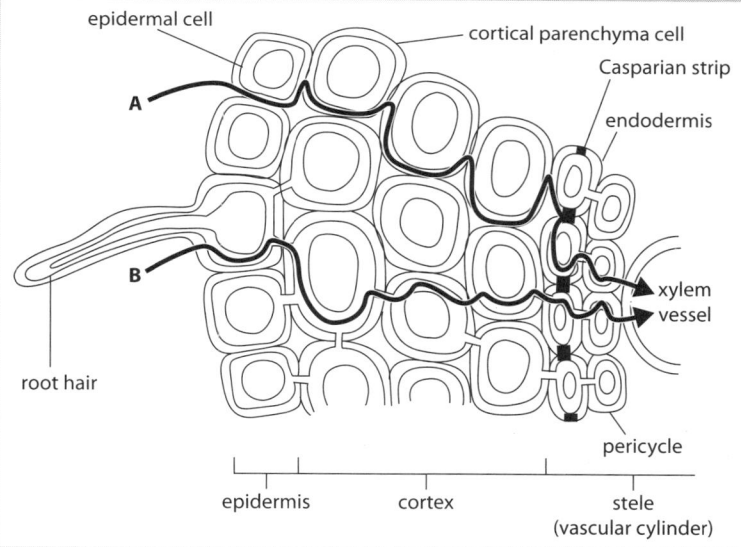

Figure 7.10

CONTINUED

a   Two of the pathways for water transport are shown on Figure 7.10 as **A** and **B**. **State** the names of these two pathways. [1]

b   The cell walls of cells in the endodermis often contain the substance suberin. **Explain** the function of this substance. [2]

Maize plants grown in heavily waterlogged soils often exhibit increased growth of a hollow root tissue called aerenchyma which is surrounded by an extensive coat of suberin. Figure 7.11 shows the development of aerenchyma and suberin in the roots.

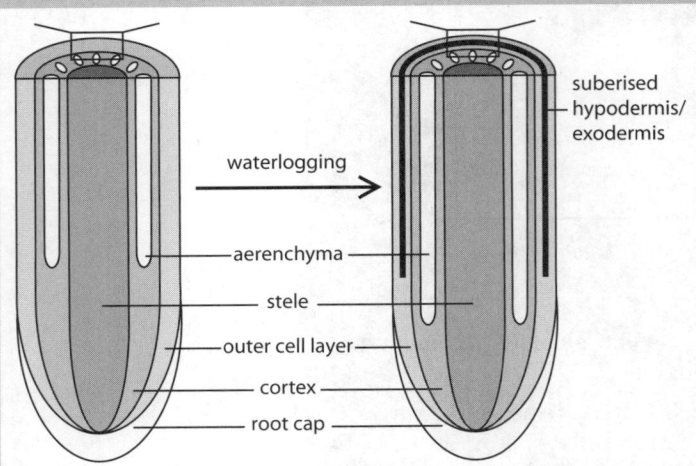

Figure 7.11

c   **Suggest** why the development of the aerenchyma tissue and suberin is advantageous to plants growing in waterlogged soils. [4]

[Total: 7]

COMMAND WORDS

State: express in clear terms.

Explain: set out purposes or reasons / make the relationships between things evident / provide why and/or how and support with relevant evidence.

Suggest: apply knowledge and understanding to situations where there is a range of valid responses, in order to make proposals / put forward considerations.

CONTINUED

3 An investigation was carried out into the rate of transpiration and width of xylem vessels of Scots pine trees during a 24-hour period. The results are shown in Figure 7.12.

  a **Describe** the composition of xylem tissue and explain how it is designed to transport water throughout the plant. **[4]**

> **COMMAND WORD**
>
> **Describe:** state the points of a topic/give characteristics and main features.

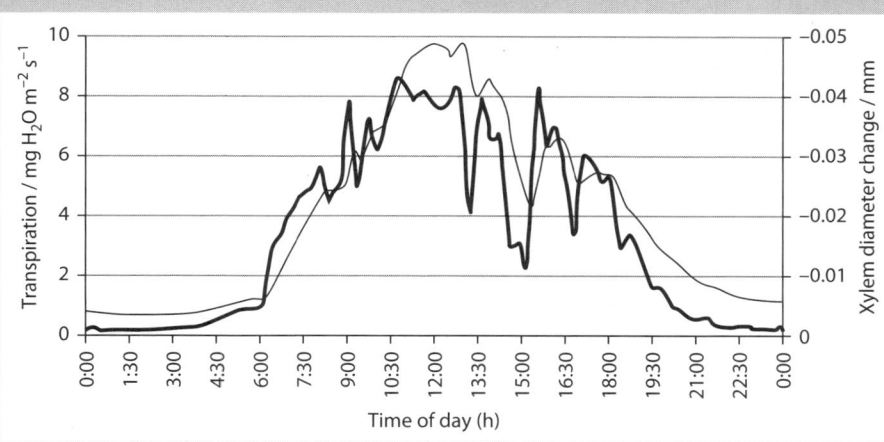

Figure 7.12

  b **i** Describe how xylem diameter and transpiration changed throughout the 24-hour period. **[2]**

     ii Using your knowledge of the cohesion–tension theory, explain the association between xylem diameter and transpiration rate shown in Figure 7.12. **[2]**

     iii Explain the changes in xylem diameter over the 24-hour period shown in Figure 7.12. **[5]**

[Total: 13]

4 An investigation into the effect of rainfall on the size of stomata (mean area) and density of stomata on the lower surface of leaves of *Eucalyptus globulus* was carried out. The results are shown in Table 7.5.

Parameter	High rainfall	Low rainfall
Mean maximum stomatal size / µm^2	169.30	125.70
Mean stomatal density/ mm^{-2}	238.80	229.60
Mean water loss / mol m^{-2} s^{-1}	1.66	1.25

Table 7.5

CONTINUED

a i Suggest how mean stomatal density could be determined in the laboratory. [4]

ii **Compare** the effect of rainfall on the development of stomata in *Eucalyptus* leaves. [1]

iii Explain the advantages to the *Eucalyptus* plant of changing stomatal patterns in response to the environment. [2]

b Figure 7.13 shows a diagram of part of a leaf from the xerophytic plant *Ammophila*.

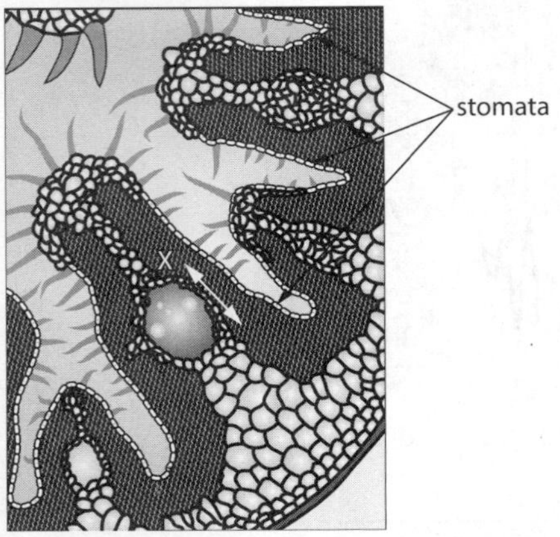

i **Calculate** the length of the vascular bundle labelled **X**. [1]

Figure 7.13

ii *Ammophila* is a xerophytic plant. Use the diagram to explain how the leaf is adapted to reduce water loss. [4]

[Total: 12]

COMMAND WORDS

Compare: identify/comment on similarities and/or differences.

Calculate: work out from given facts, figures or information.

5 An experiment was carried out into the absorption of mineral ions by the roots of barley seedlings. Root tissue was placed into two solutions of radioactive potassium ions, one of which also contained a respiratory inhibitor, potassium cyanide. The amount of potassium absorbed by the roots was measured at several intervals. The experiment was repeated with calcium ions. The results are shown in Table 7.6.

Time / hours	Quantity of ions absorbed / μmol			
	Potassium	Potassium + potassium cyanide	Calcium	Calcium + potassium cyanide
0	0	0	0	0
1	5	3	5	5
3	14	3	7	7
6	22	4	10	7
12	40	4	10	7

Table 7.6

CONTINUED

a Describe the effect of adding potassium cyanide on the absorption of potassium and calcium ions into the root tissue. [2]

b Suggest and explain the processes by which potassium and calcium ions are absorbed by root tissue. [4]

c Explain how the mineral ions are transported to the leaves. [4]

[Total: 10]

6 Figure 7.14 shows a diagram of a light micrograph of sieve tube elements.

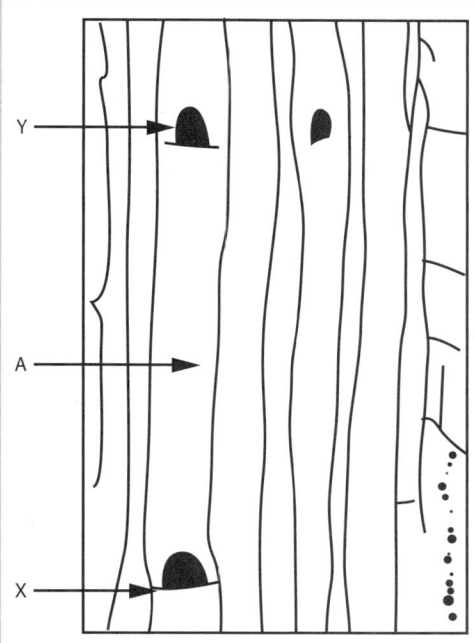

Figure 7.14

a i Calculate the length, between X and Y in micrometres of the sieve tube element labelled **A**. Magnification ×145. [2]

ii Explain the differences between the structure of xylem vessels and sieve tubes. [3]

b An investigation was carried out into the effect of the presence of fruit on the speed of **translocation** of sucrose from tomato leaves.

A leaf on a tomato plant that had the fruit removed was given a pulse of radiolabelled $^{14}CO_2$ and the amount of radioactivity remaining in the leaf monitored over a 10-hour time period.

This was repeated with plants bearing fruits at 20 °C and 30 °C.

KEYWORD

Translocation: the transport of assimilates such as sucrose through a plant, in phloem tissue

> **CONTINUED**

Figure 7.15 shows the results.

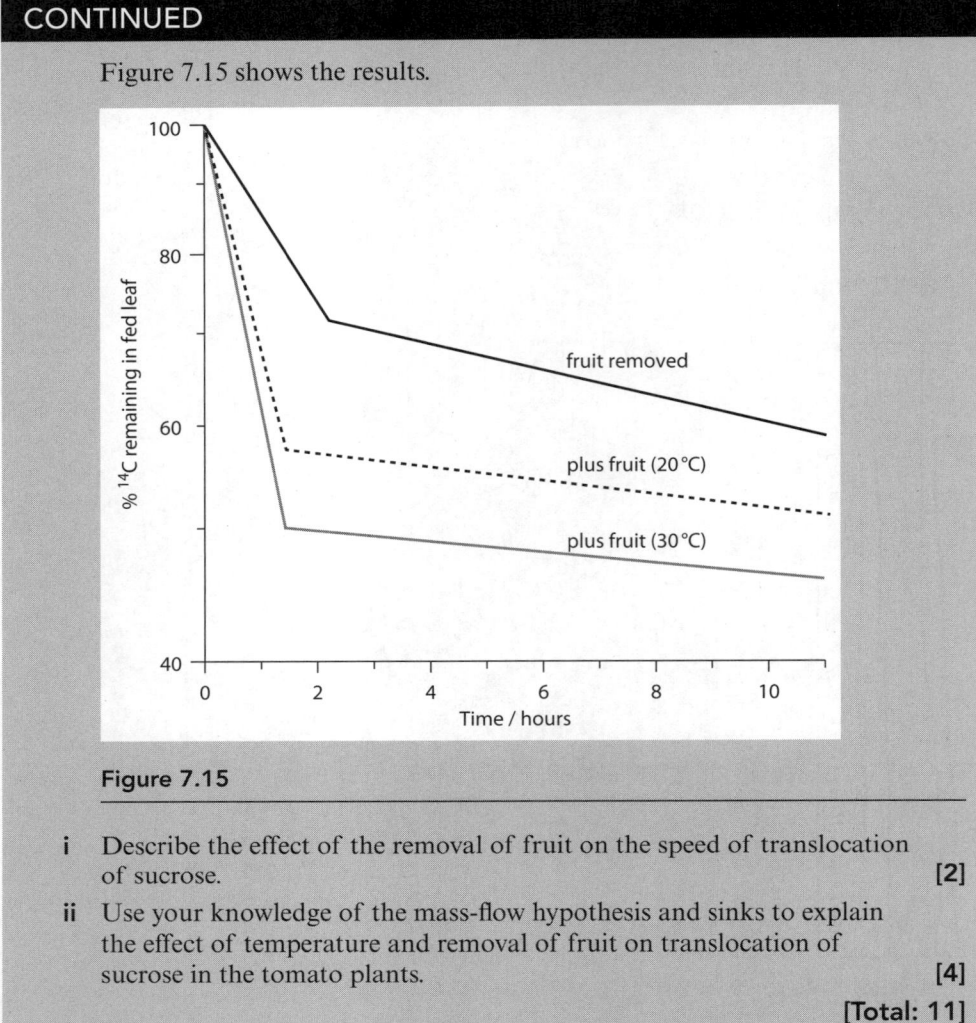

Figure 7.15

i Describe the effect of the removal of fruit on the speed of translocation of sucrose. [2]

ii Use your knowledge of the mass-flow hypothesis and sinks to explain the effect of temperature and removal of fruit on translocation of sucrose in the tomato plants. [4]

[Total: 11]

> Chapter 8
Transport in mammals

CHAPTER OUTLINE

The questions in this chapter cover the following topics:
- the structure of the mammalian circulatory system
- how the structures of arteries, arterioles, veins, venules and capillaries are related to their functions
- the structure and functions of blood, including the transport of oxygen and carbon dioxide
- drawing the blood vessels and blood cells from slides, photomicrographs or electron micrographs
- the formation and functions of tissue fluid
- the structure and function of the heart
- the cardiac cycle and its control.

Exercise 8.1 Comparing the strength and elasticity of arteries and veins

You need to understand the different properties of arteries and veins and relate them to their functions. Data presentation is an important practical skill and you need to be able to draw appropriate tables, graphs and plan diagrams of structures.

In this exercise, you will:
- develop your understanding of the different properties of arteries and veins
- develop general skills in the drawing of results tables and line graphs
- build on the drawing skills that you developed in Chapters 1, 5 and 7.

1   A student carried out an experiment to investigate the effect of hanging increasing masses from **arteries** and **veins** on their length and elasticity.

   The width of an artery taken from a sheep was measured using a ruler. A 50 g mass was then hung on the artery and the stretched length measured. The mass was removed, and the recoiled length measured. This was then repeated with increasing masses until the artery snapped.

   The experiment was then repeated for the vein.

   The student's results are shown in Figure 8.1.

KEY WORDS

artery (plural: arteries): a blood vessel that carries blood away from the heart; it has a relatively thick cell wall and contains large amounts of elastic fibres

vein: a blood vessel that carries blood back towards the heart; it has relatively thin walls and contains valves

<u>Aorta</u>
0 g – 15 mm
50 g – 17 mm, 15 mm
100 g – 19 mm, 15 mm
150 g – 20 mm, 15 mm
200 g – 21 mm, 15 mm
250 g – 23 mm, 15 mm
300 g – 26 mm, 15 mm
350 g – 31 mm, 15 mm
400 g – 35 mm, 15 mm
450 g – 42 mm, 15 mm
750 g – 46 mm, 15 mm
1 kg – 51 mm, 16 mm
1.5 kg – 58 mm, 18 mm
2.0 kg – 63 mm, 23 mm
3.0 kg – snapped!

<u>Vein</u>
0 g – 19 mm
50 g – 23 mm, 19 mm
100 g – 25 mm, 19 mm
150 g – 27 mm, 24 mm
200 g – 28 mm, 25 mm
250 g – 31 mm, 28 mm
300 g – 35 mm, 31 mm
350 g – 38 mm, 35 mm
400 g – 44 mm, 41 mm
450 g – snapped!

Figure 8.1: The lengths of arteries and veins with different masses.

a   Design a results table to display the student's data.

b   Plot a line graph to show the effect of increasing mass on the length of the artery and vein.

c   Calculate the maximum percentage increase in length for the artery: ((greatest stretched length − starting length) ÷ starting length)) × 100

Repeat this for the vein. Explain why comparing percentage increase is more valid than comparing the increase in length.

d   The student proposed that the maximum weight that the artery can withstand is 2.0 kg. Is this a correct assumption and how could the student improve the experiment to identify the breaking point?

e   Figure 8.2 is a photograph of an artery and a vein.

   i   Make a plan diagram of the section, clearly labelling: the artery, vein and lumen.

   ii   Calculate the actual maximum diameters of the lumen of both blood vessels.

   iii   Compare the structure of the artery and vein. Copy out and complete Table 8.1.

Structural feature	Artery	Vein
overall shape		
lumen		
tunica media (middle layer)		
tunica externa (outer layer)		

Table 8.1: Comparisons of the structures of arteries and veins.

> **TIP**
> You need to plot four lines using the same axes for both the stretched lengths and recoiled lengths. We cannot be confident that intermediate points will fit the line, so join your points with straight lines.

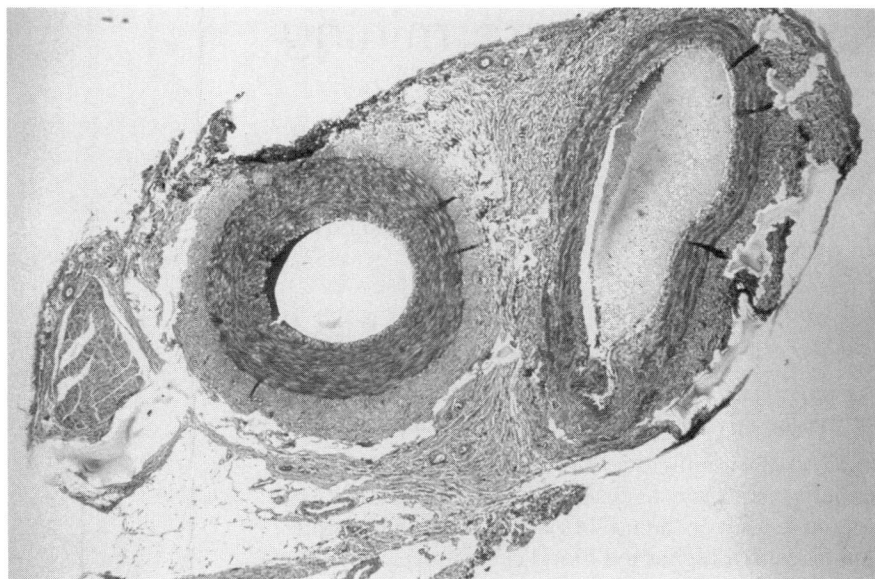

Figure 8.2: Artery and vein. Magnification ×60.

f Use Figure 8.2 and your knowledge of blood vessel structure to explain the student's results.

g Figure 8.3 shows a longitudinal section through an artery. Draw a plan diagram of the artery. Label the tunica externa (outer layer) and tunica media (middle layer) and endothelium.

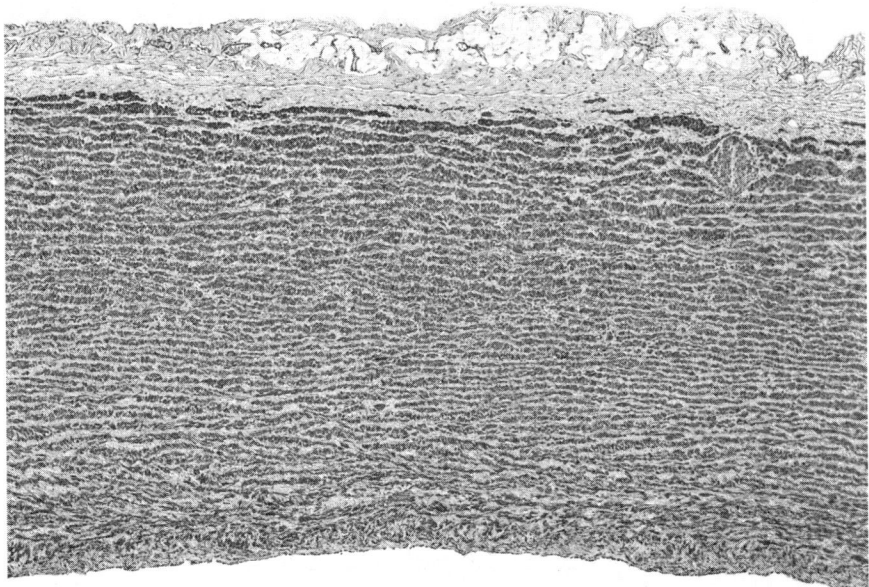

Figure 8.3: Longitudinal section through an artery.

Exercise 8.2 Methods for determining red blood cell counts

You need to understand the functions of the main components of blood, especially the role of red blood cells in the transport of oxygen. You also need to able to analyse unfamiliar data and experiments in order to reach conclusions and evaluate how valid a method is.

In this exercise, you will:

- develop your understanding of the functions of blood
- develop your analytical skills when using unfamiliar data
- develop your understanding of practical skills, including a method used to count cells using a haemocytometer, writing a risk assessment and evaluating results.

The measurement of the percentage (%) of an individual's blood that is composed of red blood cells is called the haematocrit test. It is often used to test for conditions such as anaemia (where red blood cell count is low) or the use of performance enhancing drugs or blood doping (which often leads to a higher red blood cell count).

A simple method for determining red blood cell percentage by volume of blood is by centrifugation. A 7.5 cm long narrow glass capillary tube is loaded up with blood. The bottom of the capillary is sealed with clay and the blood centrifuged for five minutes at 10 000 rpm. An example of this is shown in Figure 8.4. The percentage of the blood that is composed of red blood cells is then determined by holding the capillary up against a haematocrit chart as shown in Figure 8.5.

Figure 8.4: Fractions of blood obtained by centrifugation.

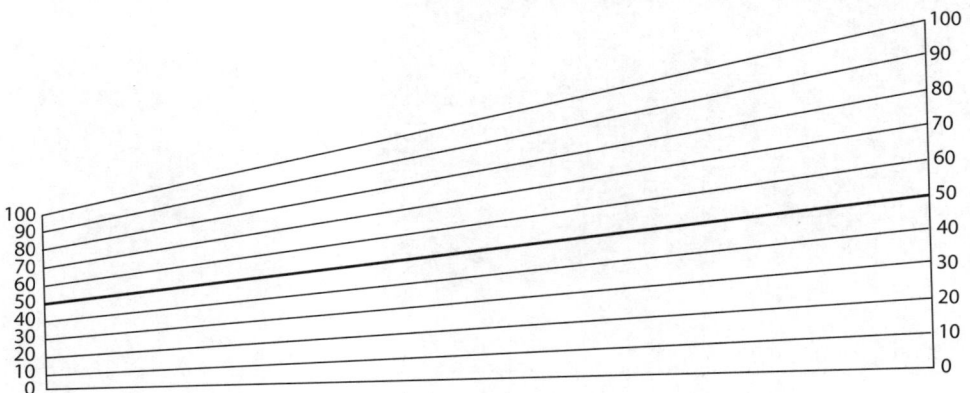

Figure 8.5: A haematocrit chart showing how the percentage of red blood cells is determined.

8 Transport in mammals

1 The three fractions that blood separates into contain:
   - mixed white blood cells and platelets
   - erythrocytes (red blood cells)
   - plasma.

   Match each fraction with the letters A, B and C as shown in Figure 8.4.

2 To determine the percentage of blood that is composed of red blood cells, the capillary tube is held against the chart so that the bottom of the sample is at the zero line and the top is at 100%. The percentage of red blood cells can then be read from the chart.

   Explain why using the chart means that the volume of blood in the capillary does not have to be a set volume.

3 Carry out a risk assessment for this method when used to determine the red blood cell concentration of human blood. Include the risks, the dangers they pose and what you would do to minimise the risk of harm.

4 The test must normally be carried out by two laboratory technicians in order to improve reliability. Eight students carried out an investigation into the reliability of the technique. They all performed haematocrit tests on five different samples of blood taken from women aged between 19 and 35. One of the women had returned from a vacation spent at high altitude. The results are shown in Table 8.2.

Blood sample	Percentage red blood cells / %							
	Student 1	Student 2	Student 3	Student 4	Student 5	Student 6	Student 7	Student 8
A	44	45	44	45	48	44	43	45
B	42	42	43	41	46	42	43	32
C	37	36	36	36	41	37	37	37
D	52	53	53	52	57	52	53	52
E	41	40	25	42	46	40	41	42

Table 8.2: Results from haematocrit tests on five different samples of blood taken from women aged between 19 and 35.

   a Identify any values in the table that you would consider to be anomalous, for example by using +/− 2 standard deviations.
   b Calculate the mean percentage of red blood cells (%) for each of the samples. Discard any readings that you consider anomalous.
   c Explain whether any of the results appear to be due to a systematic error.
   d Suggest what factors could be responsible for the anomalous values you have identified.

   Possible sources of error in the technique include:
   - improper sealing of the capillary tube before centrifugation
   - reading from the top of the meniscus of red blood cells
   - centrifugation at the wrong speed (either too high or too low)

- leaving the sample for too long after centrifuging so that the red blood cells are not as tightly packed
- including the 'buffy coat' (white blood cell and platelets fraction) in the reading.

    e The partial pressure of oxygen at high altitude is reduced. Time spent at high altitude causes the body to synthesise more red blood cells. Identify the sample that was from the woman who had spent time at altitude, and explain the advantages and possible dangers of the change in her blood.

    f Use the results in the table to suggest and explain what the minimum number of repeats should be when testing clinical samples. Would the same technician carrying out the repeats produce valid results?

5 There are many factors that can affect the haematocrit test.

Predict and explain whether the following events would cause a rise or fall in haematocrit:

    a loss of a large amount of plasma due to capillary damage from severe burns

    b over-hydration by drinking large volumes of water

    c anaemia caused by a lack of iron in the diet

    d immediately after a major trauma leading to extensive bleeding

    e one day after a major trauma after bleeding has stopped.

6 One of the students read that the size of individual red blood cells can vary and decided to investigate whether the percentage of red blood cells determined by the haematocrit test is correlated to the actual number of red blood cells.

    a Explain why the red blood cell size could affect the value determined by the haematocrit.

    b Reorder the following steps below into the correct sequence:

        A The number of cells inside five of the 25 squares of the central grid are counted.

        B Blood is placed into the space between coverslip and slide using a pipette.

        C The haemocytometer is placed on the microscope stage and the central squares focused on using the ×40 objective.

        D A coverslip is placed over the counting chamber and the sides pressed down firmly until Newtonian rings appear. The gap between the slide and coverslip is exactly 0.1 mm.

    c Figure 8.6 shows five squares of a haemocytometer grid.

    Count the total number of cells in all five squares shown in Figure 8.6.

    d The length of one side of each of these squares is 0.2 mm and the height of solution above the slide is 0.1 mm. Calculate the ***total*** volume of cell suspension over all five squares.

    e Use the equation below to calculate your cell density: cell density = number of cells in five squares ÷ volume of suspension in mm^3.

    Now convert this answer into cells per cubic decimetre by multiplying by 1 000 000, expressing your answer in standard form.

> **TIP**
>
> To get an actual count of red blood cells, we must use a haemocytometer. A haemocytometer is a special microscope slide with two sets of ruled grids in the centre, and deep grooves either side.

> **TIP**
>
> When counting red blood cells using a haemocytometer, you need to count the number of cells in five of the 25 medium-sized squares. When counting, any cells that touch the right and bottom edges are not included, while those that touch the top and left edges are included.

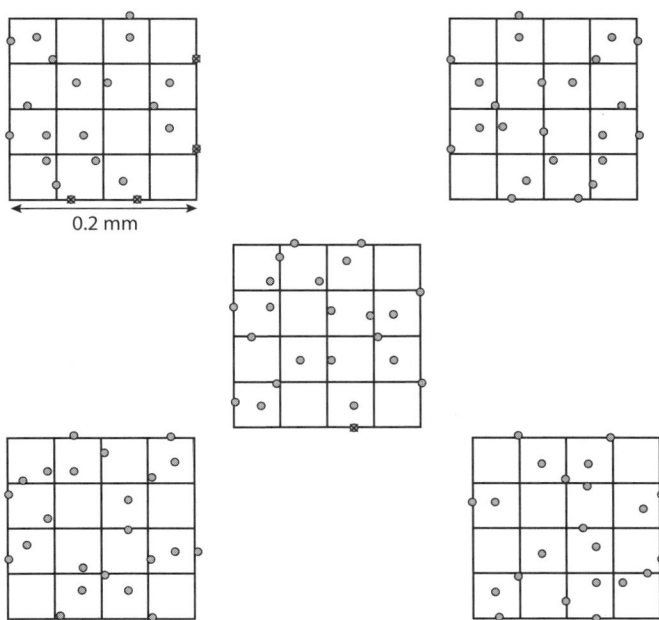

Figure 8.6: Five squares from haemocytometer central grid.

f Calculate the red blood cell count in cells per cubic decimetre for the following cell numbers counted from five squares:
   a 76
   b 125
   c 52.

g The student took 30 different blood samples and carried out both a haematocrit test and cell counts on each sample. Explain what graph the student should plot in order to determine whether the two methods show a correlation with each other.

Exercise 8.3 Interpreting graphs of heart rate and blood pressure

Understanding how graphs relate to the topic under investigation and using them to extract numerical data are important skills. The topic of mammalian transport uses numerous graphical representations. You need to be able to apply your knowledge of mammalian transport to unfamiliar graphs.

In this exercise, you will:

- develop your understanding of the **heart (cardiac) cycle**, pressure changes in the heart and oxygen dissociation curves of **haemoglobin**
- develop your ability to process unfamiliar data to reach conclusions.

1 Figure 8.7 shows a typical electrocardiogram (ECG) from a healthy human. An ECG shows the electrical activity that occurs in the heart each time it contracts. Although you do not need to understand what each part of an ECG

> **KEY WORDS**
>
> **heart (cardiac) cycle:** the sequence of events taking place during one heart beat
>
> **haemoglobin:** the red pigment found in red blood cells, whose molecules contain four iron atoms within a globular protein made up of four polypeptides, and that combines reversibly with oxygen

trace represents, you should be able to apply your knowledge of mammalian transport to interpret it. Each of the waves labelled PQRST represents one heart contraction. We can use ECG traces to calculate the heart rates.

To calculate the heart rate for the ECG shown in Figure 8.7, carry out the following steps.

a   First, calculate the time that one *small* square represents in Figure 8.6.

b   Now, determine how many small squares one heart (cardiac) cycle takes (good places to use are the tops of the R waves).

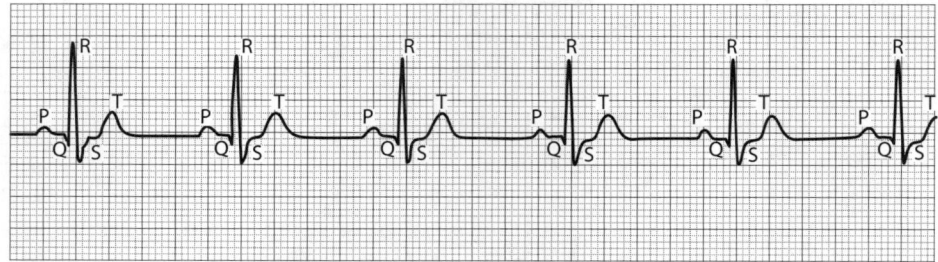

1 large square = 0.2 s

Figure 8.7: Electrocardiogram trace from a normal, healthy human.

c   You should now calculate the time taken for one heart cycle by multiplying the answers to parts **a** and **b** together.

d   To determine the heart rate, determine how many heart cycles can fit into 60 s. Heart rate = 60 ÷ time taken for one heart cycle

e   Now calculate the heart rates of the ECG traces **A** and **B** shown in Figure 8.8.

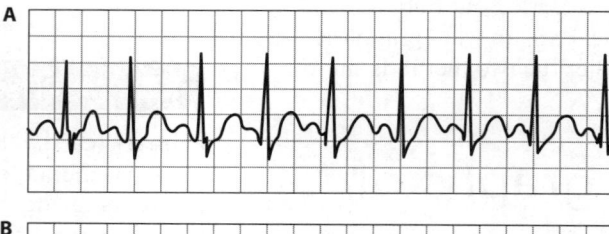

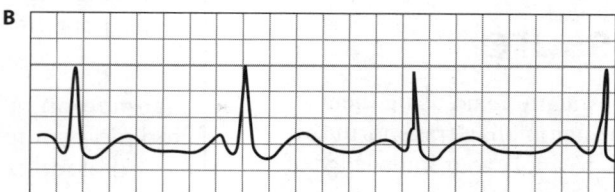

Figure 8.8: Two ECG traces. **A** Fast heart rate. **B** Slow heart rate.

2   The transport of blood occurs by mass flow. This means that blood will always flow from a higher pressure to a lower pressure. We can see how the pressures change in the different chambers of a heart during a heart cycle and predict when the blood is entering and exiting. We can also predict when the valves open and close.

Figure 8.9 shows the pressure and volume changes occurring during two cardiac cycles in the left ventricle.

8 Transport in mammals

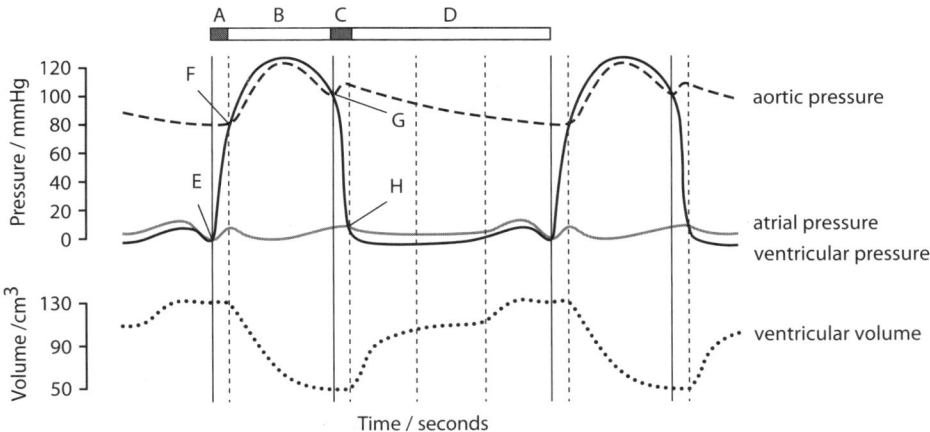

Figure 8.9: Pressure changes in aorta, left ventricle and atrium.

a   Identify on the chart the period when the ventricle is filling. Look for an area where pressure is greater in the atrium than in the ventricle.

b   Identify on the chart the period when the ventricle is emptying. Look for an area where the pressure in the ventricle is higher than in the aorta.

c   Identify on the chart the period when the ventricle is undergoing a contraction but its volume is not changing. Use the graph to explain why the ventricle does not empty.

d   Identify the letter where the bicuspid valve closes. Explain your answer.

e   Identify the letter where the bicuspid valve opens. Explain your answer.

f   Identify the letter where the semilunar valve closes. Explain your answer.

g   Identify the letter where the semilunar valve opens. Explain your answer.

h   Describe how the pressure generated by the right ventricle would differ from the pressure generated by the left ventricle. Explain why the two ventricles generate different pressures.

TIP

When interpreting these graphs, do not try to commit every detail to memory. Remember the following two rules:
- Blood will try to flow from higher to lower pressure.
- Valves will prevent the blood flowing backwards by closing.

3   You need to able to understand and use oxygen dissociation curves for haemoglobin both quantitatively and qualitatively. Do not try to learn every possible curve position for every situation – it is easier to work out what is happening.

Figure 8.10 shows oxygen dissociation curves for three mammals.

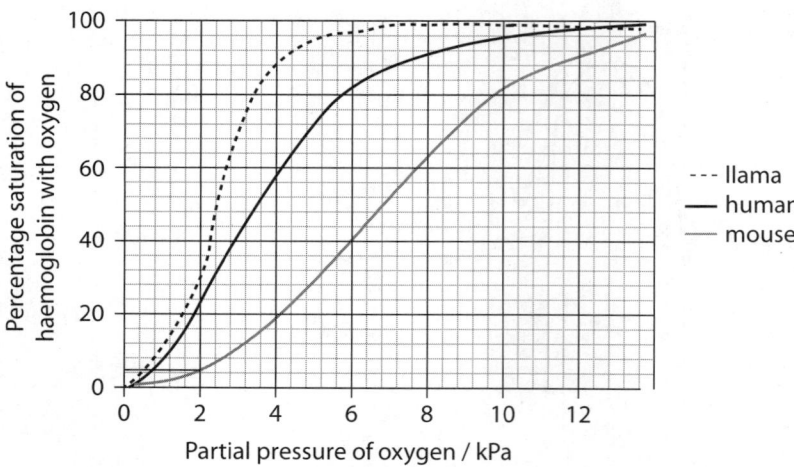

Figure 8.10: The oxygen dissociation curves for A llama, B human and C mouse haemoglobin.

a At normal sea level, the partial pressure of oxygen in the lungs is approximately 12 kPa. Follow the steps below to determine the oxygen saturation of human haemoglobin in the alveolar capillaries at this value.

Using Figure 8.10:

Step 1 Place a ruler on the x-axis at 12 kPa and draw a line up to the human curve.

Step 2 Use the ruler to draw a line from this point on the curve to the y-axis. This is the oxygen saturation value at 12 kPa.

Step 3 Copy out Table 8.3 below and enter this value in the table.

Species	Oxygen saturation of haemoglobin / %			Volume of oxygen bound to 1 g of haemoglobin / cm³			Volume of oxygen released to tissues / cm³	
	At 12 kPa	At 2 kPa	At 6 kPa	At 12 kPa	At 2 kPa	At 6 kPa	At atmospheric pressure (12 kPa)	At high altitude (6 kPa)
human								
llama								
mouse								

Table 8.3: Results table.

b Now do the same to determine the oxygen saturation of human haemoglobin in capillaries in respiring tissues (2 kPa) and alveoli at high altitude (6 kPa). Fill in the data in the table.

c 1 g of fully saturated haemoglobin will carry 1.3 cm³ of oxygen. Calculate the volume of oxygen carried by human haemoglobin at 12 kPa, 2 kPa and 6 kPa. Enter your values in your table.

d Now compare how much oxygen is delivered to tissues at normal and high altitudes.

For normal altitude:

Volume of oxygen released in tissue = volume of oxygen bound at 12 kPa − volume of oxygen bound at 2 kPa

For high altitude:

Volume of oxygen released in tissue = volume of oxygen bound at 6 kPa − volume of oxygen bound at 2 kPa

Enter your results in your table.

e Repeat steps **a** through to **d** for llama and mouse haemoglobin.

f Llamas live at high altitude. Mice are small mammals that maintain a constant body temperature. Explain the advantages to both these animals of their haemoglobin dissociation curves.

Exercise 8.4 Planning an experiment to investigate the effect of exercise on heart rate

Planning valid experiments that clearly identify the independent, dependent and standardised variables is an essential skill in A Level Biology.

In this exercise, you will:

- plan a method to investigate the effects of exercise on heart rate
- develop your understanding of how to plan valid experiments
- develop your understanding of independent, dependent and standardised variables.

1 **Independent variable**

   The independent variable is the variable that you will be changing and investigating in your investigation.

   a State what the independent variable is.
   b State how you will change it.
   c State the range and how many different values will you use.
   d State how will you ensure it is standardised between repeats.

2 **Dependent variable**

   The dependent variable is the variable that you will measure in the experiment.

   a State what the dependent variable is.
   b State how you will measure it accurately.
   c State how you will decide whether you have enough repeats (sample size) to ensure reliability.

CAMBRIDGE INTERNATIONAL AS & A LEVEL BIOLOGY: WORKBOOK

3   **Standardised variables**

The standardised variables are all other variables that could affect the experiment. They must be kept constant to ensure that the results are valid.

   a   Give a list of all variables that could affect the dependent variable.

   b   State how you will ensure that they do not change.

4   **Experimental plan**

When you have completed questions 1–3, use your answers to write a full experimental method for investigating the effects of exercise on heart rate, giving all relevant practical details and precautions.

EXAM-STYLE QUESTIONS

1   Figure 8.11 shows the oxygen dissociation curves of human haemoglobin in the presence of a range of concentrations of carbon dioxide.

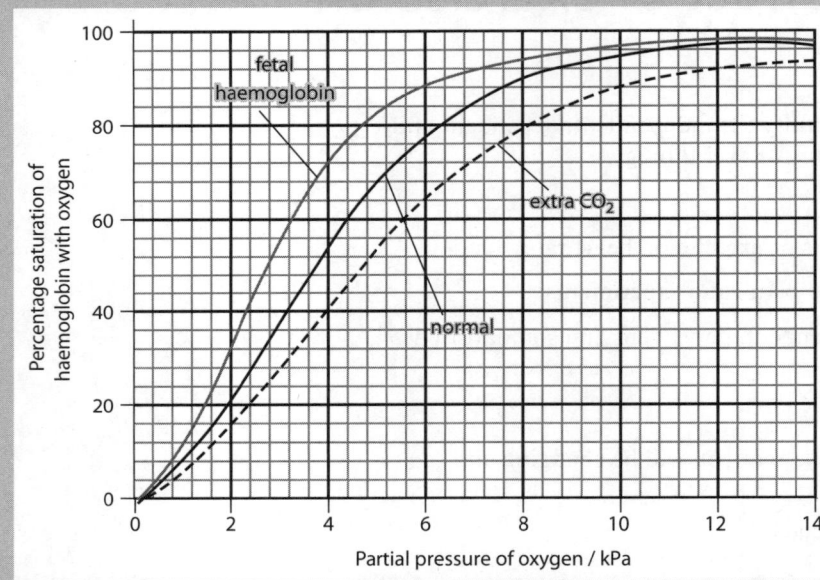

Figure 8.11

   a   **Explain** how the structure of the haemoglobin molecule generates an oxygen dissociation curve that is sigmoidal (S-shaped). [4]

   b   i   The partial pressure of oxygen in the alveoli is 12 kPa. Use Figure 8.11 to determine the percentage saturation of normal haemoglobin in the alveoli. [1]

       ii 1 g of fully saturated haemoglobin carries 1.3 cm³ of oxygen. **Calculate** the volume of oxygen bound to 1 g of haemoglobin at a partial pressure of 12 kPa. [1]

COMMAND WORDS

Explain: set out purposes or reasons / make the relationships between things evident / provide why and/or how and support with relevant evidence.

Calculate: work out from given facts, figures or information.

iii Use Figure 8.11 and your answer to (ii) to determine the volume of oxygen released to a rapidly respiring muscle with a partial pressure of oxygen of 2 kPa. [3]

iv Explain the advantage of the effect of CO_2 on the oxygen haemoglobin saturation curve. [3]

c Explain why the fetal oxygen dissociation curve is different from that of normal adult haemoglobin. [2]

[Total: 14]

2 Table 8.4 shows the pressures inside the left ventricle and left atrium of a heart during one cardiac cycle.

Time / seconds	Atrial pressure / kPa	Ventricle pressure / kPa	Aortic pressure / kPa
0.0	1.4	1.2	11.7
0.1	1.8	3.9	11.6
0.2	1.4	11.9	11.5
0.3	1.5	16.0	15.8
0.4	1.5	6.7	14.7
0.5	1.4	1.2	14.5
0.6	1.5	1.3	12.5
0.7	1.4	1.3	12.1
0.8	1.4	1.3	11.7

Table 8.4

a i One complete cardiac cycle takes 0.8 s. Calculate the pulse rate in beats per minute. [2]
ii **State** between which times the ventricle is emptying. [1]
iii State between which times the bicuspid valve is closed. [1]
iv State the time at which the ventricle is relaxing but not refilling. [1]

b **Describe** how the beating of a resting heart is controlled by electrical impulses. [5]

[Total: 10]

> **COMMAND WORDS**
>
> **State:** express in clear terms.
>
> **Describe:** state the points of a topic / give characteristics and main features.

CONTINUED

3 Figure 8.12 is a light micrograph of human blood cells.

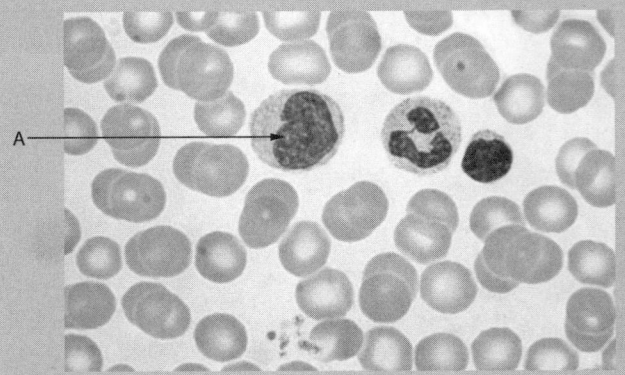

Figure 8.12

a i Make an accurate drawing of four red blood cells and all the white blood cells. Clearly label: red blood cells, monocyte, neutrophil and lymphocyte. [4]

ii Calculate the diameter of the cell labelled A. [2]

b Explain how the structure of red blood cells maximises the transport of oxygen to tissues. [4]

[Total: 10]

4 Figure 8.13 shows the mechanisms for transporting carbon dioxide in the blood.

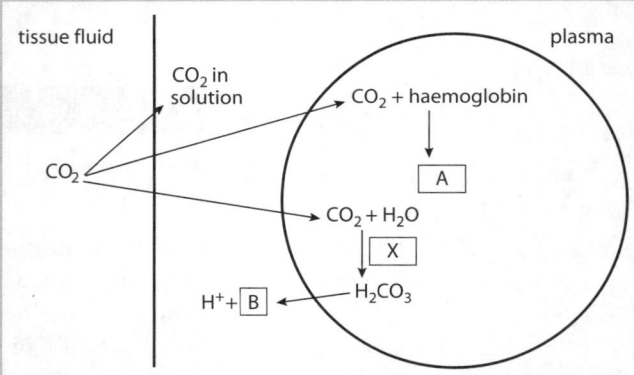

Figure 8.13

a State the names of substances A and B. [2]

b State the name of the enzyme that catalyses reaction X on the diagram. [1]

[Total: 3]

CONTINUED

5 Figure 8.14 shows how blood pressure, cross-sectional area and blood velocity change in different blood vessels.

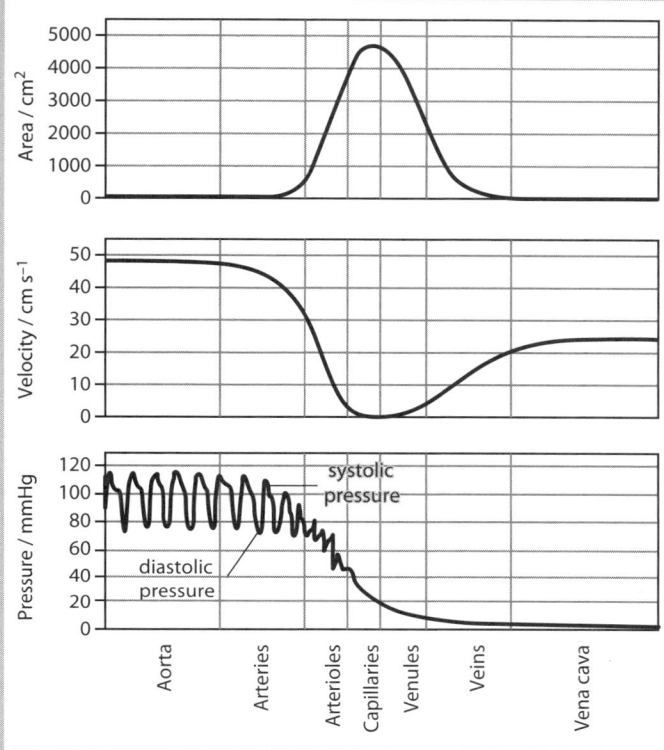

Figure 8.14

- **a** **i** Describe and explain how the blood pressures change as the blood moves from the aorta to the vena cava. [4]
  - **ii** Describe and explain how the velocity changes as the blood moves from the aorta to the vena cava. [3]
- **b** Varicose veins are veins that appear bloated and swollen with blood. They are often due to damaged valves in the veins. **Suggest** how damaged valves could lead to the appearance of varicose veins. [2]

[Total: 9]

> **COMMAND WORD**
>
> **Suggest:** apply knowledge and understanding to situations where there is a range of valid responses, in order to make proposals / put forward considerations.

Chapter 9
Gas exchange

> **CHAPTER OUTLINE**
>
> The questions in this chapter cover the following topics:
> - the structure of the human gas exchange system
> - the distribution of tissues and cells within the gas exchange system
> - the functions of these tissues and cells
> - how gases are exchanged in the lungs.

Exercise 9.1 Comparative analysis of breathing

A spirometer is a piece of apparatus that can be used for investigating the volumes of air exchanged during breathing. A range of experiments can be performed with it, such as comparing the rates of oxygen uptake of individuals under different conditions. In this exercise, you will:

- develop your understanding of gas exchange in humans
- develop your ability to analyse data and interpret graphs.

1 A team of students set out to investigate lung volumes, breathing rates and the rate of oxygen consumption.

They used a spirometer, which is shown in Figure 9.1. A spirometer can be filled with atmospheric air or pure oxygen. When using the carbon dioxide absorber, extreme care should be taken as it is very difficult to detect whether oxygen is running out. The spirometer can be used with or without the carbon dioxide absorber attached. The spirometer was filled with air.

In the first experiment, the spirometer was set up without the carbon dioxide absorber. The students attached the mouthpiece and nose clip. They breathed in and out normally, took a full breath and then exhaled all the air. They then returned to normal breathing.

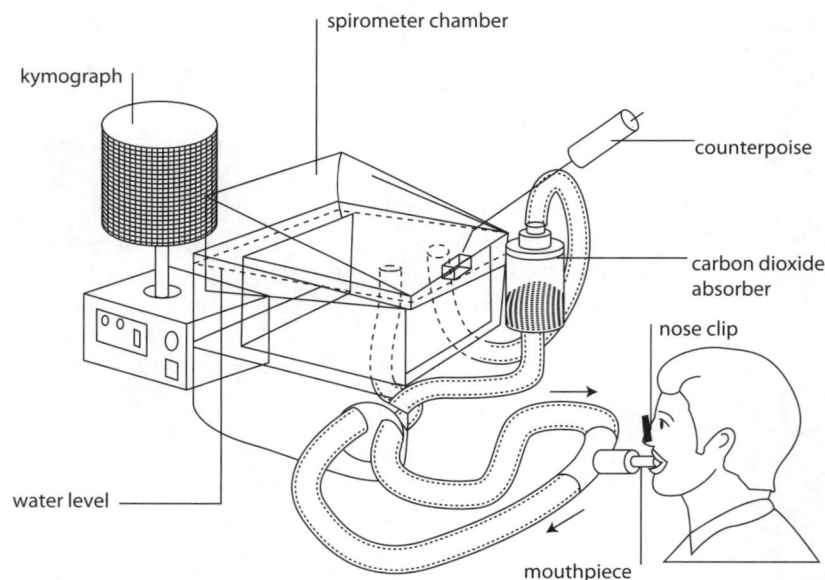

Figure 9.1: A spirometer set up to measure breathing volumes and rates.

A graph showing the changes in lung volume is shown in Figure 9.2.

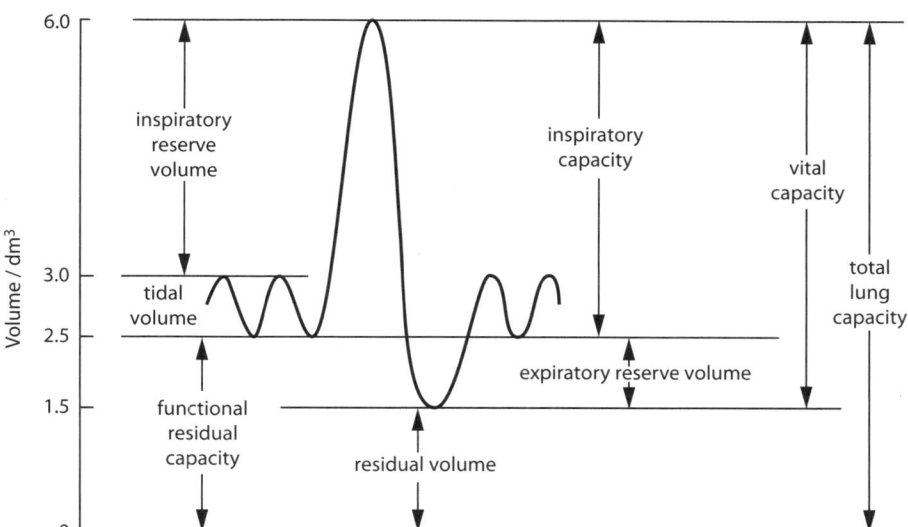

Figure 9.2: Graph showing changes in lung volume.

a The tidal volume (TV) is the amount of air exchanged during quiet breathing. Calculate the tidal volume.

b The inspiratory reserve volume (IRV) is the extra volume of air that can be inhaled, and the expiratory reserve volume (ERV) is the extra volume of air that can be exhaled. Calculate the expiratory and inspiratory volumes.

c The vital capacity (VC) is the maximum amount of air that can be exchanged in one breath. It is equal to the IRV + ERV + TV. Calculate the vital capacity.

d The residual volume is the air that remains in the respiratory system after maximum exhalation. Calculate the residual volume.

e Determine the amount of air that is exchanged during one minute if the student is performing tidal breathing at a rate of 17 breaths per minute.

2 The students then inserted the carbon dioxide absorber filled with potassium hydroxide crystals and breathed in and out of the spirometer for two minutes. The resultant trace is shown in Figure 9.3.

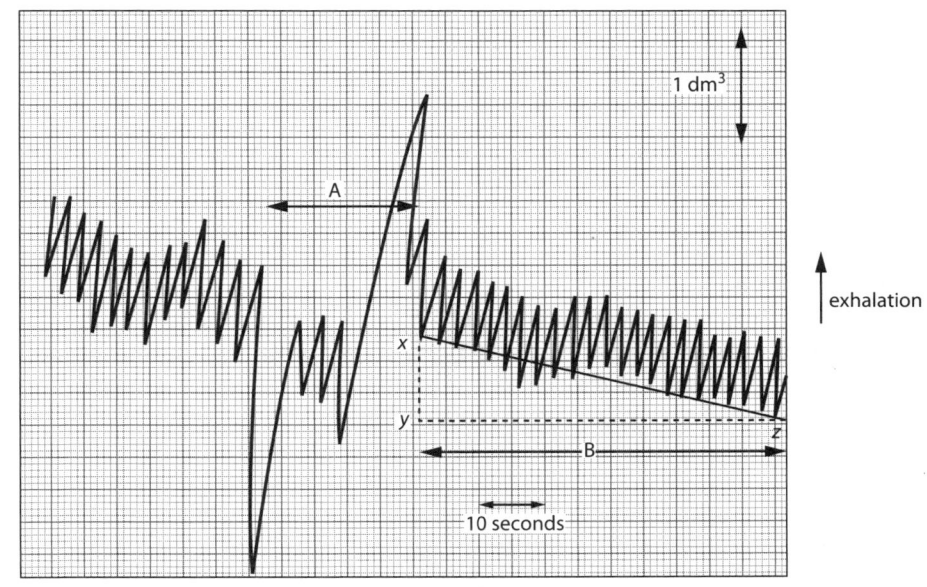

Figure 9.3: Spirometer trace obtained when potassium hydroxide is included in the carbon dioxide absorber.

a Use the graph to determine the approximate vital capacity from section **A** of the graph. To do this, follow these steps:

Step 1 Measure the length of the 1 dm³ scale bar in mm.

Step 2 Determine how much volume in dm³ that 1 mm represents (1 ÷ length in mm).

Step 3 Use a ruler to measure the approximate length of the ERV + VC + IRV.

Step 4 Multiply the length in mm by the amount that 1 mm represents.

b Now work out the breathing rate of the patient over section **B** on the diagram:

Step 1 Count the number of breaths between points y and z.

Step 2 Use the 10 s scale bar to determine how long period **B** lasts in minutes.

Step 3 Calculate the rate in breaths per min (number of breaths ÷ time taken).

c You can now use the graph to calculate the mean volume of air exchanged per minute during tidal breathing:

Step 1 Over section **B**, determine the volume of five different breaths and then calculate the mean tidal volume by adding them together and dividing by five.

Step 2 Now multiply the mean tidal volume by the breathing rate.

d The spirometer was set up with potassium hydroxide in the carbon dioxide absorber. This removes exhaled carbon dioxide so that the trace falls. The amount that the trace falls represents the volume of oxygen used.

We can thus calculate the volume of oxygen used over section **B** by drawing a gradient line $(x-z)$:

Step 1 Calculate the volume of oxygen used in dm³ by measuring line $x-y$.

Step 2 Now calculate the rate of oxygen use. This is actually the gradient of the line $x-y$. Gradients in graphs are calculated by dividing the height (y-axis) by the length (x-axis).

> **TIP**
>
> Be careful converting from seconds to minutes. Remember that 0.5 of a minute is 30 s.

3 Many medical conditions are known to affect the ability to exchange air. Peak flow meters are used to measure the forced vital capacity (FVC), which is the maximum amount of air that can be exhaled, and the forced expiratory volume in one second (FEV_1), which is the maximum amount of air that can be exhaled in one second.

Table 9.1 below shows the FVC and FEV_1 values for an unaffected adult human and patients suffering with asthma, pulmonary fibrosis and chronic obstructive pulmonary disease (COPD).

Patient	Volume of air exhaled / dm³								
	Breath 1		Breath 2		Breath 3		Mean		
	FVC	FEV$_1$	FVC	FEV$_1$	FVC	FEV$_1$	FVC	FEV$_1$	FEV$_1$/FVC ratio
Unaffected	4.80	3.85	4.75	3.78	4.78	8.82			
COPD	4.32	2.25	4.51	2.45	4.28	2.16			
Pulmonary fibrosis	3.78	3.01	3.67	3.02	3.75	3.10			
Asthma	4.75	3.09	3.45	3.01	4.73	2.95			

Table 9.1: FVC and FEV$_1$ values for an unaffected adult human and patients suffering with asthma, pulmonary fibrosis and chronic obstructive pulmonary disease (COPD).

Copy Table 9.1.

a Calculate the mean FVC and FEV$_1$ values for each condition, excluding any anomalous values, and enter your answers in the table.

b Calculate the ratio of FEV$_1$/FVC and enter your values in the table.

c It has been suggested that the FEV$_1$/FVC ratio could be used to diagnose respiratory disorders. Evaluate this suggestion.

d Asthma is a condition in which the bronchi and trachea become narrowed. Pulmonary fibrosis is a condition where the **alveoli** develop thick scar tissue. COPD is a condition where there is a severe loss of elasticity in the alveoli and narrowing of the airways.

Use this information to suggest reasons for the FEV$_1$/FVC ratio of each of the patients.

> **KEY WORD**
>
> **alveolus (plural alveoli):** one of the tiny air sacs of the lungs which allow for rapid gaseous exchange

Exercise 9.2 Comparing the structure of different parts of the airways

You need to understand the structure of the different parts of the airways and the functional roles of different cell types and tissues. In this exercise, you will:

- develop your understanding of the airway structures
- develop your understanding of gas exchange by comparing the composition of inhaled, exhaled and alveolar air.

1 Copy Table 9.2 below and fill in the blank spaces using your own knowledge. Write in 'yes' or 'no' to show presence or absence.

Region of airway	Cartilage	Smooth muscle	Ciliated epithelial cells	Goblet cells
trachea				
bronchus				
terminal bronchus				
respiratory bronchiole				
alveolar duct				
alveoli				

Table 9.2: Results table.

2 Copy Table 9.3 below and write in the functions of the different cell and tissue types.

Cell / tissue type	Function
cartilage	
smooth muscle	
ciliated epithelia	
goblet cells	

Table 9.3: Results table.

> **KEY WORDS**
>
> **smooth muscle:** type of muscle tissue found in walls of blood vessels (except capillaries), trachea, bronchi and bronchioles, alimentary canal and ureter; the muscle cells are not striated

3 You can compare the composition of gases at different points in the airway (see Table 9.4).

The value of atmospheric pressure at sea level is 100 kPa. This pressure is made up of the partial pressures of all the gases present in the air:

total atmospheric pressure = $pN_2 + pO_2 + pCO_2 + p$ (other gases)

100 kPa = 78 kPa + 21 kPa + 0.03 kPa + 0.97 kPa

(100% = 78 % + 21 % + 0.03% + 0.97%)

Region	pN_2 / kPa	pO_2 / kPa	pCO_2 / kPa	pH_2O / kPa	p(others) kPa
inhaled	78.60	20.90	0.04	0.06	0.40
alveolar	74.90	13.60	5.30	6.00	0.20
exhaled	74.50	15.70	3.60	6.00	0.20
pulmonary artery	76.39	5.33	6.00	N/A	N/A
pulmonary vein	76.39	12.67	5.33	N/A	N/A

Table 9.4: Composition of gases at different points in the airway.

a Compare the composition of inhaled, alveolar and exhaled air.
b Use your knowledge of gas exchange to explain the differences in the partial pressures of oxygen and carbon dioxide in the different airway regions.
c Explain the differences in the partial pressure of nitrogen gas in the airway regions. At atmospheric pressure, little nitrogen gas will dissolve into the blood.

Exercise 9.3 Analysing health data

Although you do not need to have a detailed understanding of how smoking tobacco affects the respiratory system, you should be aware that it is one of the leading causes of respiratory system diseases. It is important that you can compare any data in a valid way to reach appropriate conclusions. In this exercise, you will:

- develop your data analysis skills by investigating data related to smoking tobacco
- develop your graph drawing skills.

Table 9.5 shows the number of smokers found in different countries. There are clearly more smokers in China than other countries – this is not a valid comparison though, as China has more people and so would naturally be expected to have more smokers. For epidemiological data that compares health statistics it is common to express findings as a fraction of the population (often in thousands) in order to give a proportional comparison.

Country	Total population	Estimated number of smokers	Number of smokers per 1000
Russia	142 467 651	78 357 208	
China	1 393 783 836	599 327 049	
Ukraine	44 941 303	19 774 173	
Vietnam	92 547 959	29 615 347	
Egypt	83 386 739	25 849 889	
Bangladesh	158 512 570	39 628 143	
Mexico	123 799 215	14 855 906	
India	1 267 401 849	101 392 148	

Table 9.5: Number of smokers found in different countries.

To calculate the number of people per 1000 population who smoke in China you must do the following:

- calculate the proportion who are smokers (599 327 049 ÷ 1 393 783 836)
- scale it up to 1000 people by multiplying your answer by 1000

1   a   Now, copy the table and complete the last column giving the number of smokers per 1000 people for each country.

To visualise data it is often easiest to look at a graph. In this case, the independent variable is a categoric variable – it has the names of the countries – so we need to plot a bar chart.

    b   Plot a bar chart of the data – make sure that you place the independent variable on the *x*-axis and leave a gap between your bars.

When we look at epidemiological data it is often necessary to compare more than one set of data, which may be measured with different units or are in different magnitudes.

Table 9.6 shows some historic data about the incidence of deaths from lung cancer and the number of cigarettes smoked per person per year during the 20th century. The two variables have different magnitudes and so it is difficult to plot the data meaningfully using one axis.

Year	Cigarettes smoked per person per year	Lung cancer deaths per 100 000 people
1900	900	N/A
1910	1100	N/A
1920	2300	20
1930	2700	25
1940	4000	40
1950	4300	100
1960	4200	175
1970	3900	185
1980	3100	175
1990	3000	150
2000	2900	135

Table 9.6: Historic data about the incidence of deaths from lung cancer and the number of cigarettes smoked per person per year during the 20th century.

2   **Step 1**   On a piece of graph paper, draw the *x*-axis, label it 'Year' and generate a continuous scale.

    Step 2   Draw the *y*-axis on the left hand side, label it 'Cigarettes smoked per person per year' and work out a continuous scale that uses most of the graph paper.

    Step 3   Now go to the right hand side of your graph and at year 2000 draw a second *y*-axis, this time labelled 'Lung cancer deaths per 100 000 people' and work out a different scale that again uses most of the graph paper.

    Step 4   Using the left hand *y*-axis, carefully plot the points for the frequency of smoking.

As we cannot safely predict the intermediate points and the data are continuous, join the points with a ruler and a sharp pencil. Make a key for the line labelled 'Cigarettes smoked per person per year'.

Step 5 Using the right-hand *y*-axis, carefully plot the points for the frequency of lung cancer. As we cannot safely predict the intermediate points and the data are continuous, join the points with a ruler and a sharp pencil. Make a key for the line labelled 'Lung cancer deaths per 100 000 people'.

Exercise 9.4 Correlations and causal factors

A correlation is where there is a link between two variables, but we cannot say for certain that one variable causes a change in the other. A positive correlation is where both variables increase together, a negative correlation is where one variable decreases as the other increases.

Correlation does not mean that one variable is causing a change in the other one. For example, over 90% of deaths occur when people are in bed. This is a positive correlation and certainly not causation – bed is not generally a dangerous place.

To show a **causal link** we need to do a full experiment where only one variable is changed compared to a **control group**. Most data about health effects on smoking tends to show correlations.

In this exercise, you will:

- develop your understanding of correlations
- develop your understanding of scattergrams.

1 Look at your graph from Exercise 9.3, question 2. We can often look for correlations when we plot graphs that have more than one set of data on them.

   a Is there a positive or negative correlation between smoking and lung cancer deaths?

   Describe your answer in as much detail as possible.

   b Does the research prove that smoking is a cause of lung cancer, and what experiment could be performed to prove that smoking is the actual cause of lung cancer?

2 As well as plotting two line graphs on the same axes to look for correlations, we can also produce a scattergram where we plot the two variables against each other. Figure 9.4 shows a scattergram where the cigarette consumption for states in the USA is plotted against the death rate from bladder cancer in that state. A line of best fit is drawn through the points.

> **KEY WORDS**
>
> **causal link (relationship):** where one factor directly brings about an effect in another
>
> **control group:** the group in an experiment or study that does not receive treatment by the researchers and is then used as a benchmark to measure how the other tested subjects do; often referred to as the placebo group

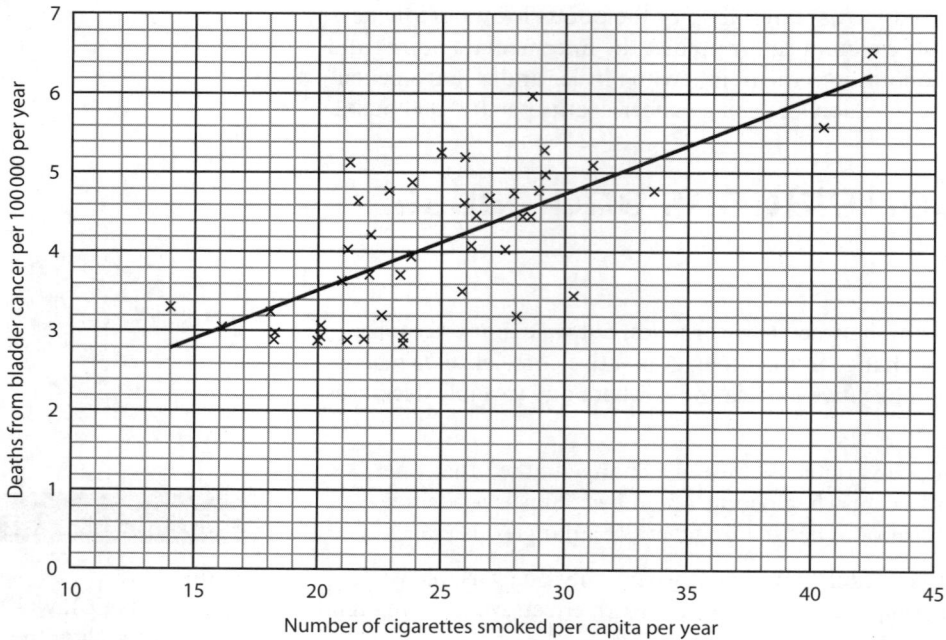

Figure 9.4: Scattergram to show number of cigarettes smoked per year per capita compared to deaths from bladder cancer per 100 000 in US states.

Each point represents the cigarette consumption plotted against the incidence of bladder cancer for that state.

a Does the graph show a positive or negative correlation between smoking and incidence of bladder cancer?

b Evaluate how strong the correlation is. For this you need to decide how much the angle of the line of best fit increases and how close all the points lie to it.

c Does the scattergram prove that smoking causes bladder cancer? What other evidence could researchers look for in order to be more secure in the link between smoking and bladder cancer?

3 Table 9.7 below shows raw data for the incidence of leukaemia and kidney cancer in the different US states. Use scattergrams to decide whether or not there is a strong correlation between cigarette consumption and kidney cancer and leukaemia. Outline your reasons for why the data fit your conclusions.

State	Cigarettes smoked per capita	Deaths per 100 000 from kidney cancer	Deaths per 100 000 from leukaemia
AK	30.34	4.32	4.90
AL	18.20	1.59	6.15
AZ	25.82	2.75	6.61
AR	18.24	2.02	6.61
CA	28.60	2.66	7.06
CT	31.10	3.35	7.20
DC	40.46	3.13	7.08
DE	33.60	3.36	6.45

State	Cigarettes smoked per capita	Deaths per 100 000 from kidney cancer	Deaths per 100 000 from leukaemia
FL	28.27	2.41	6.07
ID	20.10	2.46	6.62
IL	27.91	2.95	7.27
IN	26.18	2.81	7.00
IO	22.12	2.90	7.69
KS	21.84	2.88	7.42
KY	23.44	2.13	6.41
LA	21.58	2.30	6.71
MA	26.92	3.03	6.89
ME	28.92	3.22	6.24
MD	25.91	2.85	6.81
MI	24.96	2.97	6.91
MN	22.06	3.54	8.28
MO	27.56	2.55	6.82
MS	16.08	3.54	8.28
MT	23.75	3.43	6.90
NB	23.32	2.92	7.80
ND	19.96	3.62	6.99
NE	42.40	2.85	6.67
NJ	28.64	3.12	7.12
NM	21.16	2.52	5.95
NY	29.14	3.10	7.23
OH	26.38	2.95	7.38
OK	23.44	2.45	7.46
PE	23.78	2.75	6.83
RI	29.18	2.84	6.35
SC	18.06	2.05	5.82
SD	20.94	3.11	8.15
TE	20.08	2.18	6.59
TX	22.57	2.69	7.02
UT	14.00	2.20	6.71
VT	25.89	3.17	6.56
WA	21.17	2.78	7.48
WI	21.25	2.34	6.73
WV	22.86	3.28	7.38
WY	28.04	2.66	5.78

Table 9.7: Incidence of leukaemia and kidney cancer in US states.

EXAM-STYLE QUESTIONS

1 Figure 9.5 shows two magnifications of a section through part of an airway.

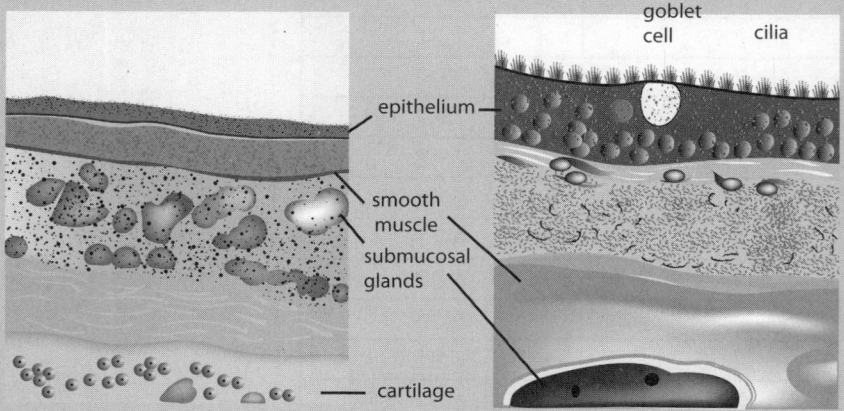

Figure 9.5

a The part of airway that the section was taken from was the:
  A bronchus
  B terminal bronchiole
  C respiratory bronchiole
  D alveolar duct. [1]

b **Explain** the functions of the goblet cells and ciliated epithelial cells. [4]

c Asthma is often caused by allergic responses and causes sufferers to have breathing difficulties. Figure 9.6 shows a cross section through a section of airway from a healthy person and a section of airway from a patient with severe asthma.

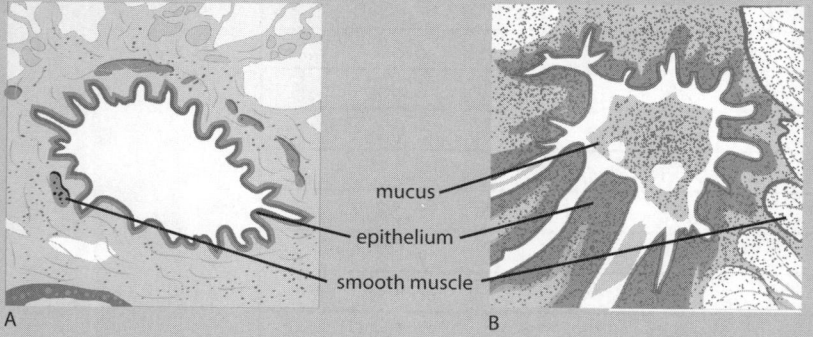

Figure 9.6

> **COMMAND WORD**
>
> **Explain:** set out purposes or reasons / make the relationships between things evident / provide why and/or how and support with relevant evidence.

CONTINUED

i Draw up a table and use it to **compare** the differences in the airways of the normal and asthmatic patients. [4]

ii Use Figure 9.6 to explain why people with asthma often have lower oxygen concentrations in their blood. [3]

iii The drug salbutamol is used to induce the smooth muscles of the airways to relax.

Figure 9.7 shows the structures of the hormone adrenaline and the drug salbutamol.

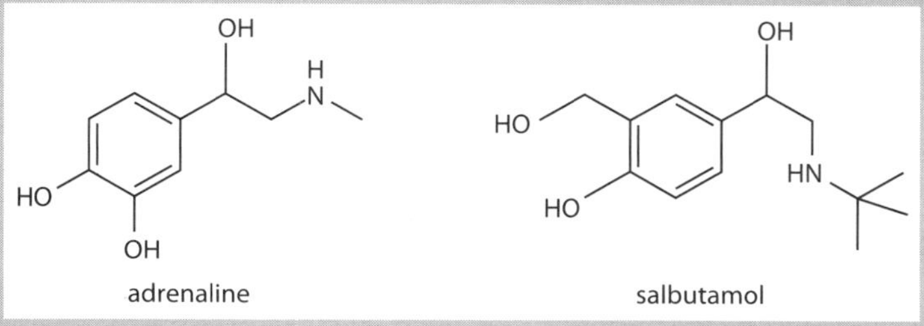

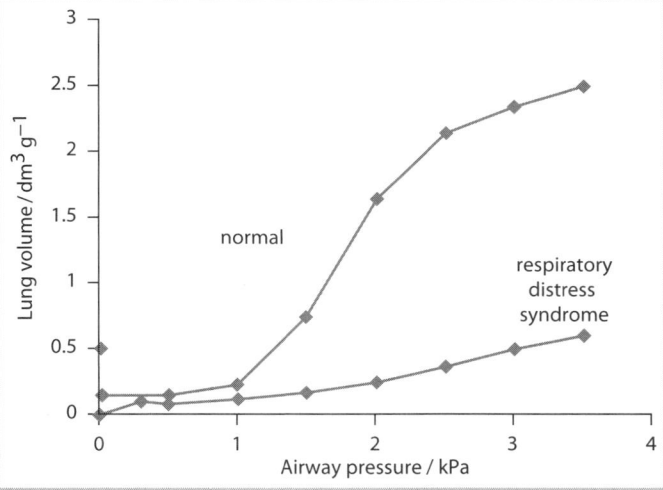

Figure 9.7

Suggest how salbutamol makes breathing easier during an asthma attack. [3]

[Total: 15]

2 Respiratory distress syndrome is a condition that occurs in young babies. It is characterised by problems in breathing and low oxygen levels in the blood. Figure 9.8 is a graph showing the amount of pressure that has to be generated in order to increase the volume of the lungs in healthy babies and those with respiratory distress syndrome.

Figure 9.8

COMMAND WORDS

Compare: identify/comment on similarities and/or differences.

Suggest: apply knowledge and understanding to situations where there is a range of valid responses, in order to make proposals / put forward considerations.

CONTINUED

a i **Describe** the effect of increasing airway pressure on the lung volume of normal babies. [2]

  ii Compare the effect of increasing airway pressure on the lung volumes of normal babies and babies with respiratory distress syndrome. [2]

b Pulmonary surfactant is a substance that is released by cells in the alveoli. It contains proteins and fatty acids and reduces the surface tension of the alveolar fluid.

Figure 9.9 shows the structure of DPPC, one of the molecules found in the surfactant fluid.

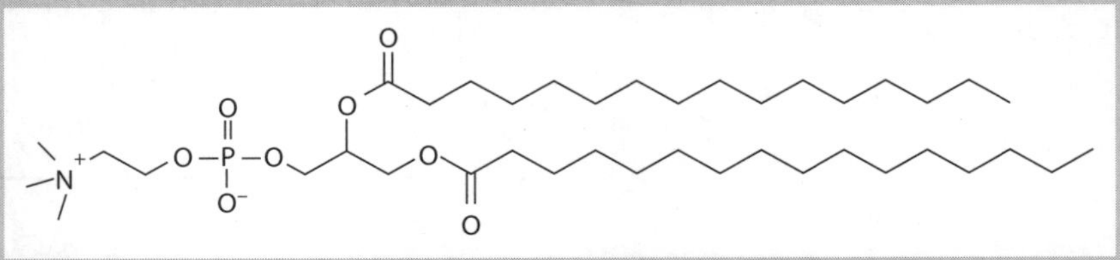

Figure 9.9

  i **State** the group of molecules to which DPPC belongs. [1]

  ii Draw out a copy of DPPC and clearly label the hydrophobic and hydrophilic regions. [1]

  iii Suggest how DPPC reduces the surface tension of aqueous solutions. [2]

c Babies who suffer from respiratory distress syndrome are often found to be short of pulmonary surfactant. Suggest and explain why these babies suffer from breathing difficulty and low oxygen concentration in the blood. [3]

[Total: 11]

3 An 'anti-smoking' vaccine has recently been trialled on smokers. Nicotine is the addictive drug found in cigarette smoke. The vaccine stimulates the body to produce **antibodies** against nicotine. When nicotine is absorbed into the blood, antibodies bind to it and prevent it passing over the blood–brain barrier and entering the brain.

In one trial, 201 smokers with no other ongoing health issues were injected with the anti-nicotine vaccine.

The smokers were divided up into two groups, half of which received a placebo (control) drug. The smokers reported how often they had smoked over the 26 weeks of the trial.

Some smokers were found to have a high response to the vaccine, producing a high concentration of antibodies, whilst others had a lower response.

COMMAND WORDS

Describe: state the points of a topic / give characteristics and main features.

State: express in clear terms.

KEY WORD

Antibody (plural antibodies): a glycoprotein (immunoglobulin) made by plasma cells derived from B-lymphocytes, secreted in response to an antigen; the variable region of the antibody molecule is complementary in shape to its specific antigen

> **CONTINUED**
>
> The percentage of patients reported as giving up smoking is shown in Table 9.8.
>
> | Level of response to vaccine | Percentage of patients receiving vaccine who did not smoke between weeks 19–26 of trial |
> |---|---|
> | high | 24.6 |
> | low | 12.6 |
> | placebo group | 13.0 |
>
> **Table 9.8**
>
> a Suggest and explain why the vaccine might help to prevent people smoking. [2]
>
> b Critics of the vaccine have suggested that the vaccine could cause some people to smoke more cigarettes. Explain why it could lead to a higher consumption of cigarettes. [2]
>
> c Explain how the placebo (control) group should be treated. [2]
>
> d Evaluate the claim that the vaccine is a success and will help to reduce smoking. [4]
>
> [Total: 10]

Chapter 10
Infectious disease

CHAPTER OUTLINE

The questions in this chapter cover the following topics:

- what is meant by infectious and non-infectious disease
- the organisms that can cause some infectious diseases
- how cholera, malaria, TB and HIV/AIDS are transmitted
- the ways in which biological, social and economic factors influence the prevention and control of cholera, malaria, TB and HIV/AIDS
- the factors that influence global patterns of malaria, TB and HIV/AIDS
- how penicillin acts on bacteria and why antibiotics do not affect viruses
- how bacteria become resistant to antibiotics
- the consequences of antibiotic resistance and how we can reduce its impact.

Exercise 10.1 Types of infectious disease

You need to know the causes and features of various infectious diseases. This exercise will:

- develop your knowledge of cholera, malaria, TB and HIV.

1   Table 10.1 summarises infectious diseases, **pathogens** and clinical features.

    Copy and complete Table 10.1 using the information provided.

Disease	Name of pathogen	Type of pathogen	Method of transmission	Clinical features

Table 10.1: Summary of infectious diseases, pathogens and their clinical features.

> **KEY WORD**
>
> **pathogen:** an organism that causes infectious disease

10 Infectious disease

Disease
- cholera
- malaria
- TB
- HIV/AIDS

Name of pathogen
- human immunodeficiency virus
- *Mycobacterium tuberculosis, M. bovis*
- *Vibrio cholerae*
- *Plasmodium vivax, P. falciparum, P. ovale, P. malariae*

Type of pathogen
- protoctist
- bacterium
- virus

Method of transmission
- semen and vaginal fluids, infected blood and blood products, across the placenta, via breast milk
- insect vector – *Anopheles* mosquito
- water borne, food borne
- airborne droplets or undercooked meat / unpasteurised milk

KEY WORDS

protoctist: a member of the Protoctista kingdom; eukaryotic organisms which are single-celled or made up of groups of similar cells

virus: very small (20–300 nm) infectious particle which can replicate only inside living cells and consists essentially of a simple basic structure of a genetic code of DNA or RNA surrounded by a protein coat

Clinical features
- fever, anaemia, nausea, headaches, shivering, sweating
- flu-like symptoms initially as the disease progresses, pneumonia, other infections, cancers, weight loss
- severe diarrhoea, weight loss, dehydration
- cough, chest pain, coughing up blood, fever, shortness of breath

Exercise 10.2 HIV/AIDS

You need to understand how HIV is transmitted. This exercise focuses on the epidemiology of HIV and AIDS. It looks at different sets of global data regarding the incidence and **prevalence** of HIV and associated diseases. In this exercise, you will:

- develop your understanding of the transmission of HIV
- develop your analytical skills by examining data on the spread of HIV in Africa.

Figure 10.1 shows the prevalence of HIV in South Africa in men and women of different age groups.

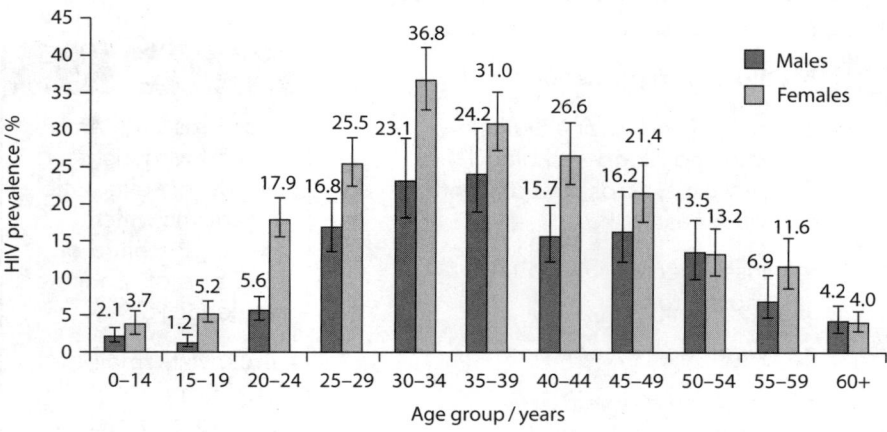

Figure 10.1: Recorded HIV cases in South Africa.

KEY WORDS

disease prevalence: the number of people who have disease at any one time

standard deviation: a measure of how widely a set of data is spread out on either side of the mean

1   a   Describe how the percentage of males *and* females with HIV changes with different age groups.

   b   Describe any 'turning points' where an increase or decrease in a trend changes direction.

   c   Now give a quantitative increase or decrease. Don't just quote data points, manipulate them. Use the graph to calculate the increase in percentage from the 0–14 age group in females to the 30–34 age group.

   d   Compare the male and female trends. Is the incidence of female HIV always higher and by approximately how much? Are there ages where the female incidence is not higher?

   e   If **standard deviation** bars are shown, use them. If the standard deviation bars (or ranges) show an overlap, the difference between male and female incidence is not significant. State which age ranges show a more significant difference in female incidence compared to male incidence.

TIP

You may be asked to describe the patterns shown by a set of data, and you may need to add more detail than just a simple trend. Consider the question 'Describe how the percentage of individuals who are HIV positive changes with age.' To answer this fully, break the answer up into several sections.

10 Infectious disease

> **TIP**
>
> After describing the data, you will often be asked to suggest an explanation. The word *suggest* means that you need to use your knowledge of the course to think of a reason that fits with the data – do not always expect to have a 'ready made' answer.
>
> Again, break up the graph into different sections:

2 **a** Suggest reasons for the general increase in incidence followed by decrease.

   b Suggest reasons for the generally higher incidence of HIV in women between the ages of 20–24. For this, think about the data. The data were gathered from blood samples that were taken when people attended doctors and hospital clinics, usually for reasons unrelated to HIV infection. The data may not be a true representation of the actual incidence of infection.

3 Figure 10.2 shows the global prevalence of HIV infection and the incidence of new cases of TB in 2003. Figure 10.3 shows the prevalence of HIV and incidence of TB in patients in South Africa between 1980 and 2000.

> **TIP**
>
> If you are asked to evaluate data, you need to look at things that support the conclusion and things that go against it.

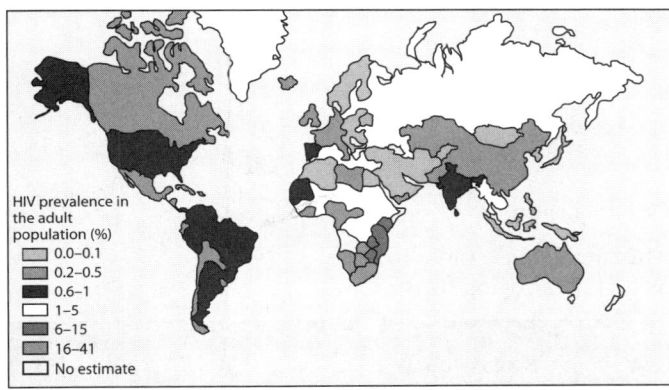
A

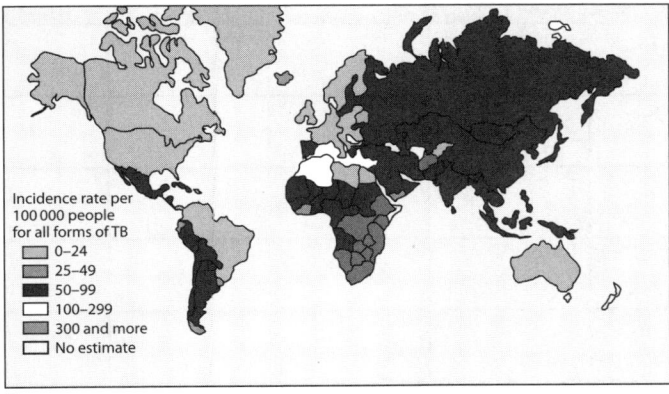
B

Figure 10.2: **A** HIV prevalence in the adult population and **B** estimated TB incidence rate in different countries.

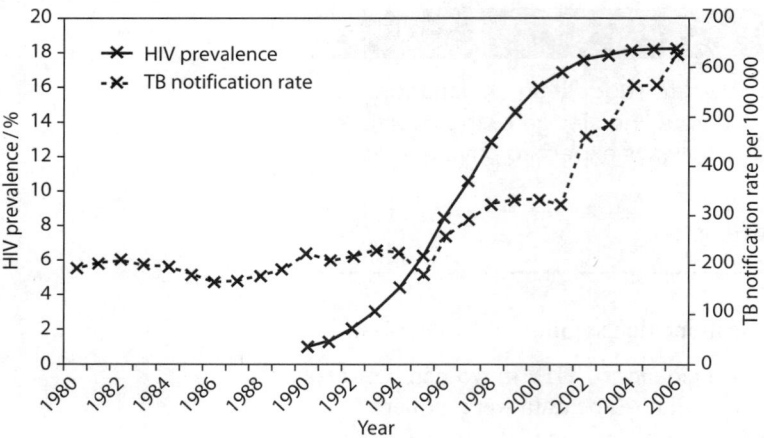

Figure 10.3: Prevalence of TB and HIV in South Africa between 1980 and 2006.

Many people have claimed that HIV is a cause of TB. Look at Figures 10.2 and 10.3 in order to evaluate this claim, and use some of your own knowledge of the diseases.

You should consider the following questions:

- Are there any correlations?
- Is there evidence for a direct causal link?
- How big is the sample size?
- Do many variables change?
- Are there any data that seem to conflict with the conclusion?

4 Pie charts are often used in order to see how proportionate data is. Table 10.2 shows a set of data for the reporting of new cases of HIV in a region of the USA.

Reported group	Number of new cases of HIV		Proportion of new cases of HIV		Angle on pie chart	
	Male	Female	Male	Female	Male	Female
homosexual contact	3436	0	0.6			
heterosexual contact	1260	1699				
injection drug use	802	352				
homosexual contact and injection drug use	172	0				
other	57	21				
total	5726	2072				

Table 10.2: Set of data for the reporting of new cases of HIV in a region of the USA.

10 Infectious disease

a Use the following steps to construct a pie chart to show the proportions of HIV in males in each of the groups:

Step 1 Copy out the table above.

Step 2 Work out the proportions of males that were HIV positive and reported homosexual contact, for example:

$(3436 \div 5726) = 0.6$

Step 3 Work out the angle that this proportion represents on a pie chart:

$0.6 \times 360° = 216°$

Step 4 Carry out calculations for all the male reported groups and enter these in the table.

Step 5 Use a compass to draw a circle of approximate radius 4 cm. Use a protractor and ruler to measure out an arc of 216° from the centre.

Step 6 Repeat this for all the categories and label each sector.

b Now use the data to plot a pie chart for female HIV cases.

c Look at your two pie charts and compare the proportions of male and female HIV cases.

d How could healthcare professionals use the data to target prevention of HIV transmission in the community?

5 Figure 10.4 shows the number of diagnoses of HIV and AIDS in the UK between 1981 and 2010 and the number of deaths from AIDS related disorders. Haart (highly active antiretroviral therapy) is the name for the treatment regime for HIV infection using combined antiretroviral drugs such as AZT.

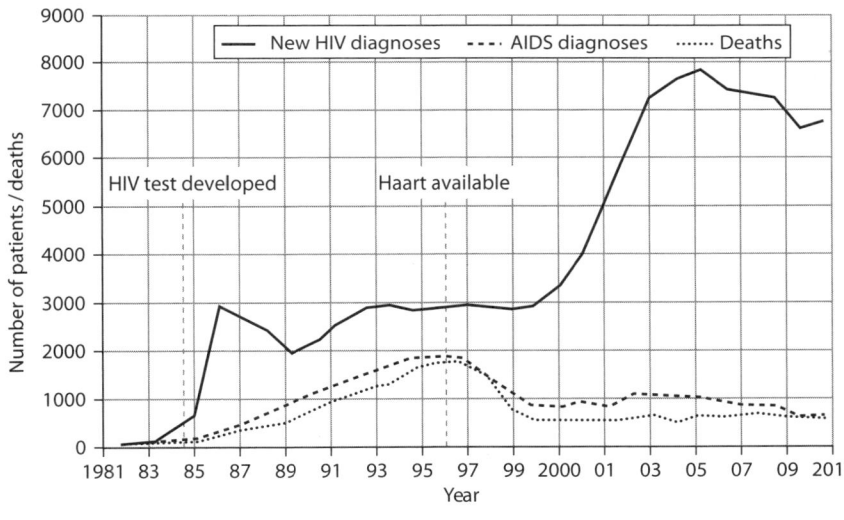

Figure 10.4: Graph of new HIV diagnoses, new AIDS diagnoses and deaths from AIDS between 1981 and 2010.

a Use the graph and your own knowledge to suggest reasons for the changes in AIDS diagnoses, new HIV diagnoses and deaths over the period 1981 to 2010.

TIP

You should look carefully at how the graphs change in relation to each other and the introductions of HIV testing and Haart treatment.

b AZT is one drug that is often used to treat HIV. The chemical structure of AZT is shown in Figure 10.5, along with the structure of the nucleotide thymidine. Use Figure 10.5 and your own knowledge of DNA structure to suggest how AZT slows down the activity of the viral enzyme reverse transcriptase.

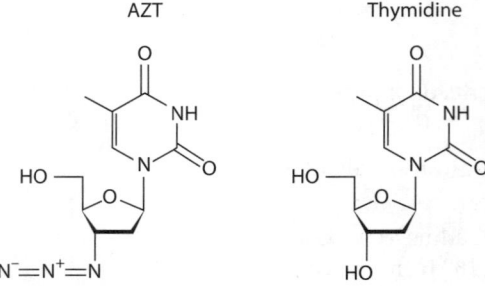

Figure 10.5: Structure of AZT and thymidine.

c Suggest why AZT can have many side-effects.

Exercise 10.3 Cholera

You need to understand how cholera is transmitted and how it can be prevented. This exercise looks at the symptoms of cholera and how the bacterium causes them. Cholera is often treated by oral rehydration therapy (ORT), and this exercise should also help you to gain an understanding of how this works and the ethics of testing new therapies. In this exercise, you will:

- develop your understanding of the causes and treatment of cholera
- develop your understanding of how to make solutions of particular concentrations
- develop your understanding of osmosis and water transport across membranes.

1 The cholera bacterium, *Vibrio cholera*, releases a toxin that causes intestinal epithelial cells to actively pump out chloride ions into the lumen of the intestine. This causes water loss from the blood and intestinal epithelial cells. This is shown in Figure 10.6.

Use your knowledge of osmosis to explain why cholera infection leads to dehydration.

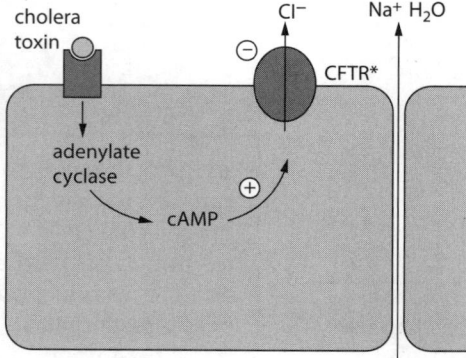

Figure 10.6: The action of cholera toxin on intestinal cells. * CFTR: A membrane protein that actively pumps chloride ions out of the cell.

2 People infected by cholera suffer from severe diarrhoea, which causes dehydration. In an effort to rehydrate patients, they are often given oral rehydration salt solutions to drink. These solutions consist of a mixture of ions and glucose measured out in carefully controlled concentrations. Figure 10.7 shows a packet of oral rehydration therapy.

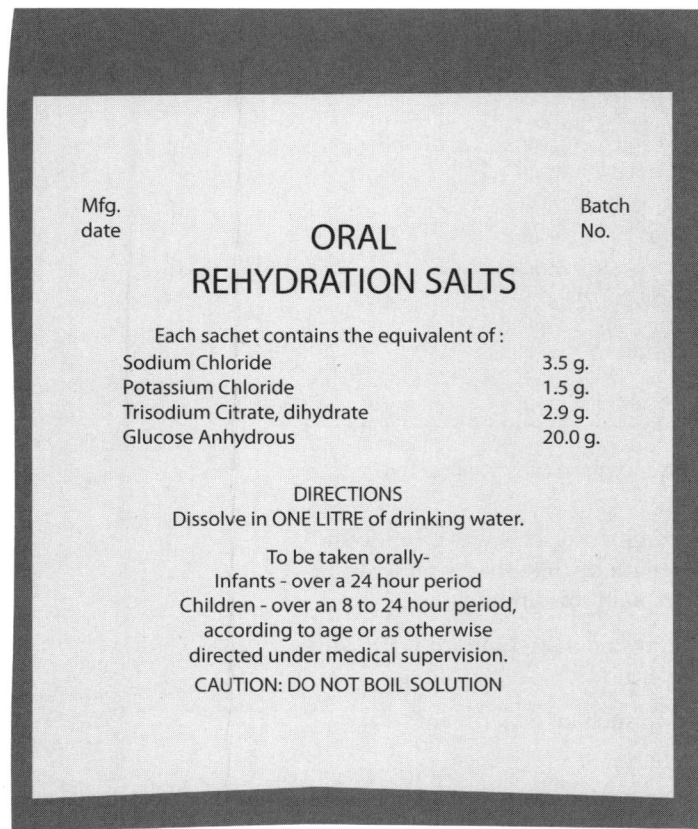

Figure 10.7: An example of oral rehydration salts.

Calculate the molar concentrations of sodium chloride, potassium chloride and glucose found in the oral rehydration salts when dissolved in 1 dm^3 of water.

3 ORT helps the intestinal epithelial cells to take up water and pass it into the blood. It is critical that the oral rehydration salts are made up to the volume shown on the label. If too much or too little water is added the diarrhoea is not improved or sometimes found to be worse.

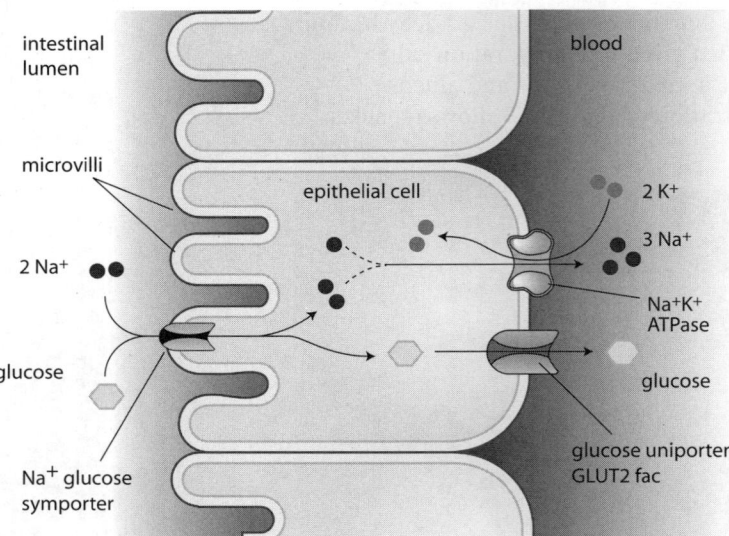

Figure 10.8: Sodium–glucose symporter in intestinal epithelial cells.

a Use Figure 10.8 to explain how ORT works and explain why it is so important to add the correct volume of water.

b Ground up rice powder is often used instead of glucose. This was often found to be more effective and worked for a longer length of time than simply using glucose. Use your knowledge of digestion to explain this finding.

4 An experiment was carried out to compare the effectiveness of standard ORS with the rice powder ORS on children infected with cholera.

a Suggest four factors that should be controlled in such an experiment.

b Suggest how the experiment could be made reliable.

c Suggest any ethical decisions that would have to be considered.

Exercise 10.4 Antibiotic resistance

Antibiotics are drugs that specifically kill bacteria. When they were initially developed in the 1940s, they were highly effective and saved millions of lives. As time has gone by more and more species of bacteria have developed resistance to them. You need to understand how bacteria have become resistant to antibiotics and the impact this is having on medicine. In this exercise, you will:

- develop your understanding of antibiotic resistance and the techniques used to test for resistant bacteria
- develop your analytical skills by investigating data from tests of the antibiotic resistance of different bacteria.

Bacteria, such as those responsible for tuberculosis, are becoming resistant to many different antibiotics. When a patient is admitted with tuberculosis, it is important to identify which antibiotics the bacteria are resistant to.

> **KEY WORD**
>
> **antibiotic:** a substance produced by a living organism that is capable of killing or inhibiting the growth of a microorganism

10 Infectious disease

1 a Copy Table 10.3 and fill in whether or not each statement is true or false.

Statement	True / false
Antibiotics cause bacteria to mutate.	
Horizontal transmission of antibiotic resistance can occur between different species of bacteria.	
Conjugation occurs in vertical transmission of antibiotic resistance.	
Antibiotics should be used to treat measles.	
Penicillin is not effective against *M. tuberculosis*.	
Overuse of antibiotics is one of the causes of MRSA.	

Table 10.3: Results table.

b Why is it important to identify the correct antibiotic to use from the point of view of the patient and the general population?

2 In order to test antibiotic resistance, there are several tests that can be performed. One of these tests is the antibiotic disc method. The stages in this are shown in Figure 10.9.

Figure 10.9: Stages in carrying out the disc method for measuring antibiotic resistance.

a State the variables that are controlled when carrying out the disc method.
b Explain why a sterile technique is essential throughout.
c Explain why the bacterial plate is examined by placing it on a black background using reflected light. Figure 10.10 shows examples of these plates.

After measuring all the zones of inhibition around the antibiotic discs, the measurements are compared with standard tables to ascertain whether the bacteria are susceptible, resistant or intermediate to a particular antibiotic. When the investigation is carried out, two quality control plates are created simultaneously using standard *E. coli* and *S. aureus* bacteria. These quality control plates are treated exactly the same as the plate with the bacteria under investigation. Table 10.4 shows the reference diameters of the zones of inhibition for different antibiotics.

Drug	Diameter of zone of inhibition / mm			Quality control strains / mm	
	susceptible	intermediate	resistant	E.coli	S. aureus
amoxicillin	≥ 17	14–16	≤ 13	16–22	27–38
chloramphenicol	≥ 18	13–17	≤ 12	21–27	19–26
ciprofloxacin	≥ 21	16–20	≤ 15	30–40	22–30
erythromycin	≥ 23	14–22	≤ 13	N/A	22–30
nalidixic acid	≥ 19	14–18	≤ 13	22–28	N/A
norfloxacin	≥ 17	13–16	≤ 12	28–35	17–28
co-trimoxazole	≥ 18	≤ 12	≤ 10	23–29	24–32
tetracycline	≥ 19	15–18	≤ 14	18–25	24–30

Table 10.4: The reference diameters of the zones of inhibition for different antibiotics.

d What is the purpose of the quality control strain plates?

e Figure 10.10 shows antibiotic disc test plates for two patients. Also shown are the quality control plates.

The magnification of the plates is ×0.75.

Use a ruler to determine the **true** diameters of the zone of inhibition for each plate. Draw out Table 10.5 to display your results and write in the final columns whether or not the bacteria are S – susceptible, I – intermediate, R – resistant, or N – not known, for both patients. Record 'N' for not known if the control plates suggest that an antibiotic was not working.

Drug	Diameter of zone of inhibition /mm				Resistance state of bacteria (R, I, S, N)	
	E. coli	S. aureus	Patient 1	Patient 2	Patient 1	Patient 2
(1) amoxicillin						
(2) chloramphenicol						
(3) ciprofloxacin						
(4) erythromycin						
(5) nalidixic acid						
(6) norfloxacin						
(7) co-trimoxazole						
(8) tetracycline						

Table 10.5: Results table.

10 Infectious disease

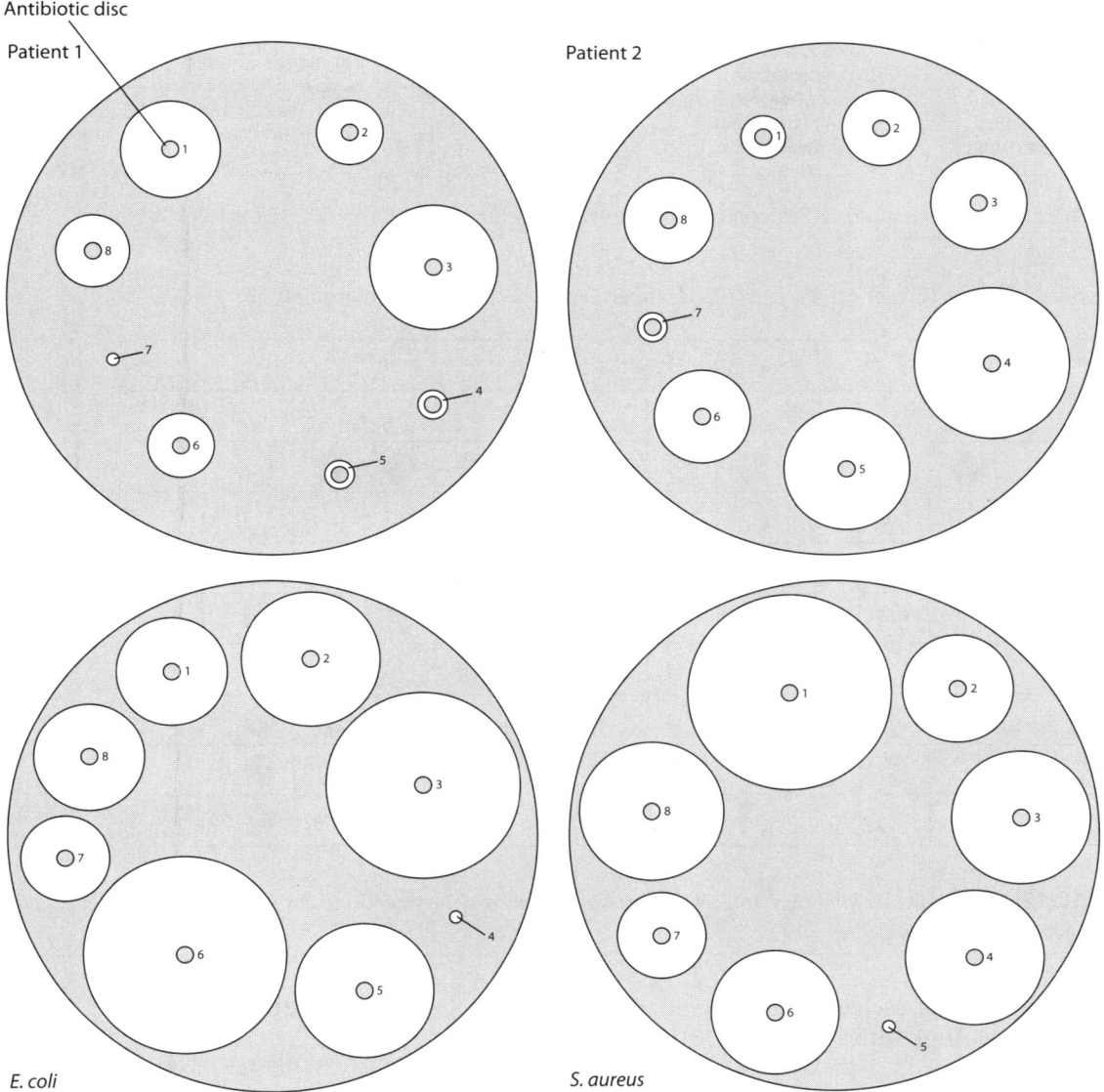

Figure 10.10: Antibiotic resistance test plates on bacteria taken from two patients infected with *M. tuberculosis*, and two quality control plates (*E. coli* and *S. aureus*). The number by each disc corresponds to the number given to each antibiotic in the table.

    f Some of the zones of inhibition for patient 2 had isolated bacterial colonies growing within them. Suggest what these colonies could be caused by.

3 The MIC or Minimal Inhibitory Concentration technique is a method for identifying the minimum concentration of antibiotic that will kill a particular strain of bacteria. It is used for identifying the level of resistance the bacteria have developed to different antibiotics. The method is shown in Figure 10.11.

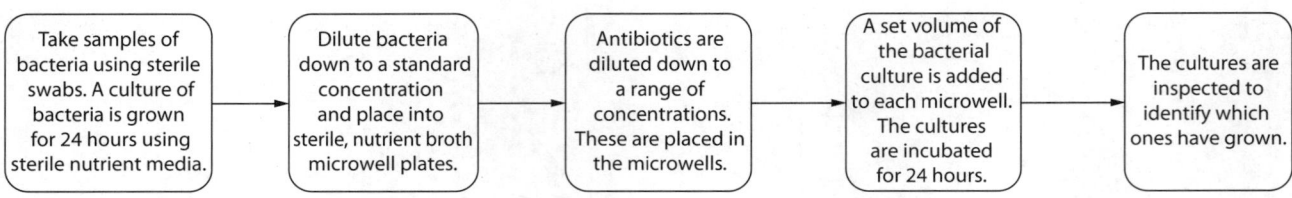

Figure 10.11: Minimal Inhibitory Concentration method for determining antibiotic resistance in bacteria.

The results of an MIC test on a strain of *M. tuberculosis* are shown in Figure 10.12.

Antibiotic	Concentration of antibiotic / µg dm⁻³											0 (sterility control)
	128	64	32	16	8	4	2	1	0.5	0.25	0	
Amoxicillin	●	●	●	●	●	●	●	●	●	●	●	○
Chloramphenicol	○	○	○	○	○	○	○	○	○	○	○	○
Ciprofloxacin	○	○	○	○	○	○	○	●	●	●	●	●
Erythromycin	○	○	○	○	○	○	○	○	●	●	●	○
Nalidixic acid	○	○	○	○	○	●	●	●	○	●	●	○
Norfloxacin	○	●	○	○	○	○	○	○	●	●	●	○
Co-trimoxazole	○	●	●	●	●	●	●	●	●	●	○	○
Tetracycline	○	○	○	○	○	○	○	○	○	●	○	○

Figure 10.12: The results of Minimum Inhibitory Concentration test on a strain of *M. tuberculosis*. Shaded ovals show where bacteria are growing.

Explain how 10 cm³ of each concentration of antibiotic solution could be made by serial dilution, starting with a 128 µg dm⁻³ stock solution.

State which antibiotic test:

a   would need repeating due to contamination of the stock broth
b   has one well which received cross-contamination
c   has a MIC of 1.0 µg dm⁻³
d   has a MIC of 0.25 µg dm⁻³
e   lacked live bacteria in all samples
f   has bacterial resistance at all concentrations tested
g   had a well missed out when bacteria were added
h   has a MIC of 128 µg dm⁻³.
i   Explain the roles of the tests with no antibiotic and the sterility control.

10 Infectious disease

EXAM-STYLE QUESTIONS

1 a Figure 10.13 shows the number of recorded cases of malaria and the rainfall in a region of South Korea over a 12-month period.

   i **Give** the name of *one* of the species of the parasite that causes malaria. [1]

   ii **Describe** the change in number of recorded cases of malaria over the 52-week period. [2]

   iii Use the data in Figure 10.13 and your own knowledge of the life cycle of the malaria parasite to **explain** the change in number of recorded cases of malaria over the 52-week period. [5]

b Artemisinin is a drug that was first used to treat malaria in the 1990s. By 2013, 392 million treatment courses were being used. In 2007, the WHO called for artemisinin to only be prescribed as part of a multiple drug package where patients received a high dose of the drug alongside another one or two anti-malaria drugs. By 2015, large numbers of cases of malaria parasite resistance against artemisinin in South-East Asia were reported. Figure 10.14 shows the proportion of cases of malaria that were resistant to artemisinin in different countries.

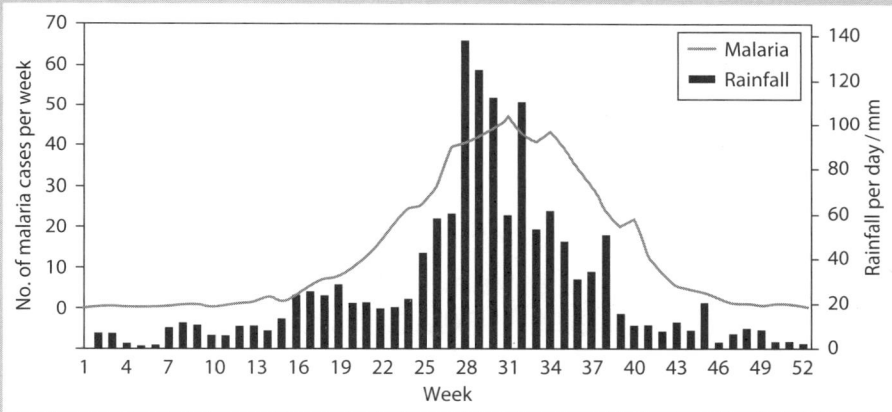

Figure 10.13

COMMAND WORDS

Give: produce an answer from a given source or recall/memory.

Describe: state the points of a topic / give characteristics and main features.

Explain: set out purposes or reasons / make the relationships between things evident / provide why and/or how and support with relevant evidence.

CONTINUED

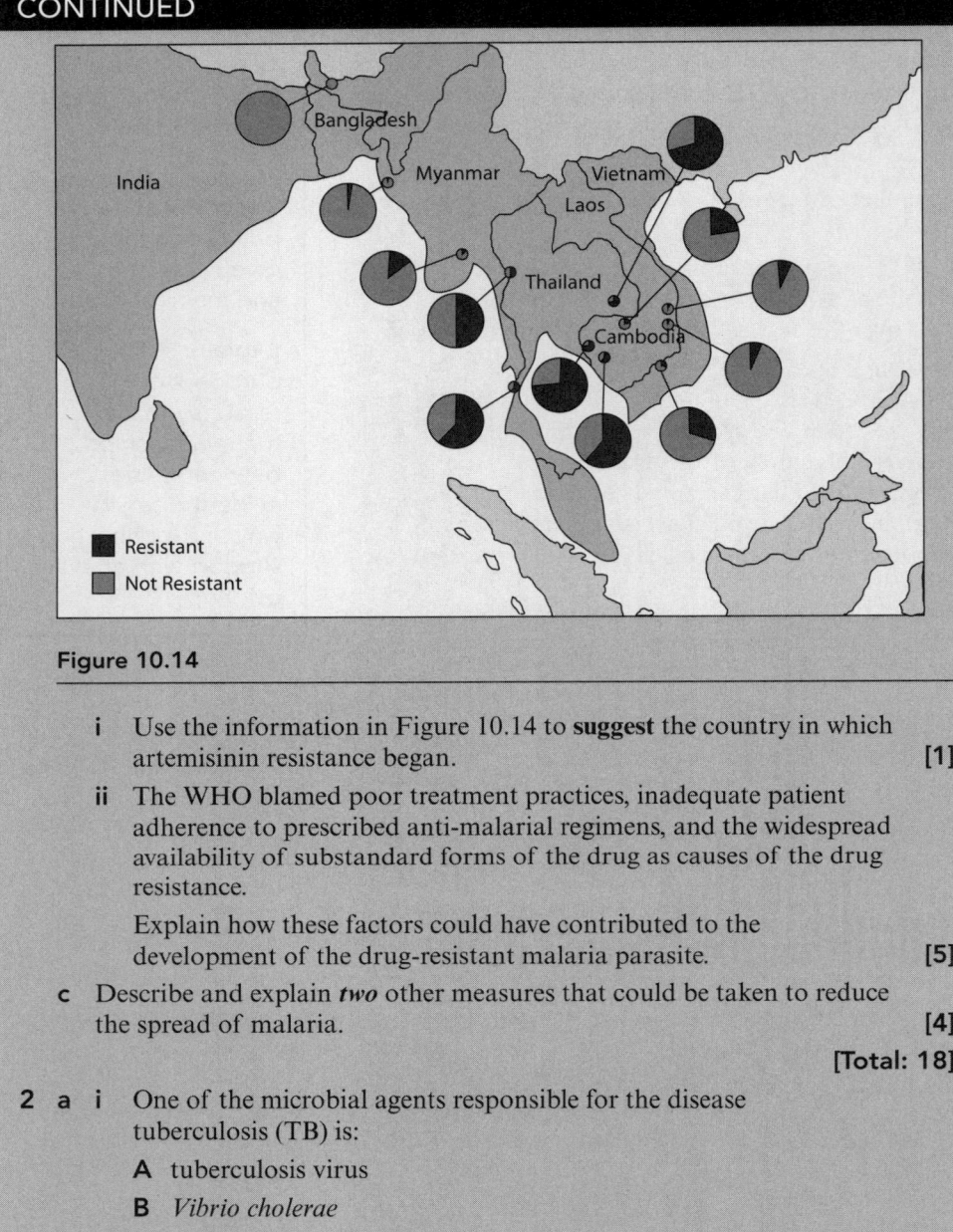

Figure 10.14

  i   Use the information in Figure 10.14 to **suggest** the country in which artemisinin resistance began. [1]

  ii The WHO blamed poor treatment practices, inadequate patient adherence to prescribed anti-malarial regimens, and the widespread availability of substandard forms of the drug as causes of the drug resistance.

   Explain how these factors could have contributed to the development of the drug-resistant malaria parasite. [5]

c Describe and explain *two* other measures that could be taken to reduce the spread of malaria. [4]

[Total: 18]

2 a i One of the microbial agents responsible for the disease tuberculosis (TB) is:

   A tuberculosis virus
   B *Vibrio cholerae*
   C *Mycobacterium bovis*
   D *Plasmodium falciparum*. [1]

  ii **Outline** the method of transmission and the clinical features of tuberculosis. [5]

COMMAND WORDS

Suggest: apply knowledge and understanding to situations where there is a range of valid responses, in order to make proposals / put forward considerations.

Outline: set out the main points.

CONTINUED

b Figure 10.15 shows the effect of GDP, a measure of national wealth, on the prevalence of TB in different European Union countries.

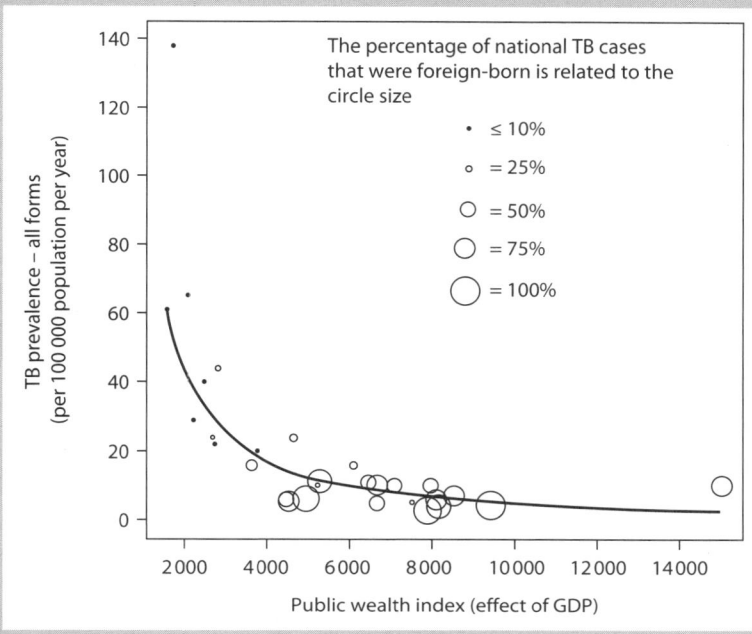

Figure 10.15

   i Use the information in Figure 10.15 to describe the effect of GDP on the prevalence of TB. [3]

   ii Explain the effect of GDP on tuberculosis prevalence. [4]

c Tuberculosis is treated with a 6–9 month course of several antibiotics. In some countries, people who refuse to receive treatment for the full course may be placed into detention and forcibly treated. Explain the reasons why it is important to ensure that treatment is completed. [6]

[Total: 19]

3 In 2010, a major earthquake occurred in Haiti causing the destruction of many buildings and much of the infrastructure of essential services, such as water and electricity supplies. In the aftermath of the earthquake, thousands of people entered refugee camps. Several weeks after the earthquake, a cholera epidemic began that lasted several years.

  a **i** Name the organism that causes cholera. [1]

     ii Outline the clinical features of cholera infection. [2]

     iii Suggest and explain why the cholera epidemic occurred after the earthquake. [6]

CONTINUED

b The number of cases of cholera in Haiti that occurred after the earthquake, between 2010 and June 2014, and the cholera fatality rate are shown in Table 10.6.

Year	(Oct–Dec) 2010	2011	2012	2013	June 2014	Total
cases	185 351	352 033	101 722	58 650	7 451	705 207
deaths	4 101	2 927	927	572	32	8 559
fatality rate / %	2.2		0.9	1.0	0.4	1.2

Table 10.6

   i **Calculate** the fatality rate for 2011. [1]
   ii Suggest and explain the WHO measures that caused the reduction in the number of cases and in the fatality rate during the years after the earthquake. [3]
   iii Explain why more economically developed nations generally have a much lower incidence of cholera. [3]

[Total: 16]

COMMAND WORD

Calculate: work out from given facts, figures or information.

4 Figure 10.16 shows a diagram of the HIV virus.

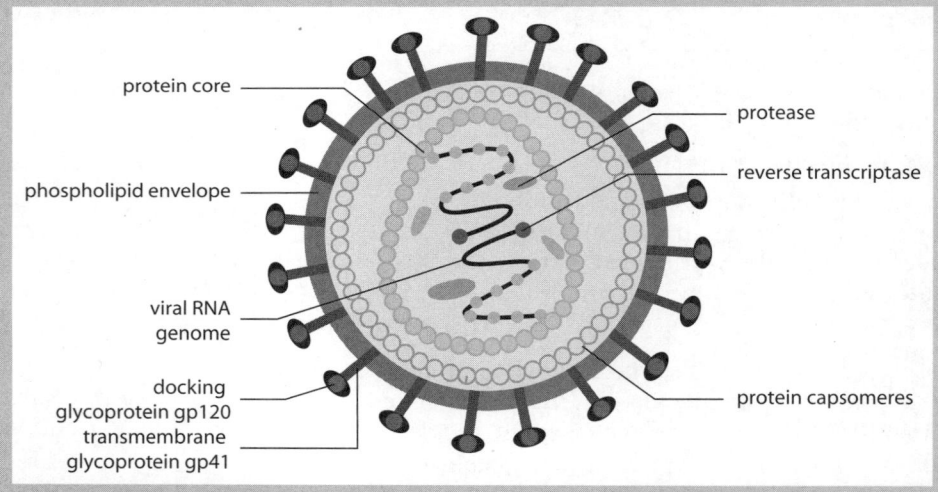

Figure 10.16

10 Infectious disease

CONTINUED

a Explain how the structure of the HIV virus enables it to infect cells. [4]

b Figure 10.17 shows the estimated global population infected with HIV, the annual increase in new HIV infections and the death rate from AIDS.

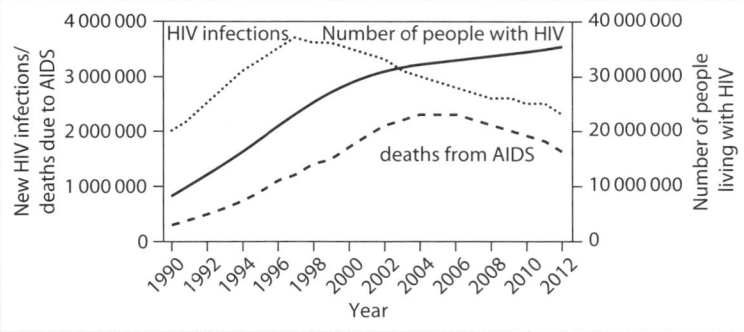

Figure 10.17

i Explain the difference between HIV and AIDS. [2]

ii **Compare** and **contrast** the trends in new HIV infections and the number of people living with HIV between 1990 and 2012. [2]

iii Suggest and explain the trend in the number of people living with HIV between 1990 and 2012. [4]

Figure 10.18 shows the projected age pyramids for the population of Botswana both with and without AIDS.

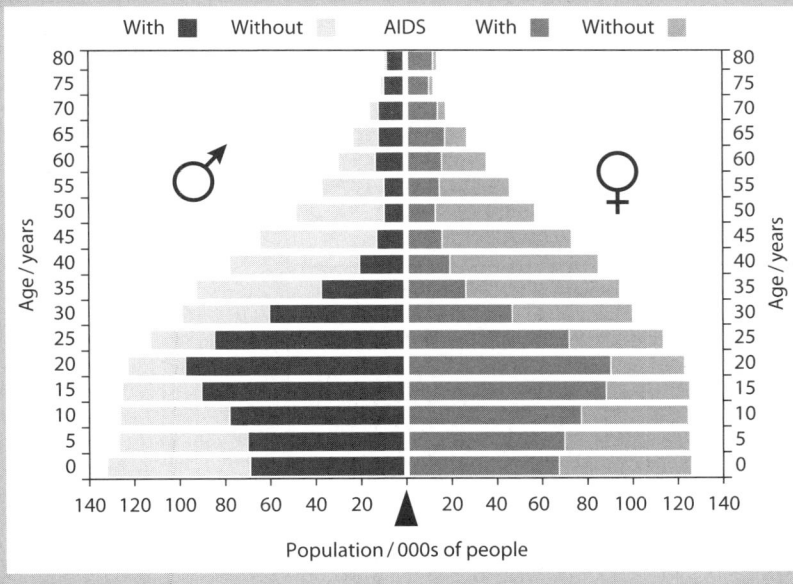

Figure 10.18

COMMAND WORDS

Compare: identify / comment on similarities and/or differences.

Contrast: comment on differences.

CONTINUED

    iv Use Figure 10.18 to describe the projected effect of AIDS on the population profile of Botswana in 2020. **[3]**

    v Explain the differences in the population age pyramids with AIDS and without AIDS. **[3]**

    vi Use Figure 10.18 to suggest and explain the socioeconomic difficulties that Botswana may face in 2020. **[4]**

    [Total: 22]

5 Puerperal fever was a common cause of death of women following childbirth in the 19th century and early 20th century. It is caused by bacterial infection.

Ignaz Semmelweis investigated the transmission of puerperal fever in two clinics in Vienna in the 19th century. In one ward, childbirth was supervised by medical students who had often carried out prior anatomical dissections. In the second ward, childbirth was supervised by midwives who did not carry out other duties.

Semmelweis reached the conclusion that the medical students were transferring bacteria to the women after pathological anatomical dissections. He formulated a hypothesis that hand washing in chlorine solutions would reduce the spread of infections.

Table 10.7 shows results of the research of Ignaz Semmelweis on the causes of puerperal fever. The numbers of maternal deaths are shown at different times.

  a **i** Calculate the death rate from puerperal fever after pathological anatomy (1823–33). **[1]**

    ii Use the table below to explain how Semmelweis reached his conclusion that medical students were transferring the infections from pathological anatomy investigations. **[2]**

    iii Evaluate the hypothesis that hand washing in chlorine solutions prevented infections. **[3]**

	Births	Maternal deaths	Death rate / %
before pathological anatomy (1784–1823)	71 395	897	1.25
after pathological anatomy (1823–33)	28 429	1 509	
Ward 1, male medical students, no chlorine washings (1841–7)	20 042	1 989	9.92
Ward 2, female midwives, no chlorine washings (1841–7)	17 791	691	3.38
Ward 1, male medical students, since chlorine washings (1891–1959)	47 938	1 712	3.57
Ward 2, female midwives, since chlorine washings (1891–1959)	40 770	1 248	3.06

Table 10.7

CONTINUED

b In the 1940s, antibiotics became available to treat bacterial infections. Figure 10.19 shows the death rate of mothers from bacterial infection following childbirth in the UK.

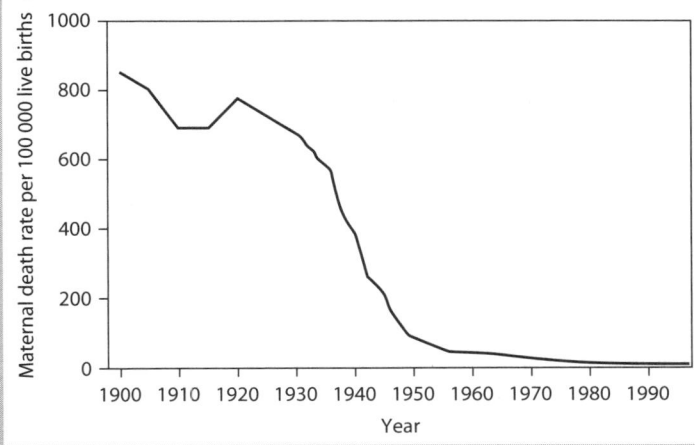

Figure 10.19

i   Explain why the data are expressed as number of deaths per 100 000 live births. [1]

ii Use Figure 10.19 to describe the impact of antibiotics on maternal death rate between 1940 and 1950. [2]

iii Describe the processes by which bacteria such as MRSA have become resistant to many different antibiotics since the widespread use of antibiotics beginning in the 1940s. [6]

iv Explain what measures should be taken to reduce the rate at which bacteria are becoming resistant to antibiotics. [3]

[Total: 18]

Chapter 11
Immunity

> **CHAPTER OUTLINE**
>
> The questions in this chapter cover the following topics:
> - the roles of phagocytes, B-lymphocytes and T-lymphocytes in the immune response
> - the structure and function of antibodies
> - active, passive, natural and artificial immunity
> - vaccination
> - monoclonal antibodies.

Exercise 11.1 Writing a good answer on the immune response

Before answering a question, it is often a good idea to take a few minutes to plan your answer before you begin to write. It is also useful to notice the number of maximum marks available. This will help you to judge how many relevant points you should try to include in your answer. Also remember to try to use appropriate subject-related terminology where you can.

1   Look at the following question:

   Explain the roles of *phagocytes* (macrophages), *B-lymphocytes* and *T-lymphocytes* in the immune response. [5]

   a Read the three sample answers.

   Example X

   T-lymphocytes start the immune response. They are activated by specific antigens. T-helper cells release cytokines which stimulate specific B-cells to divide into plasma cells and produce antibodies and also stimulate macrophages to carry out phagocytosis. T-killer cells patrol the tissues and if they encounter non-self antigens on the surface of host cells they bind to the cell and release toxins like hydrogen peroxide to kill the host cell and the pathogen inside it. They also cause the lysis of these cells. B-lymphocytes are activated when they recognise complementary specific antigens from a pathogen which they recognise as non-self. They divide by mitosis (clonal expansion) into plasma cells and memory cells. Plasma cells produce antibodies which cause pathogens to agglutinate, neutralise toxins etc. Memory cells remain in the body for a second encounter with the same antigen.

> **KEY WORDS**
>
> **phagocyte:** a type of cell that ingests (eats) and destroys pathogens or damaged body cells by the process of phagocytosis; some phagocytes are white blood cells
>
> **B-lymphocyte:** a type of lymphocyte that gives rise to plasma cells and secretes antibodies
>
> **T-lymphocyte:** a lymphocyte that does not secrete antibodies; T helper lymphocytes stimulate the immune system to respond during an infection, and killer T lymphocytes destroy human cells that are infected with pathogens such as bacteria and viruses

Phagocytes carry out phagocytosis, engulf harmful bacteria by endocytosis, fuse with the pathogen, and enter body in a phagocytic vacuole and digest them with hydrolytic enzymes of lysosomes. Also, they cut the pathogens out to display their antigen on their surface for antibodies (humoral response).

Example Y

Macrophages are phagocytes that have as a function to uncover the antigen (foreign cell) receptor, for lymphocytes to know how to attack the foreign cell. T-lymphocytes can be either helper, which stimulate B-lymphocytes to produce antibodies or killer, which attach to pathogens and secrete toxins. They normally use the receptor uncovered by macrophages.

Example Z

Macrophages, B-lymphocytes and T-lymphocytes are used to detect and destroy foreign bodies/antigens that come into contact with the body's immune system. Receptors on antibodies detect these antigens and cause B-lymphocytes to be secreted (memory and plasma cells) and T-helper and killer cells. The memory cells detect the non-self matter during the first immune response and plasma cells are used to secrete more antibodies to destroy the antigens. T-lymphocytes continue the process by destroying, and antigens and the memory cells remain in the event of these antigens entering again so that a rapid secondary immune response is met.

    i   Rank the three answers in order, with the one that you think is the best first.

    ii   Write a brief commentary on each answer, stating clearly what is good about each one, and what could be improved.

b Write your own answer to the question.

Exercise 11.2 Choosing the appropriate type of graph

Graphs are drawn to display data in a form that helps people to understand the relationship between the variables on the x-axis and the y-axis. You will already have drawn many bar charts, line graphs and histograms (frequency diagrams), and this exercise gives you practice in deciding when to use each of these methods of displaying information.

Line graphs are used when there is a continuous variable on both the x-axis and the y-axis. A continuous variable is one that goes up steadily – there is a continuous range of numbers, and the data can be of any value within that range. Each point is plotted as a small cross (or a dot with a circle around it), and the points are then joined with ruled straight lines between points, or a best fit line is drawn. See example in Figure 11.1.

> **KEY WORDS**
>
> **line graph:** a graph drawn when both the x-axis variable and the y-axis variable are continuous

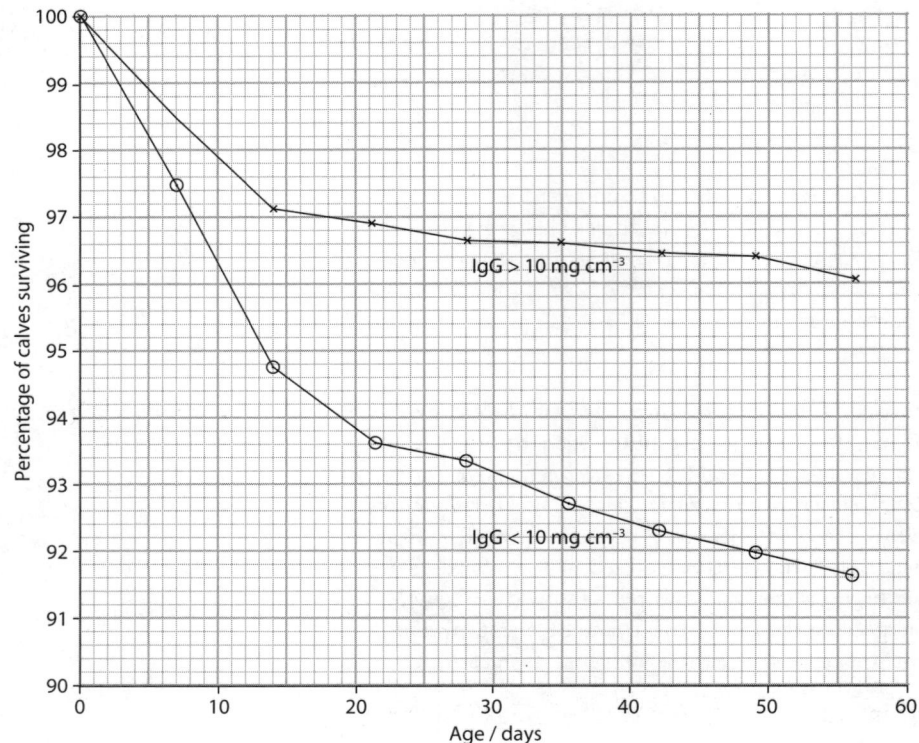

Figure 11.1: Line graph showing the survival rates of calves with two levels of IgG antibodies in their blood in the first 56 days after birth.

Bar charts are used when there is a discontinuous variable on the x-axis and a continuous variable on the y-axis. A discontinuous variable is one where there are distinct categories that are separated from one another. The bars are drawn so that they do not touch. See example in Figure 11.2.

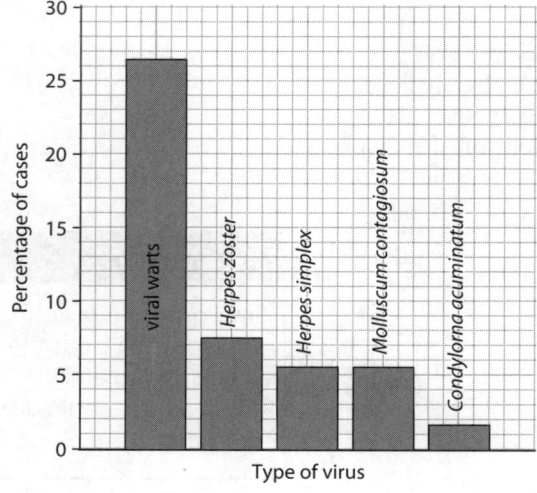

Figure 11.2: Bar chart showing the frequency of different viral infections of the skin, in patients treated with immunosuppressant drugs after a kidney transplant.

KEY WORDS

bar chart: a graph drawn when the x-axis variable is discontinuous and the y-axis variable is continuous; bars do not touch

Histograms, also called **frequency diagrams**, are used when the variable on the *x*-axis is a series of categories that are continuous, and the *y*-axis shows the frequency of each category. The bars are drawn touching one another. See example in Figure 11.3.

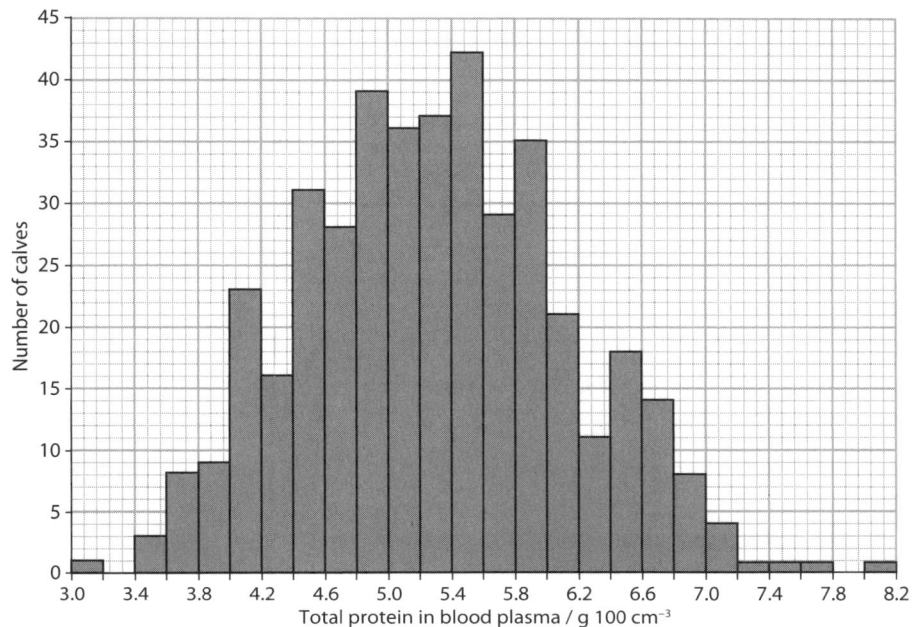

Figure 11.3: Histogram showing the number of calves with different levels of total protein (including antibodies) in their blood 24 hours after birth.

In the example in Figure 11.3, the numbers on the *x*-axis scale are shown on the lines between the bars. These numbers show where the range for the next bar begins. So, for example, the histogram shows that there are 16 calves with a level between 4.20 and 4.39 g 100 cm^{-3}. You may also see histograms where the numbers are placed between the lines, so that they align with the centre of each bar; this is also acceptable.

KEY WORDS

histogram (frequency diagram): a graph drawn when the variable on the *x*-axis is a series of categories that are continuous, and the *y*-axis shows the frequency of each category; bars touch each other

TIP

There is not always a sharp dividing line between these types of graph. There are occasions when either a line graph or a histogram would be suitable, and others when either a bar chart or a histogram would be suitable. This is the case in some of the questions below, so don't worry too much if you are having trouble deciding between two alternative types of graph. Consider the question carefully, make your decision and then justify it.

1   A study was carried out to find out if a person's sociability affected how likely they were to get a cold. A sample of 334 people were given tests to determine how sociable they were. Each person was assigned to one of five categories, 1 to 5, with category 1 having the highest sociability and category 5 having the lowest sociability.

The percentage of people in each category getting a cold during the experimental period was recorded. The results are shown in Table 11.1.

Sociability score	Percentage of people getting a cold
1	34
2	28
3	27
4	22
5	18

Table 11.1: Percentage of people by category getting a cold during the experimental period.

a   Decide on the best way to display these results graphically, and give your reasons.
b   Draw the graph to display the results.

2 Researchers measured the numbers of a type of lymphocyte called an NK cell in different parts of the bodies of mice that had been injected with an antigen. They recorded the NK cells as the percentage of the lymphocytes in that part of the body. The results are shown in Table 11.2.

Part of body	Percentage of lymphocytes that were NK cells
spleen	2.1
blood	4.0
lung	5.4
liver	3.2
bone marrow	0.8

Table 11.2: Percentage of NK cells in different parts of the bodies of mice that had been injected with an antigen.

a   Decide on the best way to display these results graphically, and give your reasons.
b   Draw the graph to display the results.

3 Table 11.3 shows the number of reported cases of *Haemophilus influenzae* (Hib) infection in Canada between 1985 and 2004.

Year	Number of cases of reported cases of Hib
1985	410
1986	543
1987	518
1988	690
1989	516
1990	407
1991	385
1992	291
1993	133
1994	80
1995	72
1996	81
1997	82
1998	58
1999	19
2000	24
2001	47
2002	46
2003	56
2004	91

Table 11.3: Number of reported cases of *Haemophilus influenzae* (Hib) infection in Canada between 1985 and 2004.

a Decide on the best way to display these results graphically, and give your reasons.

b Draw the graph to display the results.

Exercise 11.3 Displaying and analysing data about colostrum

This exercise gives you more practice in choosing the right kind of graph to draw. There is a calculation to do, a question that asks you to remember earlier work on movement across cell membranes, and other questions that check your understanding of how antibodies confer immunity.

1 Colostrum is a type of milk produced by female mammals just after giving birth. Newborn mammals normally have very poorly developed immune systems, and are unable to make sufficient quantities of antibodies to protect themselves against infections. Colostrum contains relatively large quantities of a type of antibody called immunoglobulin G, or IgG.

   a What type of immunity is provided by the antibodies in colostrum? Choose from:

   - active, natural
   - active, artificial
   - passive, natural
   - passive, artificial

   b The ability of the small intestine of a newborn calf to absorb antibodies from the milk it drinks from its mother decreases rapidly after birth. Table 11.4 shows the percentage of antibodies that can be absorbed at birth and up to 30 hours afterwards.

Time after birth / hours	Percentage of antibodies in milk that are absorbed
0	32
6	15
12	7
18	3
24	1
30	0

Table 11.4: Percentage of antibodies that can be absorbed at birth and up to 30 hours afterwards.

   i Display these results, using the most appropriate type of graph.

   ii Antibodies are proteins. Using your knowledge of how molecules are transported across cell membranes, suggest how antibodies are absorbed into the cells lining the small intestine, in a very young calf.

iii Farmers are advised to give calves colostrum from a bottle feeder if the calf has not drunk from its mother within three hours of birth. Use the information in the text and table above to explain the reasons for this.

c The milk of cows was analysed just after giving birth (colostrum) and on the second day afterwards. The results are shown in Table 11.5.

Component of milk	Time after birth / days	
	0	2
fat / %	6.7	3.9
protein / %	14.0	5.1
antibodies / %	6.0	2.4
lactose / %	2.7	4.4
minerals / %	1.11	0.87
vitamin A / µg 100 cm^{-3}	295	113

Table 11.5: Composition of the milk of cows analysed just after giving birth (colostrum) and on the second day afterwards.

i Display these data as a bar chart, with 'Component of milk' on the x-axis.

Draw two bars touching one another for each component, representing the values for the two different times after birth. You will need to use a key to identify the two types of bar. You may also like to use two different y-axes.

ii Explain why these results should be shown as a bar chart, not a histogram or a line graph.

iii A young calf may drink as much as 2.5 dm^3 of milk in one feeding session.

Calculate the percentage difference between the quantity of antibodies it would receive in one feeding session immediately after birth and two days after birth. Show all of your working.

Exercise 11.4 Determining the effectiveness of a vaccine

This exercise, like the previous ones, asks you to choose the right kind of graph to draw. You will also need to use your knowledge about how the immune system responds to infection, and – in part (d) – apply knowledge about the effectiveness of vaccination programmes in a new context.

1 Infection with a rotavirus is a major cause of childhood death worldwide. Some children with this infection have mild diarrhoea for a short period of time, but others have severe diarrhoea, accompanied by vomiting and fever, which often

results in death. More than 80% of children are infected with this virus before they are 5 years old, but severe and potentially fatal symptoms are normally seen only in children aged between 3 and 35 months.

a A child who has had a rotavirus infection is not totally protected against further infections, but these will only be very mild.

   i Using your knowledge of the immune system, explain how surviving a first infection with a rotavirus gives protection against future infections.

   ii Name the type of immunity that you have described in your answer to **i**.

   iii Suggest why children who have survived a rotavirus infection may still be infected again, even though only mildly.

b Research to develop a safe, effective vaccine against rotavirus began in the 1970s in the USA. Trials determined that a vaccine containing attenuated viruses was safe and effective, particularly if given in three oral doses at the ages of 2 months, 4 months and 6 months.

   i Use your knowledge of the immune response to suggest why three doses of the vaccine were given.

   ii Suggest why the vaccine was given at these ages.

c Some parents reported that their children suffered mild illness after being given the vaccine. Research was carried out to test whether these claims were true. Children were given either the vaccine or a placebo (a substance that did not contain the vaccine). The children, the parents and the researchers did not know whether a particular child had been given the vaccine or the placebo. The percentage of children developing a fever during the next five days was recorded.

The results are shown in Table 11.6.

Dose	Percentage with temperature above 38 °C	
	Given vaccine	Given placebo
1 (at 2 months)	22	6
2 (at 4 months)	11	8
3 (at 6 months)	13	12

Table 11.6: Percentage of children developing a fever during five days after receiving a vaccination dose or placebo.

   i Display these data using the type of graph you think is most appropriate. Justify your choice.

   ii Discuss the extent to which these data support the parents' concern that the vaccine caused fever.

d New versions of the rotavirus vaccine have now been developed. Only two doses are now required, and symptoms of fever after vaccination are much milder and much rarer. The vaccine is now given routinely to many children in developed and developing countries all over the world. However, rotavirus continues to be a common infection in small children.

Suggest why vaccination programmes for rotavirus have not succeeded in eradicating the associated disease.

CAMBRIDGE INTERNATIONAL AS & A LEVEL BIOLOGY: WORKBOOK

EXAM-STYLE QUESTIONS

1 Figure 11.4 shows two types of white blood cell (leucocyte).

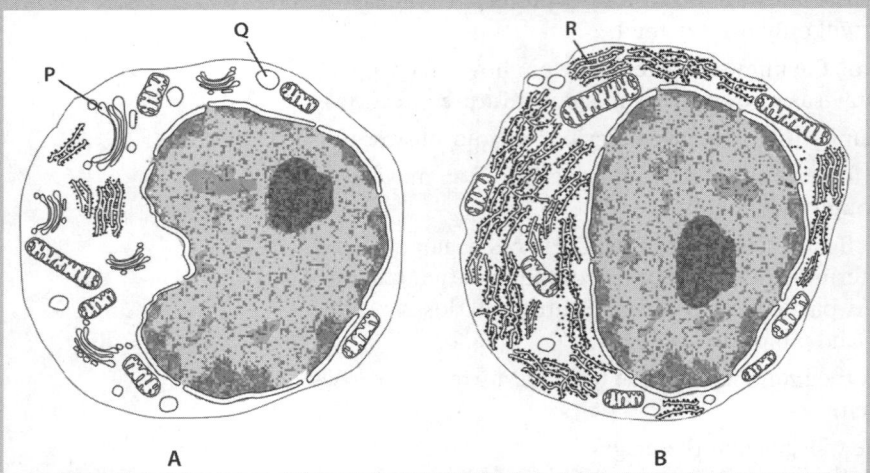

Figure 11.4

   a **Identify** cells **A** and **B**. [2]
   b Name organelles **P**, **Q** and **R**. [3]
   c **Explain** why cell **A** contains a relatively large number of organelle **P**. [3]
   d Explain why cell **B** contains a large quantity of **R**. [3]
   [Total: 11]

2 a Figure 11.5 shows how some of the different types of blood cell are formed.

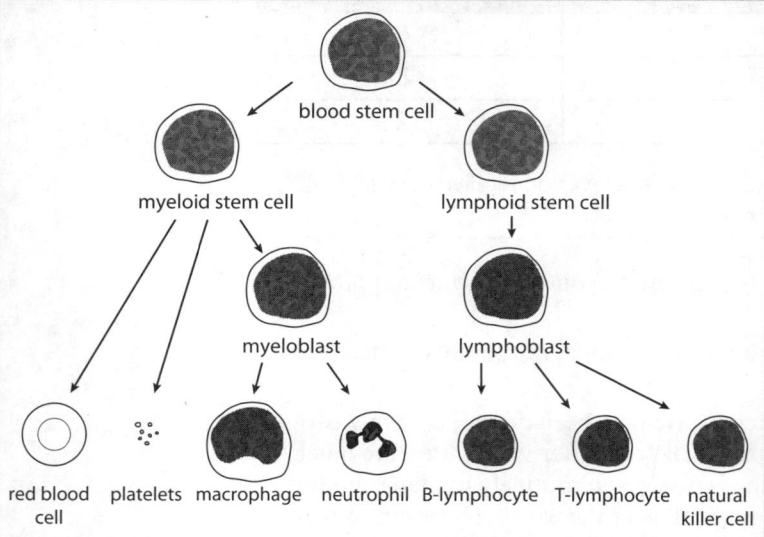

Figure 11.5

TIP

As you near the end of your course, it is important to keep revisiting earlier work. In this question, you are asked to link knowledge you have gained in your study of the immune system with knowledge relating to cell structure and function.

COMMAND WORDS

Identify: name / select / recognise.

Explain: set out purposes or reasons / make the relationships between things evident / provide why and/or how and support with relevant evidence.

TIP

This question tests your understanding of two different parts of the syllabus. For section a, you will need to think back to what you have learnt about mitosis and stem cells. It is worth taking time to plan your answer to b iii before starting to write.

CONTINUED

    i Name the type of nuclear division that takes place during the formation of these cells. [1]

    ii With reference to Figure 11.5, explain the meaning of the term *stem cell*. [2]

b For the production of monoclonal antibodies, plasma cells derived from B-lymphocytes are fused with a cancer cell.

    i Explain why neither B-lymphocytes nor plasma cells alone could be used to produce monoclonal antibodies. [2]

    ii Name the type of cell that is produced by the fusion of a plasma cell and a cancer cell. [1]

    iii **Describe** how monoclonal antibodies can be used in the diagnosis of disease. [4]

[Total: 10]

3 Chronic lymphocytic leukaemia (CLL) is a type of cancer that occurs when the processes that produce blood cells, in the bone marrow, result in the formation of large numbers of abnormal B-lymphocytes. These lymphocytes do not work as they should, and there are so many of them that not enough functioning blood cells are produced.

Table 11.7 shows information about the numbers of lymphocytes in a healthy person and a patient with CLL.

Person	Lymphocyte count /number of cells dm^{-3}
healthy person	3.5×10^9
patient with CLL	6.1×10^{10}

Table 11.7

a **Calculate** the percentage difference between the lymphocyte count of the healthy person and the patient with CLL. Show your working. [3]

b B-lymphocytes, but not other cells, have a protein called CD20 in their cell surface membranes (see Figure 11.6). This protein is known as the B-lymphocyte antigen. The function of the protein is not known, but it is thought that it may function as a Ca^{2+} channel.

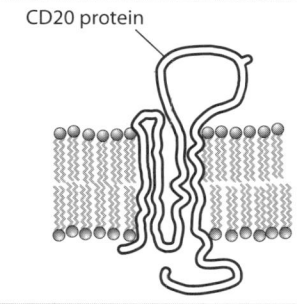

Figure 11.6

COMMAND WORDS

Describe: state the points of a topic / give characteristics and main features.

TIP

This question touches on many different parts of the syllabus. In **a**, you can practise doing a calculation with numbers in standard form. Take care in **c ii**, where there are two command words – make sure that you respond to each of these in your answer.

COMMAND WORDS

Calculate: work out from given facts, figures or information.

CONTINUED

      i Explain how a protein such as CD20 could function as a Ca^{2+} channel. [3]

      ii With reference to the CD20 protein, explain what is meant by the term *antigen*. [2]

      iii It was found that the percentage of lymphocytes with the CD20 protein in their membranes was 95.7% in the patient with CLL, and 7% in the healthy person.

         Explain this difference. [2]

  c A monoclonal antibody, rituximab, is used in the treatment of CLL. Rituximab binds with the CD20 protein.

    Figure 11.7 shows the structure of rituximab.

      i Explain how the structure of an antibody such as rituximab is related to its function. [3]

      ii A patient with CLL was given rituximab intravenously. The initial dose was 50 mg, followed by a further 150 mg after 24 hours, and a further 500 mg after 48 hours.

         The total lymphocyte count in blood samples taken at intervals during and after this treatment was recorded and then calculated as the percentage change in count compared with the count at time 0.

         The results are shown in Figure 11.8.

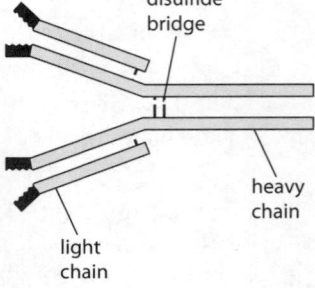

Figure 11.7

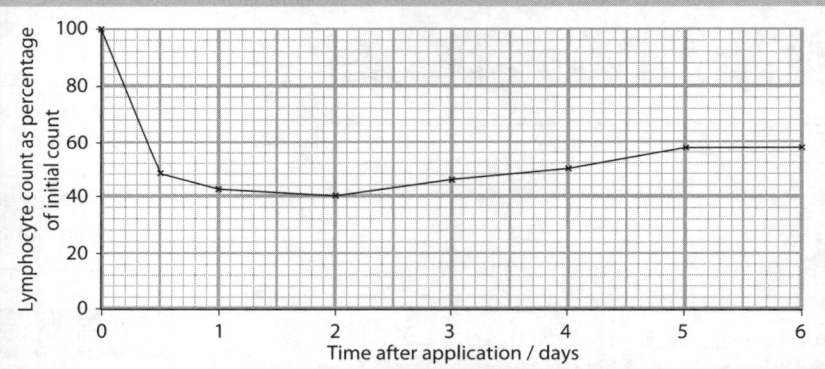

Figure 11.8

Describe **and** explain the changes in the total lymphocyte count during and after treatment with rituximab. [4]

[Total: 17]

> Chapter 12
Energy and respiration

CHAPTER OUTLINE

The questions in this chapter cover the following topics:

- the need for energy in living organisms
- the role of ATP as the universal energy currency
- how energy from complex molecules is transferred to ATP in the process of respiration
- the different energy values of respiratory substrates
- how to use simple respirometers
- how to use a redox dye to investigate respiration in yeast
- how the structure of a mitochondrion is adapted for its functions
- respiration in aerobic and anaerobic conditions
- how rice plants are adapted for growing with their roots submerged in water.

Exercise 12.1 Anaerobic respiration in yeast

You need to understand the process of **anaerobic respiration** in yeast. In this exercise, you will:

- develop your understanding of how to compare data, identifying anomalous values and comparing means whilst taking into account how the raw data vary
- develop your understanding of how to draw results tables
- develop your understanding of how to measure anaerobic respiration in yeast
- develop your understanding of how to use significant figures
- develop your understanding of how to calculate and use standard deviation.

KEY WORDS
anaerobic respiration: the enzymatic release of energy from organic compounds in living cells in the absence of oxygen from ethanol or lactate fermentation

1 Below is the method used by a student carrying out an investigation into the effect of different sugars on the anaerobic respiration of yeast.

Hypothesis: Different sugars will cause the yeast to respire anaerobically at different rates.

Apparatus:

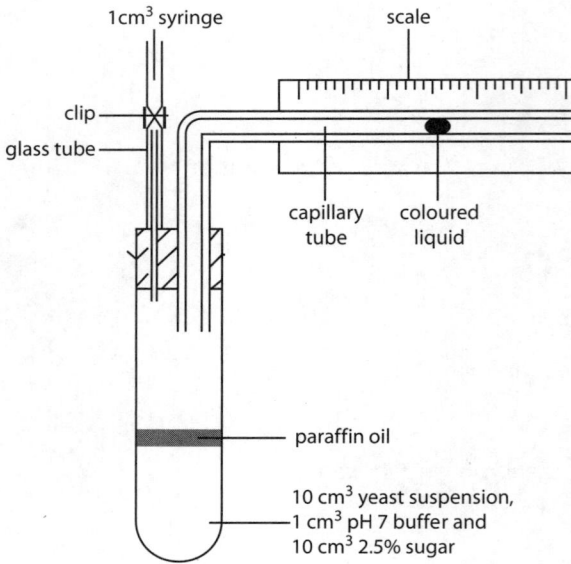

Figure 12.1: Apparatus used to investigate the effect of different substrates on anaerobic respiration in yeast.

Method: The apparatus was set up as shown in the diagram. Using a 10 cm³ pipette, 10 cm³ of 10% yeast suspension in water was placed into a boiling tube. Using a separate 10 cm³ pipette, 10 cm³ of 2.5% glucose solution was placed into a separate tube and 1 cm³ of pH 7 buffer added. All solutions were made up with boiled water that had been cooled. Both boiling tubes were placed into a water bath at 30°C for ten minutes. The solutions were then mixed and the bung placed into the boiling tube, which was then placed into the water bath. The coloured liquid was set to the start of the scale. The distance moved by the liquid in five minutes was then measured. This was repeated ten times with glucose.

The experiment was repeated with 2.5% sucrose solution, 2.5% maltose solution and with distilled water.

a i Copy Table 12.1. Identify the independent, dependent and controlled variables and the methods used to change, measure or control them in the experiment, and write them in Table 12.1. There is more than one controlled variable.

Variable type	Named variable	Method of changing, measuring or controlling
independent		
dependent		
controlled		

Table 12.1: Results table.

ii Explain the purpose of the following:
- the syringe
- the paraffin oil
- using boiled water to make up all solutions
- placing the separate boiling tubes of sugar and yeast in the water bath for ten minutes before mixing.

iii Explain why the coloured dye will move along the scale.

iv Explain why an experiment is set up with water instead of a sugar.

It is often difficult to decide how many significant figures that numbers should be recorded to in scientific experiments. The general rule is that for raw data, the number of significant figures used should be appropriate to the least accurate piece of measuring equipment.

For example, a 1 cm^3 pipette has a scale on it divided up into 0.01 cm^3 increments. Measurements should not be given to more or fewer significant figures. So, 1 cm^3 of a solution measured using this pipette would be written as 1.00 cm^3.

b i A 10 cm^3 pipette has a graduated scale that can measure to 0.1 cm^3. Using the correct number of significant figures, write down the correct measurement of 5 cm^3 of water when measured using this pipette.

When raw data are processed, you can use one more significant figure than the least accurate piece of measuring equipment. For example, three mitochondria were found to be of diameters 0.77 μm, 0.84 μm and 0.93 μm. The mean diameter of the mitochondria is:

(0.77 + 0.84 + 0.93) ÷ 3 = 0.8466666666666 recurring

The accuracy of the scale used was 0.01 μm, so we state the mean diameter as 0.847 μm.

ii The masses of six separate samples of starch from different potatoes were found to be 25.27 g, 28.43 g, 22.46 g, 21.00 g, 23.45 g and 27.45 g. The balance was accurate to 0.01 g. Which of the following gives the mean mass to the correct number of significant figures?

A 24.67666 B 24.677 C 24.676 D 24.67 E 24.7

2 Table 12.2 shows the student's table of results for an experiment.

Number of repeated experiment	Distance moved by coloured dye in 5 minutes			
	Glucose	Sucrose	Maltose	Water
1	6.9 cm	7.42 cm	3.22 cm	0.22 cm
2	7.34 cm	7.38 cm	3.4 cm	0.10 cm
3	7.2 cm	7.54 cm	3.46 cm	0.12 cm
4	8.1 cm	8.5 cm	2.9 cm	0.3 cm
5	7.54 cm	8.32 cm	2.65 cm	0.31 cm
6	0.2 cm	7.65 cm	2.91 cm	0.2 cm

Number of repeated experiment	Distance moved by coloured dye in 5 minutes			
	Glucose	Sucrose	Maltose	Water
7	7.1 cm	19.21 cm	3.1 cm	0.01 cm
8	7.8 cm	6.43 cm	2.77 cm	0.22 cm
9	7.98 cm	6.5 cm	2.8 cm	0.1 cm
10	7.52 cm	7.3 cm	3.01 cm	0.3 cm
mean	6.768 cm	8.62 cm	3.01 cm	0.19 cm

Table 12.2: Student's experimental results.

When drawing a table, you should follow certain rules:

- Ensure that all raw data are shown in a single table with ruled lines and border.
- Place the independent variable in the first column.
- Place the dependent variable in columns to the right (for quantitative observations) or descriptive comments in columns to the right (for qualitative observations).
- Place processed data (e.g. means, rates, standard deviations) in columns to the far right.
- Do not place calculations in the table, only calculated values.
- Ensure that each column is headed with an informative description (for qualitative data) or physical quantity and correct SI units (for quantitative data); units should be separated from the physical quantity using a solidus (slash).
- Do not place units in the body of the table, only in the column headings.
- Ensure that raw data are recorded to the number of decimal places and significant figures appropriate to the least accurate piece of equipment used to measure them.
- Ensure that all raw data are recorded to the same number of decimal places and significant figures.
- You can add processed data that are recorded to up to one decimal place more than the raw data.

a   The scale used to measure the distance the coloured oil travelled was accurate to 0.1 cm. Copy Table 12.3 and identify which features of the student's table are correct.

Table feature	Correct (Y/N)	Description of errors
all raw data in a single table with ruled lines and border		
independent variable in the first column		

Table feature	Correct (Y/N)	Description of errors
dependent variable in columns to the right (for quantitative observations) or descriptive comments in columns to the right (for qualitative observations)		
processed data (e.g. means, rates, standard deviations) in columns to the far right		
no calculations in the table, only calculated values		
each column headed with informative description (for qualitative data) or physical quantity and correct SI units (for quantitative data); units separated from physical quantity using either brackets or a solidus (slash)		
no units in the body of the table, only in the column headings		
raw data recorded to a number of decimal places and significant figures appropriate to the least accurate piece of equipment used to measure them		
all raw data recorded to the same number of decimal places and significant figures		
processed data recorded to up to one decimal place more than the raw data		

Table 12.3: Analysis of student's results.

b Draw a new table that takes into account all the features listed above. Do not add the mean values yet.

When processing the data, it is often useful to calculate a mean value. To generate an accurate mean value, however, we need to look closely at the raw data and identify any data points that appear not to fit the general pattern and are, in other words, anomalous. In biological experiments, it is important to note that many outlying data points may just be due to natural variation, and we must be absolutely certain that a point is not due to natural variation before we class it as an anomaly. In the student's table, the values of 0.2 cm in the glucose column and 19.21 in the sucrose column appear to be a long way from the trend of other results and therefore anomalous. You should not include these or any other anomalous values that you have identified.

c Now calculate the mean distance moved by the coloured dye in five minutes when glucose is the substrate. Do not include the value of 0.2 cm in your calculation. Write this in your table.

d Now calculate the mean distances for the coloured dye movement for all the other substrates, taking into account any clearly anomalous values.

We can now compare the mean values of the substrates. Mean values do not, however, tell the whole story. The mean distance moved by the dye when the substrate is glucose is higher than the mean distance moved when the substrate is sucrose. This does not necessarily mean, however, that every time that the yeast has glucose as a substrate the dye moves further than when sucrose is used. To help compare different means and get an idea of how reliable a set of data is, it is useful to see how much variation around the mean the raw data shows.

The range is simply the minimum value to maximum value; in the case of the distance for glucose this would be from 6.9 cm to 8.1 cm, if we discount the anomaly. The range can, however, include outliers, and a better measure is standard deviation, which tells us how much the majority of the data is spread about the mean.

Many sets of data will produce a symmetrical pattern when plotted as a frequency diagram; this is called a **normal distribution**. Examples of this are shown in Figure 12.2. The larger the standard deviation, the wider the spread of data about the mean. When looking at repeats in experimental data, a low standard deviation suggests that the data points are all close to the mean and that the mean is reliable. If the standard deviation is high, it shows that there is a lot of variation about the mean and our data are less reliable.

> **KEY WORDS**
>
> **normal distribution:** a set of data in which a graph shows a symmetrical distribution around the central value

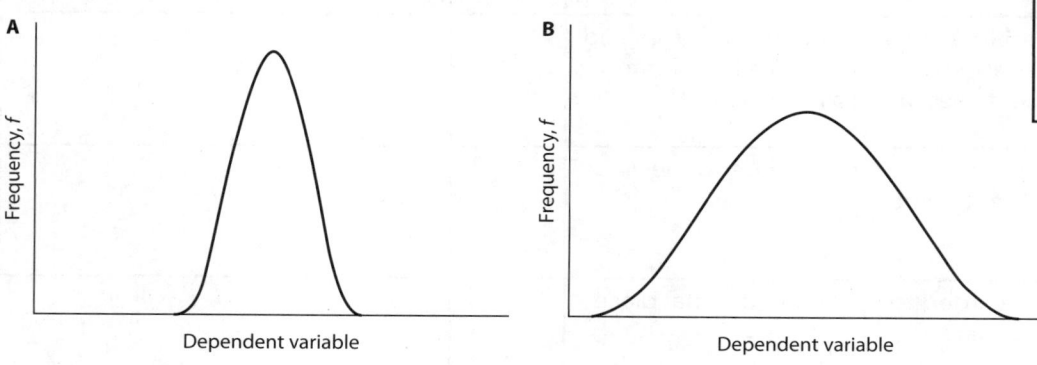

Figure 12.2: Normal distributions showing **A** small and **B** large standard deviations.

To calculate standard deviation you can either use your calculator, or a spreadsheet, or do it manually. If you do choose to do it using a calculator, however, you still need to know how to perform the calculation!

3 a Follow the steps below to calculate the standard deviation for the distance moved by the coloured dye when glucose is the substrate.

Distance moved by coloured dye (x) / cm	Distance moved by coloured dye minus the mean distance (x − x̄) / cm	(x − x̄²)
6.9	(6.9 − 7.49) = −0.59	0.35

Table 12.4: Results table.

Step 1 Copy Table 12.4.

Step 2 Write down all the distances (x) that the dye moved when glucose was used as a substrate, in the left-hand column.

Step 3 For each distance, subtract the mean distance ($x - \bar{x}$) and enter that number in the second column. Don't worry if some are negative.

Step 4 Square each number in the second column and write that in the third column $(x - \bar{x})^2$. There should be no negative numbers in this column.

Step 5 Calculate $\Sigma(x - \bar{x})^2$. To do this, add together all the $(x - \bar{x})^2$ values. The Greek capital letter sigma, Σ, means 'sum of'.

Step 6 Calculate the standard deviation: use the formula below:

$$standard\ deviation = \sqrt{\frac{\Sigma(x-\bar{x})^2}{n-1}}$$

where n is the number of samples.

Divide your value for $\Sigma(x - \bar{x})^2$ by $(n - 1)$. In this case $n = 9$. Then find the square root of this number. You now have the standard deviation.

Step 7 We can now state the mean distance moved by the coloured dye when glucose was used as a substrate, along with its standard deviation in the form:

Mean ± 1 standard deviation

b   Now calculate the standard deviations for the movement of the coloured dye when sucrose, maltose and water were used as substrates. Copy Table 12.5 and write in your answers.

Substrate	Mean distance moved by coloured dye / cm	Standard deviation / cm
glucose		
sucrose		
maltose		
water		

Table 12.2: Results table.

c   Describe the differences between the mean values of each substrate. Take into account the size of the standard deviations for each and explain whether or not the mean values are reliable.

Exercise 12.2 Energy density and calorimetry

Different substrates can be used to generate ATP through respiration. Most of the energy released from aerobic respiration comes from oxidation of hydrogen to water via the Krebs cycle and electron transport chain. The more hydrogen present per molecule, the more energy that molecule will have available per gram of substance. Calorimetry experiments aim to quantify the energy density of different substrates.

In this exercise, you will:

- develop your understanding of how calorimeters work, how experiments are designed to give accurate readings and how we can calculate the energy densities of different respiratory substrates.

Figure 12.3 shows a simple method for demonstrating the energy content of a substrate.

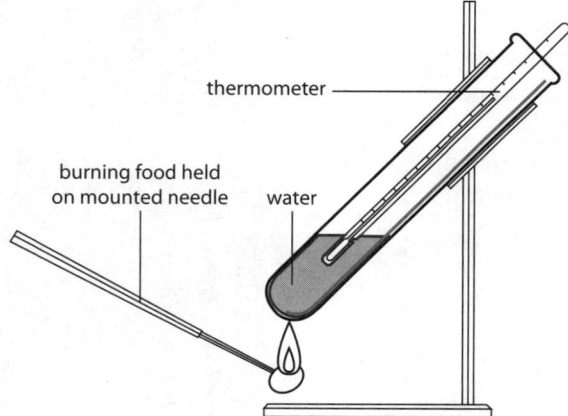

Figure 12.3: A simple calorimeter.

A known mass of substrate is placed on a metal mounted needle. A volume of 25 cm³ of water is placed into a clamped boiling tube and a thermometer placed in the water. The temperature of the water is measured (t_1). The food is ignited using a Bunsen burner and then moved under the boiling tube so that the heat energy from the burning food is used to heat the water. The maximum temperature that the water reaches is then recorded (t_2).

The energy density of the substrate is calculated using the following formula:

$$energy\ density = \frac{temperature\ rise\ (t_2 - t_1) \times 4.2 \times mass\ of\ water}{1000 \times mass\ of\ substrate}$$

The specific heat capacity of water is 4.2. It takes 4.2 J to raise the temperature of 1 g of water by 1°C. The number '1000' is present to convert from J to kJ. As water at normal temperature has a density of 1 g cm⁻³, we normally assume that the volume in cm³ is the same as the mass in g.

The units should be:

- energy density: kJ g⁻¹
- temperature rise: °C
- mass of water: g
- mass of substrate: g
- specific heat capacity of water: kJ g⁻¹ °C⁻¹.

1 Table 12.6 shows some calorimetry results that compare the energy density of starch, casein and stearic acid.

Substrate	Mass of substrate / g	Start temperature / °C	Max temperature / °C	Energy density / kJ g⁻¹
starch (carbohydrate)	1.3	5	67	
casein (protein)	1.2	7	72	
stearic acid (lipid)	0.8	4	92	

Table 12.6: Comparison of the energy density of starch, casein and stearic acid.

    a Use the formula to calculate the energy density of each substrate. Ensure that you state the units for energy density.

    b The published energy densities for carbohydrates, proteins and lipids are 15.8 kJ g⁻¹, 17.0 kJ g⁻¹ and 39.4 kJ g⁻¹, respectively. Compare the published data with the data obtained using the equipment shown above. List and explain all the sources of inaccuracy that could occur when using this experiment that might cause the differences in the results from published data.

    c State why saturated fats such as stearic acid have higher energy densities than unsaturated fats such as oleic acid.

Figure 12.4 shows a different calorimeter which is designed to give a more accurate value of energy content than the apparatus shown in Figure 12.3.

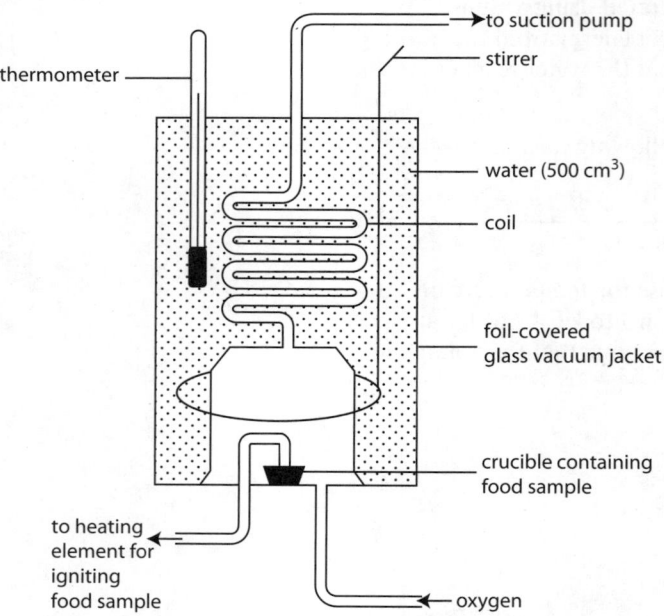

Figure 12.4: A more accurate calorimeter.

2 Copy Table 12.7 to explain how each listed feature of the calorimeter helps to ensure that the energy value for each substrate is close to the true value.

Feature	How feature improves accuracy
foil-covered glass vacuum jacket	
stirrer	
heating element inside the apparatus	
oxygen supply	
500 cm³ water volume	
coiled tubing through which exhaust gas passes	

Table 12.7: Comparison of the features of a calorimeter to ensure accuracy.

Exercise 12.3 Mitochondria experiments

There are a lot of biochemical and physiological data regarding mitochondria, which can show us how processes such as the **electron transport chain** occur. In this exercise, you will:

- develop your understanding of the mechanisms of **oxidative phosphorylation**
- apply your knowledge of respiration to unfamiliar experiments.

1 Mitochondria are isolated by homogenising liver tissue in an isotonic pH 7.4 buffer solution at 4 °C. They are separated from other cell organelles by centrifugation and then re-suspended in the same isotonic buffer at 4 °C. The suspension is placed into a small cavity slide and an oxygen probe lowered into the suspension. The oxygen probe measures the concentration of oxygen in the suspension. Different substances can then be added to the suspension and the effect of them on the oxygen concentration examined.

   a Explain why the mitochondria are placed in isotonic pH 7.4 buffer at 4 °C.

   b In the first experiment, pyruvate was added as a substrate, and adenosine diphosphate (ADP) and inorganic phosphate (P_i) added at repeated intervals. The oxygen levels were monitored, and the results are shown in Figure 12.5 below.

> **KEY WORDS**
>
> **electron transport chain:** a chain of adjacently arranged carrier molecules in the inner mitochondrial membrane, along which electrons pass by redox reactions
>
> **oxidative phosphorylation:** the synthesis of ATP from ADP and Pi using energy from oxidation reactions in aerobic respiration (compare photophosphorylation)

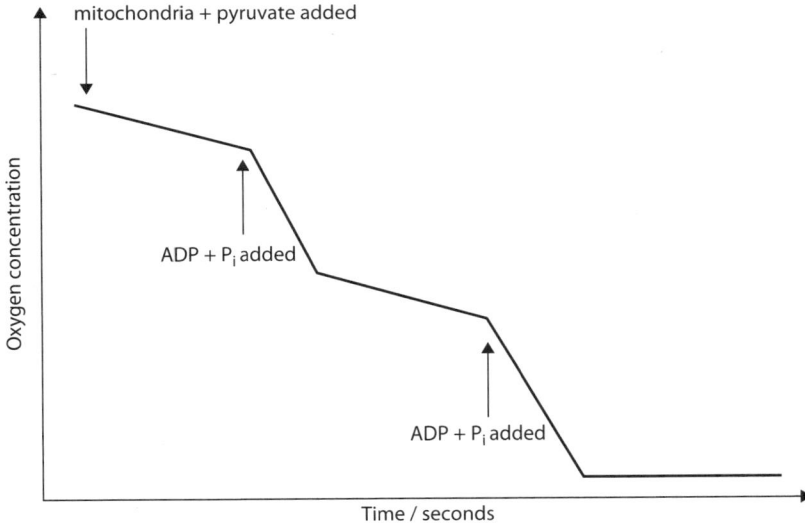

Figure 12.5: The effect of addition of ADP and inorganic phosphate (P_i) on oxygen concentration.

   i   Explain why pyruvate was added rather than glucose.
   ii Describe and explain the effects of addition of ADP + P_i on the concentration of oxygen.

2　In a further experiment, the effects of addition of the respiratory inhibitor rotenone were investigated. Rotenone inhibits the activity of the first set of electron carriers in the inner mitochondrial membrane, as shown in Figure 12.6.

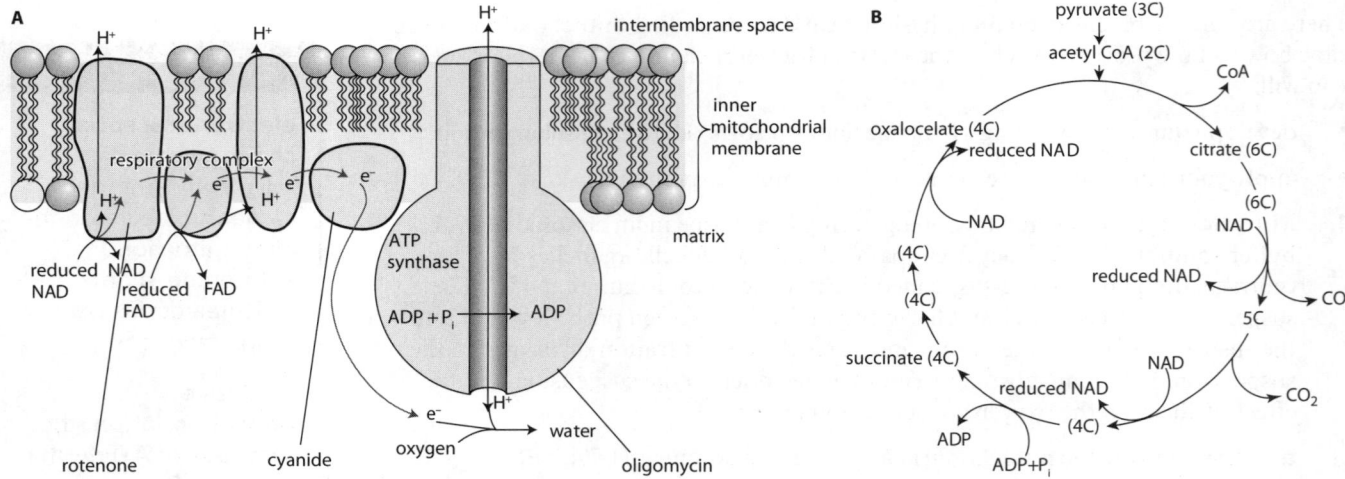

Figure 12.6: Diagrams of the **A** electron transport chain showing the sites of action of the inhibitors rotenone, cyanide and oligomycin, and **B** Krebs cycle showing the role of the intermediate sugar, succinate.

As in the first experiment, ADP and P_i were added to the mitochondrial suspension followed by a mixture of ADP, P_i and rotenone. The Krebs cycle intermediate sugar, succinate, was then added. The results are shown in Figure 12.7.

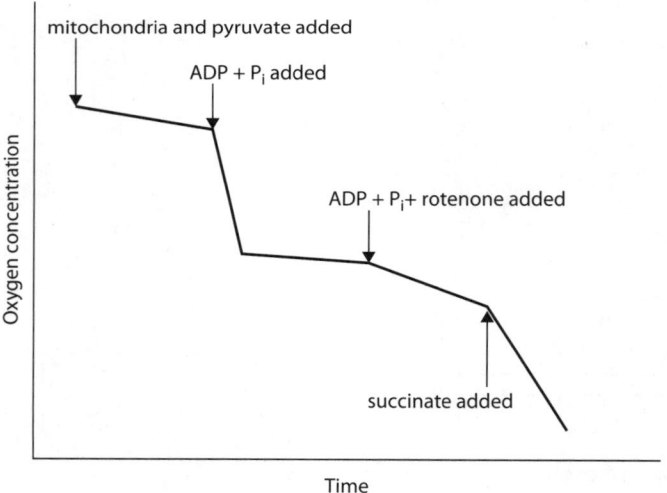

Figure 12.7: Effect of addition of ADP, Pi, rotenone and succinate on oxygen concentration.

> **KEY WORDS**
>
> **Krebs cycle:** a cycle of reactions in aerobic respiration in the matrix of a mitochondrion in which hydrogens pass to hydrogen carriers for subsequent ATP synthesis and some ATP is synthesised directly; also known as the citric acid cycle

　　a　Explain why the first addition of ADP + P_i was carried out.

　　b　Use Figures 12.6 and 12.7 to explain the effect of rotenone on oxygen uptake and the effect of succinate addition.

3 In two final experiments, the effects of the respiratory inhibitors oligomycin and cyanide were investigated. Oligomycin inhibits the movement of H⁺ ions through the ATPase channel. Cyanide inhibits the enzyme cytochrome oxidase at the end of the electron transport chain, preventing it from releasing electrons on to oxygen.

DNP is a substance that makes the inner mitochondrial membrane permeable to H⁺ ions so that they will diffuse into the matrix and dissipate their energy as heat. It 'decouples' the movement of H+ ions into the matrix from ATP synthesis.

Mitochondria were treated as in previous experiments. ADP and P_i were added to the mitochondrial suspensions followed by either ADP, P_i and oligomycin or ADP, P_i and cyanide. In both cases, DNP was then added to the mitochondrial suspension. The results are shown in Figure 12.8.

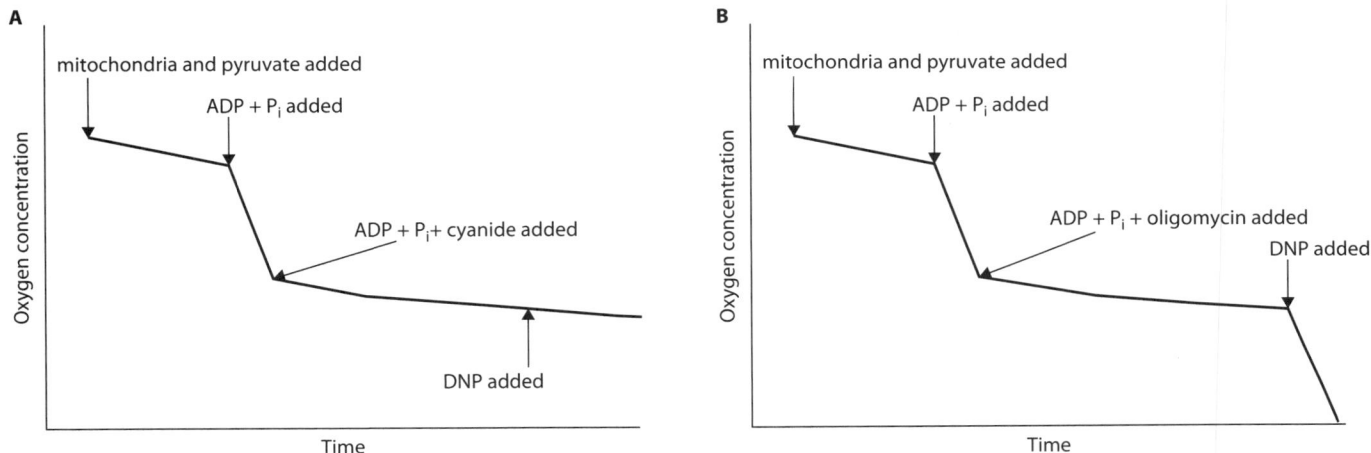

Figure 12.8: The effect of addition of **A** oligomycin and **B** cyanide with DNP on oxygen concentration.

Compare and explain the effects of oligomycin and cyanide addition with and without DNP.

Exercise 12.4 Measuring respiratory quotient (RQ) and using a respirometer

The **respiratory quotient** (RQ) is a ratio of the volume of oxygen used by an organism compared to the amount of carbon dioxide released. The RQ for the complete aerobic respiration of glucose is 1, since from the equation for aerobic respiration, 6 molecules of oxygen are used for every 6 molecules of carbon dioxide produced.

$$6O_2 + C_6H_{12}O_6 \rightarrow 6CO_2 + 6H_2O$$

$$RQ = \frac{CO_2}{O_2} = \frac{6}{6} = 1$$

KEY WORDS

respiratory quotient (RQ): the ratio of the volume of carbon dioxide released to the volume of oxygen used

Other substrates, however, do not have an RQ of 1. In this exercise, you will:

- develop your understanding how we can carry out practical work to determine the RQ of respiring organisms.
- develop your understanding of the RQ values of different substrates.

1 An investigation was carried out into how respiratory substrates change during the germination of barley seeds over a seven-day period.

Figure 12.9 shows a diagram of a **respirometer**.

> **KEY WORD**
>
> **respirometer:** a piece of apparatus that can be used to measure the rate of oxygen uptake by respiring organisms

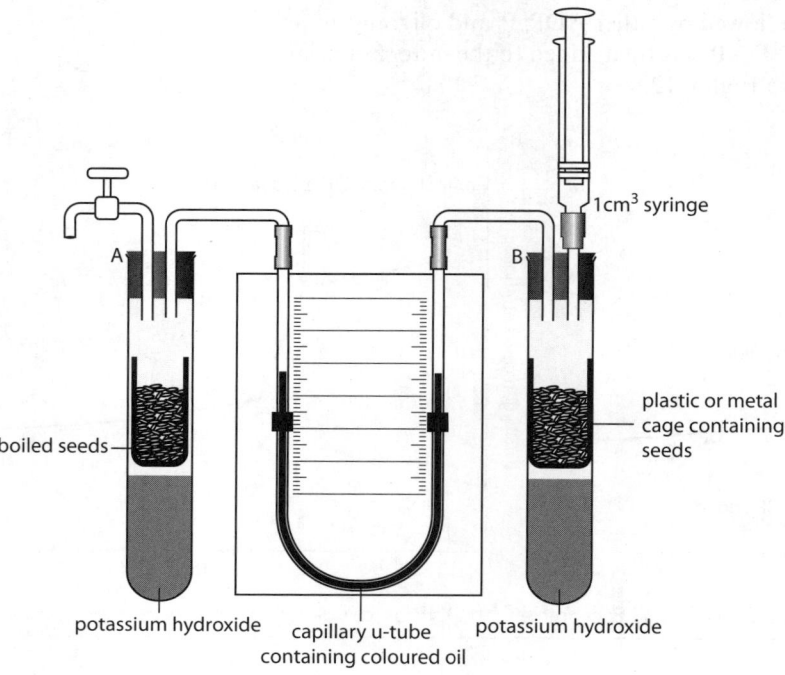

Figure 12.9: A typical respirometer.

- 5 g of barley seeds were soaked in water for 12 hours.
- After 12 hours of soaking in water the barley seeds were washed with a solution of sodium hypochlorite bleach. All glassware was also soaked with the sodium hypochlorite. The barley seeds were placed into the metal cradle in boiling tube B. A sample of 5 g of identically treated boiled barley seeds was placed into the metal cradle in boiling tube A. A volume of 10 cm³ of potassium hydroxide solution was placed into both boiling tubes as shown in Figure 12.9.
- The coloured oil in the U-tube was set so that it was at an equal height on both arms of the tube.
- The tap on boiling tube A was closed and the distance moved by the coloured liquid recorded after 15 minutes.
- This was repeated five times in total.
- The potassium hydroxide was then replaced with water and the experiment repeated five more times.
- This was repeated for seven days in total.

12 Energy and respiration

- After three days the seeds were seen to be germinating. At this point, all respirometer work was carried out, with the seeds in the dark.

  a Explain why the seeds and apparatus were washed with sodium hypochlorite bleach.

  b Explain the purpose of the 1 cm³ syringe.

  c List the variables that should be controlled.

  d Explain why, after day 3, the respirometer experiment was carried out in the dark.

  e Explain why an equal mass of boiled barley seeds was placed into boiling tube A.

  f Explain the purpose of the potassium hydroxide.

2 The volumes of oxygen used are calculated from the distances the dyes moved in the presence of potassium hydroxide. Oxygen is removed by the barley seeds, and carbon dioxide is released. The potassium hydroxide removes the carbon dioxide, so we see a reduction in volume which is the volume of oxygen removed.

Day	Mean distance moved by dye / cm		Volume of oxygen used in 15 min / mm³	Volume of carbon dioxide released in 15 min / mm³	RQ
	In presence of sodium hydroxide	In presence of water			
0	6.3	−4.1			
1	7.3	0.1			
2	7.7	0.9			
3	6.7	1.1			
4	6.5	1.6			
5	6.7	1.8			
6	7.1	0.2			
7	7.8	0.0			

Table 12.8: Results table.

a Follow the steps below to calculate the volumes of oxygen used on different days.

   Step 1 Copy Table 12.8. To calculate the volume of oxygen used, we need to assume that the U-tube manometer is a cylinder where the height is the distance moved by the coloured dye.

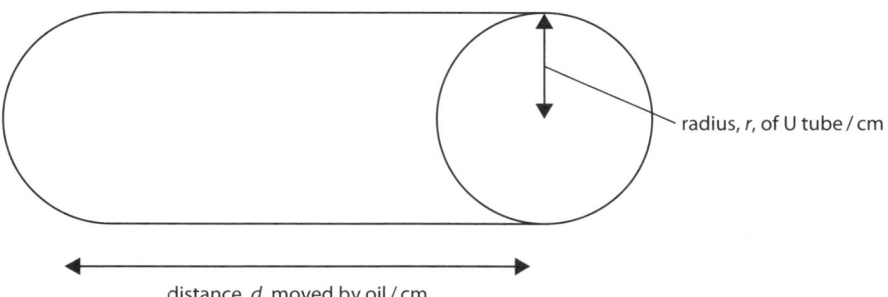

Figure 12.10: Diagram showing dimensions of a cylinder.

The formula for the volume of a cylinder is: $\pi r^2 d$

Step 2 For day 0, the distance, d, moved in the presence of potassium hydroxide was 6.3 cm. Convert this to millimetres: 6.3 cm = 63 mm

Step 3 The radius of the U tube was 0.1 cm. Convert this to millimetres: 0.1 cm = 1 mm

Step 4 Now calculate the volume of the cylinder of oxygen:

$$\text{volume} = \pi r^2 d$$

$$\text{volume} = 3.14 \times (1)^2 \times 63 = 197.82 \, mm^3$$

b Now calculate the volumes of oxygen used by the respiring barley seeds on all the days.

Write your answers in your table.

Determining the volume of carbon dioxide released by the respiring barley seeds is slightly harder. We have to look at the movement of the dye when the barley seeds are in the presence of water (with no potassium hydroxide to remove the carbon dioxide) and compare this with the movement of the dye when the barley seeds are in the presence of potassium hydroxide.

If the coloured dye has moved towards the respiring barley seeds in tube B

In Figure 12.11, the distance x is the distance moved by the coloured dye in the presence of sodium hydroxide and represents the volume of oxygen. Distance y is the distance moved by the coloured dye in the presence of water and represents the combined effect of oxygen removal and carbon dioxide release. The fluid has moved towards the respiring seeds in tube B, so that means more oxygen has been removed by the seeds than carbon dioxide has been released. The volume of carbon dioxide is the distance $(x - y)$.

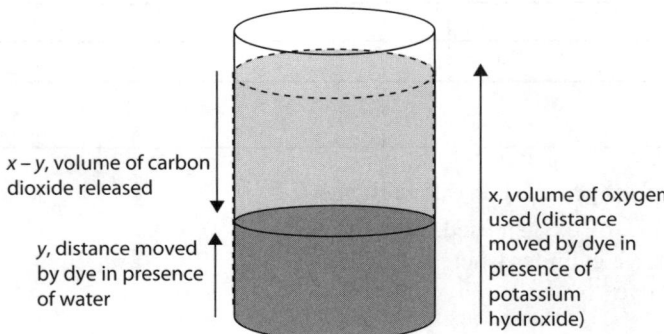

Figure 12.11: Calculating volume of carbon dioxide released when the coloured dye moves towards the respiring seeds in tube B.

c Calculate the volumes of carbon dioxide released by the seeds by following the steps below:

Step 1 On day 1, the distance moved in the presence of potassium hydroxide, x, is 7.3 cm. Convert this to millimetres: 7.3 cm = 73 mm

Step 2 The distance moved in the presence of water, y, is 0.1 cm. Convert this to millimetres: 0.1 cm = 1 mm

Step 3 Distance moved due to carbon dioxide release $x - y$ = 73 mm − 1 mm = 72 mm

Step 4 Now calculate the volume of carbon dioxide released:

volume of cylinder = $\pi r^2 d$

volume of cylinder = $3.14 \times 72 \times (1)^2 = 226.08\,mm^3$

d Now calculate the volumes of carbon dioxide released on days 2–7 and write your answers in your table.

If the coloured dye has moved away from the respiring barley seeds in tube B

In Figure 12.12, the distance x is the distance moved by the coloured dye in the presence of sodium hydroxide and represents the volume of oxygen. Distance y is the distance moved by the coloured dye in the presence of water and represents the combined effect of oxygen removal by the respiring seeds and carbon dioxide release. The fluid has moved down, so that means more carbon dioxide has been released than oxygen removed. The volume of carbon dioxide is the distance $(x + y)$.

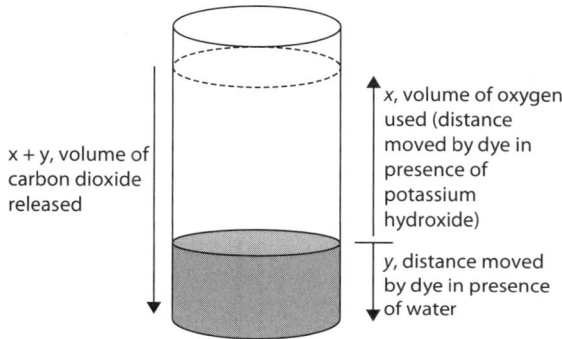

Figure 12.12: Calculating volume of carbon dioxide released when the coloured dye moves away from the respiring seeds in tube B.

e Calculate the volume of carbon dioxide released on day 0.

The change in volume for day 0 is listed as −4.1. The negative sign is to show the opposite direction of movement – we ignore this when carrying out the calculation.

Step 1 Distance x moved by dye in the presence of potassium hydroxide = 6.3 cm. Convert this to millimetres.

Step 2 Distance y moved by the dye in the presence of water = 4.1 cm. Convert this to millimetres.

Step 3 Calculate the distance moved due to carbon dioxide release, $x + y$.

Step 4 Use this distance to calculate the volume of carbon dioxide released on day 0.

Volume of a cylinder = $\pi r^2 d$

Step 5 Add the value to the table.

3 a Now use the formula given at the start of this exercise to calculate the respiratory quotients for each day and write your answers in your table.

   b Plot a suitable graph to show how the RQ values change over the seven-day period.

   c Describe the changes in RQ values over the seven-day period and suggest explanations for the changes. Typical RQ values are: aerobic respiration of carbohydrate 1.0, lipids 0.7, protein 0.9; anaerobic respiration of carbohydrate < 1.

EXAM-STYLE QUESTIONS

1 a Table 12.9 lists four different stages of aerobic respiration and some statements about aerobic respiration. Place ticks in the boxes if the statements are correct, and crosses for those that are wrong.

Statement	Glycolysis	Link reaction	Krebs cycle	Oxidative phosphorylation
NAD is reduced				
FAD is oxidised				
decarboxylation occurs				
occurs in the cytoplasm				
occurs in the matrix of the mitochondria				

Table 12.9

[5]

b **Explain** why ATP is frequently described as the 'energy currency of the cell'.

[4]

[Total: 9]

2 Figure 12.13 shows an electron micrograph of a mitochondrion.

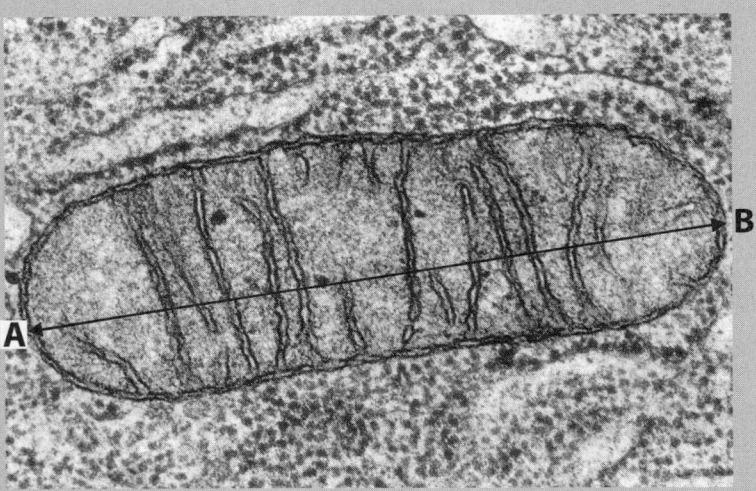

Figure 12.13

COMMAND WORD

Explain: set out purposes or reasons / make the relationships between things evident / provide why and/or how and support with relevant evidence.

CONTINUED

a **Calculate** the length in μm, between **A** and **B**, of the mitochondrion in Figure 12.13. [2]

b In a study into the effect of endurance training on muscle cells, the density of the cristae within the mitochondria was measured. It was found that after six months of endurance training, the density had increased by 43% compared to control groups. Use your knowledge of aerobic respiration to explain how this change could benefit endurance athletes. [4]

[Total: 6]

3 Mitochondria were extracted from liver tissue and placed into ice-cold isotonic solution with excess quantities of pyruvate and inorganic phosphate (P_i). ADP was added and the change in oxygen concentration of the solution measured. This was repeated until ADP addition had no more effect. The results are shown in Figure 12.14.

> **COMMAND WORD**
>
> **Calculate:** work out from given facts, figures or information.

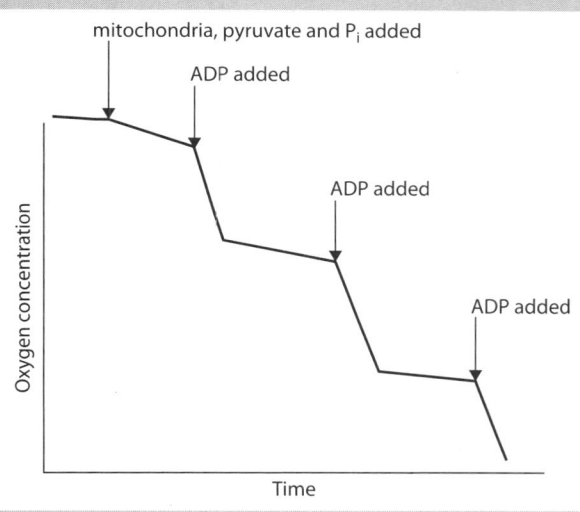

Figure 12.14

a Using your knowledge of chemiosmosis, explain the change in oxygen concentration after addition of ADP. [4]

b In a further experiment, mitochondria from a specialised adipose tissue called 'brown fat' were investigated. This tissue is found to be prevalent in hibernating animals. The results are shown in Figure 12.15.

CONTINUED

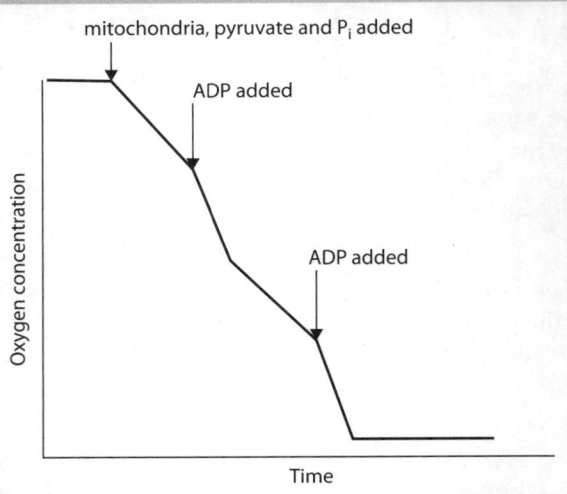

Figure 12.15

i **Compare** the results from the mitochondria extracted from brown fat with the liver mitochondria. [2]

ii The inner membranes of mitochondria from brown fat cells possess high concentrations of an H⁺ protein channel called UCP-1. UCP-1 is not attached to ATP synthase. **Suggest** and explain how the UCP-1 protein helps hibernating animals survive cold conditions. [5]

[Total: 11]

4 The respiratory quotient (RQ) is a measure of the amount of oxygen consumed by an organism compared to the amount of carbon dioxide released.

It is calculated using the following formula:

$$RQ = \frac{\text{volume of carbon dioxide released}}{\text{volume of oxygen consumed}}$$

The RQ for the aerobic respiration of glucose is 1.

The equation for the oxidation of the fatty acid oleic acid is:

$$C_{18}H_{34}O_2 + 25.5\ O_2 \rightarrow 18CO_2 + 17H_2O$$

a Calculate the RQ for the aerobic respiration of oleic acid. [1]

b An investigation was carried out into the effect of fasting on the RQ of two students. One of the students was given no food (fasting) for 12 hours whilst the other ate a normal balanced diet. The students were then made to exercise at increasing intensities in rooms set at the same temperature and humidity. The volumes of oxygen used in their bodies and carbon dioxide produced were measured and the RQ calculated for the different exercise intensities. Intensity was measured as power output. The results are shown in Figure 12.16.

COMMAND WORDS

Compare: identify / comment on similarities and/or differences.

Suggest: apply knowledge and understanding to situations where there is a range of valid responses, in order to make proposals / put forward considerations.

CONTINUED

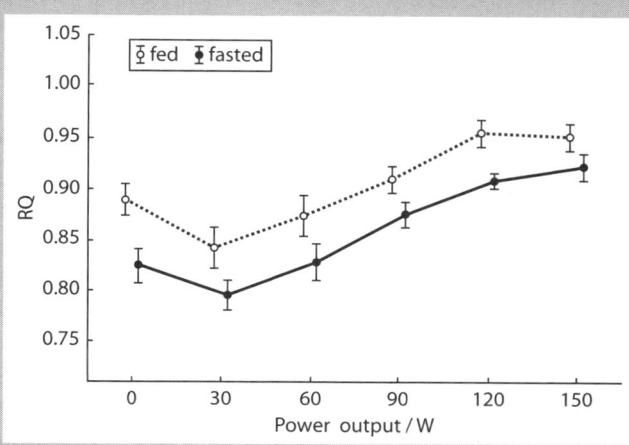

Figure 12.16

i The temperature and humidity of the room were kept the same. Suggest *two* other factors that would need to be controlled. [1]

ii Compare the effects of increasing exercise intensity on the RQ of the two students. [3]

iii Using your knowledge of RQ values for different substrates explain the results in Figure 12.16. [4]

iv Migratory birds such as Canada geese have a much higher ratio of saturated fatty acid to glycogen for their energy reserves than flightless birds such as ostriches. Using your knowledge of the energy density of different substrates, suggest and explain reasons for this difference. [3]

[Total: 12]

5 Black mangrove trees are found in muddy coastal regions. The soil is waterlogged and the roots spend time submerged. Figure 12.17 shows the root system of a black mangrove tree.

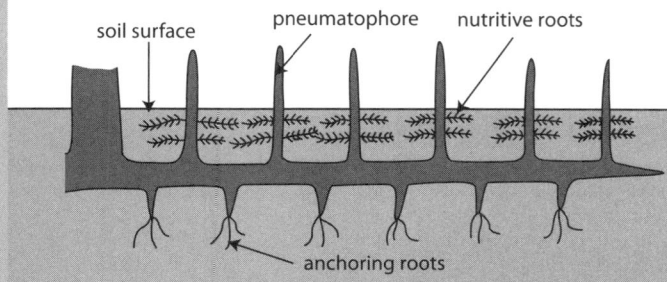

Figure 12.17

CONTINUED

The mangrove trees develop three types of root:
- anchoring roots, which help stabilise the tree
- nutritive roots, which absorb mineral ions from the soil
- pneumatophores, which are roots with large numbers of lenticels that protrude out of the mud and water.

  a Explain how the root system of the black mangrove enables it to grow in the shifting, waterlogged soil. [5]

  b **Describe** and explain how rice plants are adapted to grow whilst partially submerged in water. [5]

  [Total: 10]

COMMAND WORD

Describe: state the points of a topic / give characteristics and main features.

6 Figure 12.18 shows an outline of some stages of respiration.

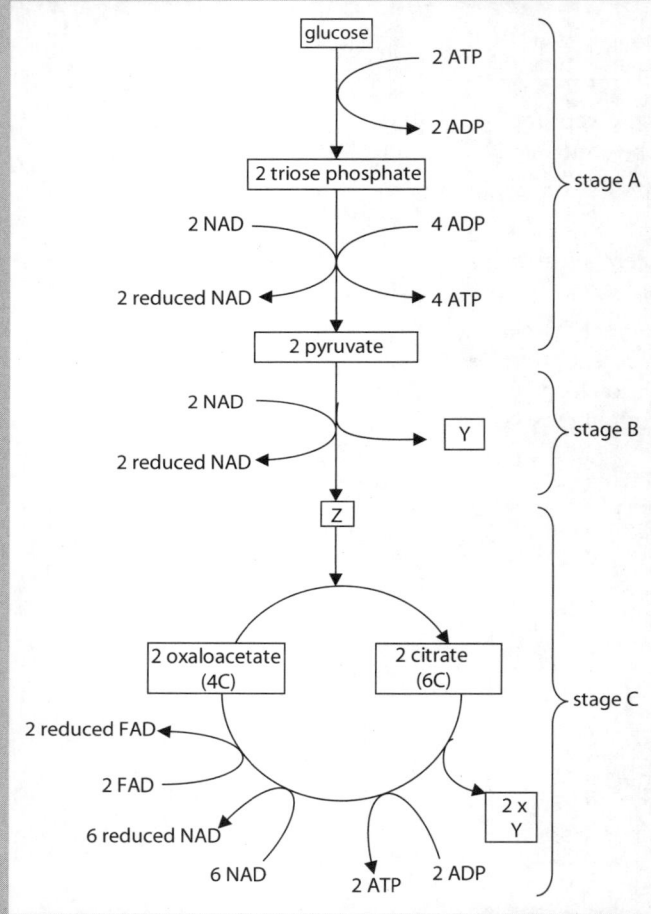

Figure 12.18

CONTINUED

a i   Name molecules **Y** and **Z**. [2]
   ii Name stages **A**, **B** and **C**. [1]
   iii **State** the exact location in the cell where stages **B** and C occur. [1]

b i   Explain how reduced NAD and reduced FAD are used to generate ATP. [5]
   ii Assuming that each reduced NAD can produce 2.5 molecules of ATP and each reduced FAD can produce 1.5 molecules of ATP, calculate the total number of ATP molecules produced from one molecule of glucose. [1]

c In the absence of oxygen, stages B and C in Figure 12.18 stop. Explain why in anaerobic conditions animals covert pyruvate into lactate. [3]

[Total: 13]

COMMAND WORD

State: express in clear terms.

> Chapter 13
Photosynthesis

> **CHAPTER OUTLINE**
>
> The questions in this chapter cover the following topics:
>
> - the absorption of light energy in the light-dependent stage of photosynthesis
> - the transfer of energy to the light-independent stage of photosynthesis and its use in the production of complex organic molecules
> - the role of chloroplast pigments in the absorption of light energy
> - how the structure of a chloroplast fits it for its functions
> - how environmental factors influence the rate of photosynthesis
> - how C4 plants are adapted for high rates of carbon fixation at high temperatures.

Exercise 13.1 The effect of different colours of light on the light-dependent reactions

The light-dependent reactions of photosynthesis include the photolysis of water to give hydrogen ions, electrons and oxygen and the synthesis of ATP. Light energy causes the excitation and loss of electrons from the primary pigments of photosynthesis. The electrons on the pigments are replaced by those from the water. The electrons pass through a range of electron carrier molecules and are eventually added to NADP along with H$^+$ ions to produce reduced NADP. The substance DCPIP and methylene blue can act as a substitute for NADP. DCPIP and methylene blue are blue in their oxidised state and colourless when reduced.

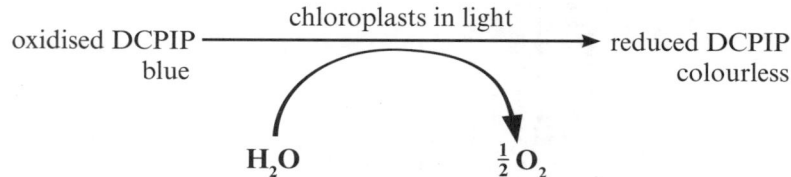

This exercise will:

- develop your understanding of how **redox indicators** such as DCPIP and methylene blue are used to investigate photosynthesis
- develop your understanding of how **standard error** can be used to compare mean values.

> **KEY WORDS**
>
> **redox indicator:** a substance that changes colour when it is oxidised or reduced
>
> **standard error:** a calculation that indicates how close the calculated mean value is likely to be to the true mean value

1   Chloroplasts were isolated from spinach leaves, suspended in an isotonic pH 7.4 buffer and kept at 4 °C. Ten test tubes were placed into an ice-cold water bath and chloroplast suspension and DCPIP added to each. The tubes were exposed to light of different colours by placing sheets of coloured cellophane in front of a lamp. The distances of the lamps to the tubes were kept constant as shown in Figure 13.1.

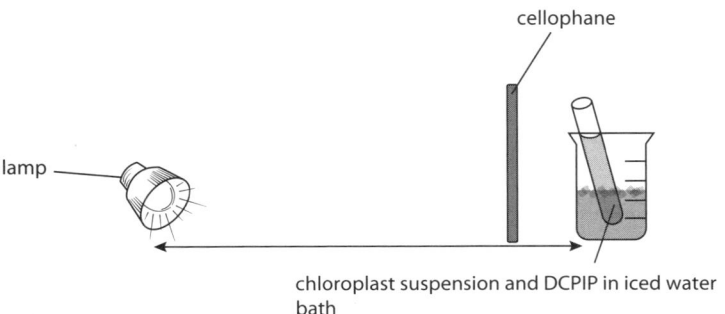

Figure 13.1: Diagram to show position of lamp and light filter.

The contents of each tube and the colour of the light to which each tube was exposed are shown in the Table 13.1. Ticks and crosses represent the presence or absence of substances.

Tube	Light colour	Wavelength range of light /nm	5 cm³ chloroplast suspension	5 cm³ DCPIP
1	white	390–780	✗	✓
2	white	390–780	✓	✗
3	none (foil wrapped)	none	✓	✓
4	white	390–780	✓	✓
5	violet	390–455	✓	✓
6	blue	455–495	✓	✓
7	green	495–575	✓	✓
8	yellow	575–600	✓	✓
9	orange	600–625	✓	✓
10	red	625–780	✓	✓

Table 13.1: Contents of each tube and the colour of the light to which each tube was exposed.

Tubes 1, 2, 3 and 4 acted as controls or were used to make comparisons.

The tubes were exposed to the appropriate light and the time taken for the DCPIP to decolourise recorded. Each of tubes 4–10 were replicated five times. The results are shown in Table 13.2.

Tube number	Colour of light	Time taken for DCPIP to decolourise / seconds					Mean	sd
		1	2	3	4	5		
1*	white	–	–	–	–	–	–	–
2**	white	–	–	–	–	–	–	–
3***	none (foil wrapped)	–	–	–	–	–	–	–
4	white	33	36	41	32	39	36.2	3.6
5	violet	72	68	79	67	68	70.8	5.0
6	blue	85	92	91	84	82	86.8	4.4
7	green	274	225	265	268	245	255.4	20.2
8	yellow	167	152	143	168	156	157.2	10.5
9	orange	96	89	102	96	97	96.0	4.6
10	red	65	59	71	73	71		

* DCPIP remained blue.
** remained green colour of chloroplast suspension.
*** remained starting colour of mixed blue DCPIP and green chloroplast suspension.

Table 13.2: Time taken for DCPIP to decolourise in different lights.

a i   Explain the purpose of each of tubes 1, 2, 3 and 4.
   ii   Calculate the mean time taken and standard deviation for DCPIP to decolourise in red light.

The experiment suggests that the mean times taken for the DCPIP to decolourise in different colours of light are different. If you repeated each colour again would the times taken to decolourise the DCPIP be the same? We can never be certain of this but we can carry out a calculation to determine whether our mean value is close to the true mean value. This calculation works out the standard error of our mean (SE). SE tells us how certain we can be that our mean is the true mean. The formula for SE is:

$$SE = \frac{s}{\sqrt{n}}$$

where SE is the standard error
s is the standard deviation
n is the sample size.

We can be 95% certain that if we carried out another experiment the result would be within two times our value of SE. This range around the mean is known as the 95% confidence limit.

Follow the steps to calculate the 95% confidence limits of the mean time taken to decolourise DCPIP in violet light:

Step 1 Calculate the standard error using the formula given above:

$$SE = \frac{5.0}{\sqrt{5}}$$

$$= \frac{5.0}{22}$$

$$SE = 2.3$$

Step 2 Now calculate 2 × SE:

$$SE = 2 \times 2.3 = 4.6$$

This means that we can be 95% confident that the mean from a second sample would be 70.8 ± 4.6, which means that the 95% confidence limits are between 66.2 and 75.4.

b Calculate the standard error and 95% confidence limits for the mean times taken for DCPIP to colourise for all the colours of light.

> **TIP**
>
> We can use the confidence limits as **error bars** on graphs. If error bars overlap, there is no significant difference in mean values. If the error bars do not overlap, we cannot definitely state that the difference is significant but we do know that there is a possibility that it is. This is shown in Figure 13.2.

> **KEY WORDS**
>
> **error bar:** a line drawn through a point or the top of bar on a graph, extending two standard errors above and below the mean value indicated by the point or bar; you can be 95% certain that the true value lies within the range indicated by the error bar

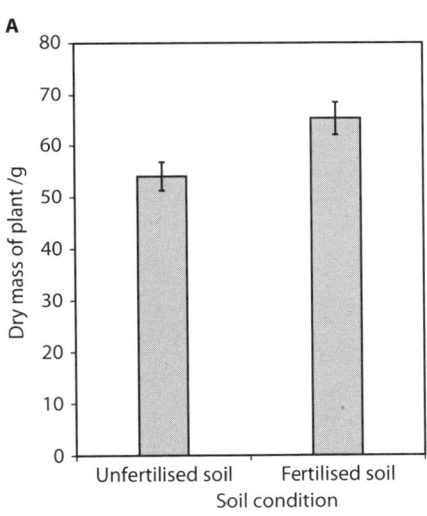
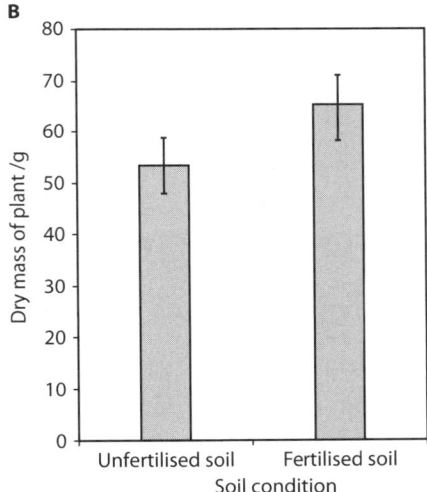

Figure 13.2: Error bars that **A** do not overlap and **B** do overlap. Where they overlap, the difference in mean is not significant.

c Plot an appropriate graph of light colour against mean time taken for DCPIP to decolourise. You will need to think what type of variables you have – categoric or continuous – before you select the correct graph. When you draw the *x*-axis, it will be helpful if the order of the colours corresponds to the increasing wavelength.

d Add error bars onto each mean plot. The upper error bar should be 2 × SE above the mean, and the lower error bar should be 2 × SE below the mean.

e Look at your graph. State which mean values show an overlap of error bars and are not significantly different from each other.

f Describe and explain the pattern shown by the graph.

2 It was suggested that the different cellophane filters not only altered the wavelength (colour) of light passing through but also the absorbance of light by the filters. The relative percentage transmission of light was measured for each colour using a light meter. The data are shown in Table 13.3. 'White' light was set at 100% transmission when passed through a clear filter.

colour of light	relative transmission of light / %
white	100
violet	23
blue	38
green	72
yellow	80
orange	75
red	58

Table 13.3: Relative percentage transmission of light measured for each colour using a light meter.

a Plot an appropriate graph to determine whether transmission of light had a significant effect on DCPIP decolourisation.

b Think carefully about the experiment. Evaluate how effective the method was at generating accurate data and identify the major sources of error. Suggest improvements that could give more accurate results.

Exercise 13.2 Chromatography and photosynthetic pigments

Identifying and isolating different substances is an important practical tool in biology. Chromatography is a method for separating different compounds on the basis of their solubility. In this exercise, you will:

- develop your understanding of how we use chromatography to identify the different pigments present in plants.

1 A scientist carried out an investigation into the effects of ageing and illumination levels on the pigments found in maple tree leaves. Three types of leaf were taken from the same maple tree: old leaves that were changing colour, young leaves from the sun-exposed upper branches and young leaves from the shaded lower parts of the tree. Pigment was extracted by cutting up the leaves and placing them into a mortar with six drops of propanone (an organic solvent) and a pinch of sand.

The leaves were then ground until a dark coloured extract was formed. A pencil line was drawn on a strip of chromatography paper 1.5 cm from the base. A fine glass capillary tube was used to place spots of extract on the pencil line. Each time a spot was placed it was allowed to dry before the next one was placed on the same spot. When a dark coloured spot had been produced, the chromatography paper was placed into a boiling tube so that the end below the dark spots was submerged in propanone solvent. The chromatogram was allowed to run until the solvent almost reached the end of the chromatography paper.

The **R_f value** of a chemical is a measure of its solubility in nonpolar solvents. For a particular solvent, the R_f should be approximately the same every time. It is calculated using the formula:

$$R_f \text{ value} = \frac{\text{distance moved by substance}}{\text{distance moved by solvent}}$$

KEY WORDS

R_f value: a number that indicates how far a substance travels during chromatography, calculated by dividing the distance travelled by the substance by the distance travelled by the solvent; R_f values can be used to identify the substance

a Calculate the R_f values for each of the pigments **A**, **B**, **C** and **D**. Always measure from the origin line to the centre of the spot. Use Table 13.4 below to identify each one.

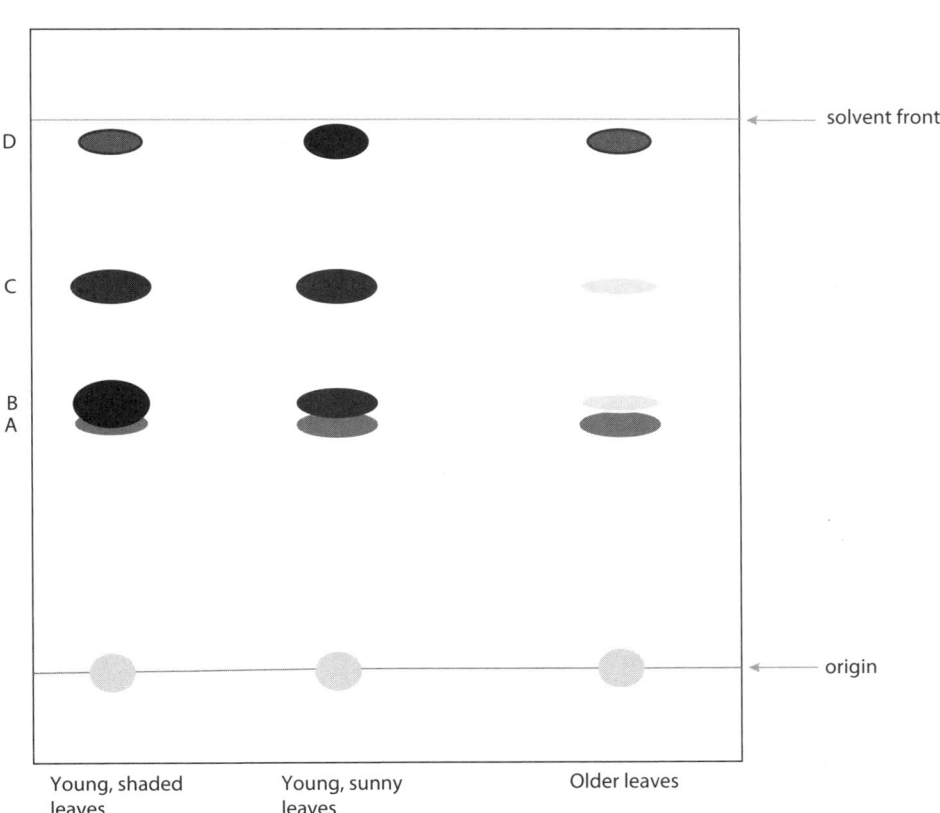

Figure 13.3: Chromatogram showing pigments in different leaves.

Pigment	R_f value
anthocyanin	0.31
carotene	0.96
chlorophyll a	0.70
chlorophyll b	0.48
phaeophytin	0.85
xanthophyll	0.44

Table 13.4: R_f values of different pigments.

b Explain why the pigment spot is dried between applications.

c Compare the differences in pigments shown on the chromatogram in Figure 13.3 for each of the leaves. Darker, larger spots suggest a higher concentration of pigment.

d There have been several hypotheses proposed to explain why different concentrations of pigments are found in leaves. Some of these hypotheses are:

- Carotene protects chlorophyll from over oxidation in bright light.
- Chlorophyll *b* absorbs wavelengths of light that reach the woodland floor.
- Xanthophyll is an **accessory pigment** that absorbs wavelengths of light not absorbed by chlorophyll.
- In autumn, or the fall, chlorophylls *a* and *b* are broken down so that leaves turn brown, red and yellow, the colours of carotene and xanthophyll.

Discuss which of these hypotheses could be consistent with the results shown in Figure 13.3.

e Explain whether or not it is acceptable to make valid, quantitative conclusions when comparing the leaves.

f Suggest how the experiment could be extended to generate quantitative data in order to compare concentrations of the pigments.

g Review the original method and carry out a risk assessment. State the risks and what strategies could be used to minimise the risks.

> **KEY WORDS**
>
> **chlorophyll:** a green pigment that absorbs energy from light, used in photosynthesis
>
> **accessory pigment:** a pigment which absorbs wavelengths of light not absorbed by chlorophyll

Exercise 13.3 Measuring the rate of photosynthesis in aquatic plants

Aquatic plants are useful for measuring photosynthesis rates as they produce oxygen gas underwater which can then be easily collected. We can use them to investigate several factors that can influence the rate of photosynthesis, such as light intensity, carbon dioxide concentration and temperature. In this exercise, you will:

- develop your understanding of the factors that can limit the rate of photosynthesis

- develop your understanding of how to investigate photosynthesis in aquatic plants
- develop your understanding of the **Spearman's rank correlation** test, a statistical test for correlations.

1 Plan an investigation into the effect of increasing light intensity on the rate of photosynthesis of *Cabomba aquatica*, a pond weed. The apparatus you have available is: a **photosynthometer** (see Figure 13.4), a glass beaker, a thermometer, a pair of sharp scissors, a paper clip, a 10 cm^3 syringe, a stop watch, a ruler graduated in millimetres, anhydrous sodium hydrogen carbonate (dry), balance, distilled water, 1 dm^3 volumetric flask, a bench lamp, a metre ruler, a clamp and stand, sources of hot and cold water.

> **KEY WORDS**
>
> **Spearman's rank correlation:** a statistical test to determine if there is a correlation between two variables when one or both of them are not normally distributed
>
> **photosynthometer:** apparatus to measure the rate of photosynthesis by collecting and measuring the volume of oxygen produced in a certain time

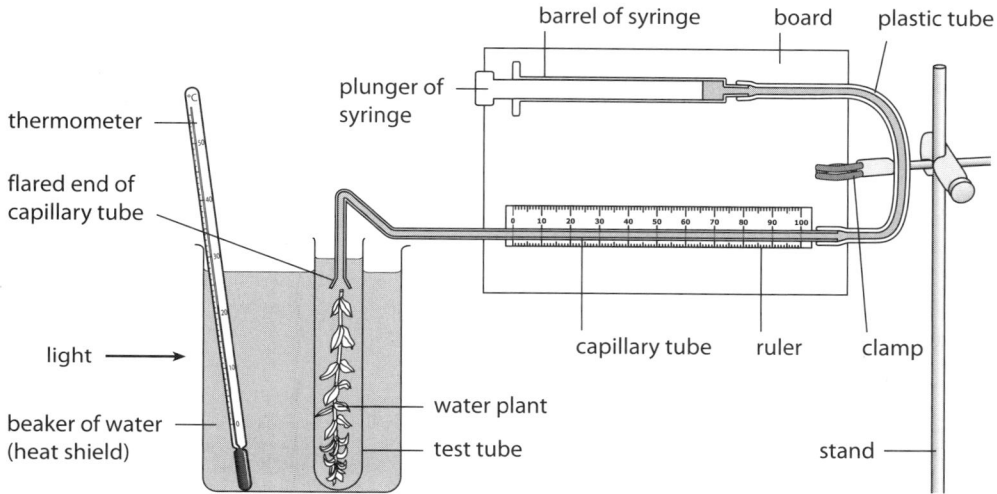

Figure 13.4: A photosynthometer set up to measure the volume of oxygen produced by an aquatic plant. The capillary and plastic tubing are filled with water at the beginning of the investigation.

Plan your experiment under the following headings.

Hypothesis

Your hypothesis should be a testable prediction of the effect of changing light intensity on rate of photosynthesis. State how you think changing light intensity will affect the rate, and draw a sketch graph of what you predict your results will be like if the hypothesis is correct.

Method

When outlining your method break it up into parts as shown here.

Independent variable: name the independent variable and give full details of how you will change it. You should state the range you will use, and the increments, and explain why you have chosen the range and increments and how you will decide on the number of repeats.

Dependent variable: name the dependent variable and give full details of how you will measure it. In this case, you will need to explain how to measure the length of a bubble of oxygen using the scale and the syringe. The method should generate accurate, quantitative results. Explain why the measure you have chosen is a measure of photosynthesis rate.

Control variables: state all the variables that you will control, explain why they need controlling and how you will control them. Do not list everything you can think of, only list relevant things.

Detailed method: in your method give practical details and the precautions you will take. For instance, 'the *C. aquatica* stem is cut using the sharp scissors at an angle in order to help it release oxygen bubbles. It is then placed cut end upwards in …'. This should include details of how you will make up any concentrations of solutions that you will need. You will also need to explain the role of the syringe.

2 The results from a similar investigation into the effect of changing light intensity on rate of photosynthesis are listed in the table below. The data suggest there is a negative correlation between the distance of the lamp from the sample and the size of the bubble of oxygen. In other words, as distance increases, the length of the oxygen bubble decreases. To find out if this correlation is significant, we need to carry out a statistical test. Spearman's rank correlation is used to find out if there is a significant correlation between two sets of variables when they are not normally distributed and may not be in a linear relationship. Follow the steps below to determine whether or not the correlation is significant.

   a Formulate a null hypothesis for the results in Table 13.5.

   In statistical tests, we make the assumption that there is no significant association or correlation between the two variables: for example, in an experiment to see if the addition of fertiliser increases plant dry mass, a null hypothesis would be 'there is no significant correlation between concentration of fertiliser and dry mass of plant.'

Distance of lamp from *C. aquatica*/cm	Rank for distance of lamp, r_1	Mean length of oxygen bubble formed after 5 minutes illumination/cm	Rank for mean length of bubble, r_2	Difference in ranks, D, $(r_1 - r_2)$	D^2
5		26.4	1		
10		20.3	3		
15		20.5	2		
20		21.6			
25		18.4			
30		18.4			
35		14.0			
40		8.2			
45	2	1.6			
50	1	1.6			

Table 13.5: Results table.

b Rank the distances of the lamp from the plant in order, with the largest as '1'. Write the ranks into your table. Two are completed for you.

c Rank the mean length of oxygen bubble, again with the largest ranked as '1'. The first three are done for you.

Two of the mean lengths of oxygen bubble are both 18.4 cm, and we will need to give them the same rank. We give them an average rank as follows:

18.4 and 18.4 would nominally rank '5' and '6'.

The average rank will be:

$(5 + 6) \div 2 = 5.5$

Both values of 18.4 are ranked 5.5, the next rank awarded will be 7.

d Calculate the difference in ranks, D, for each pair of results and write your answer in your table.

e Calculate the square of the difference in rank of each pair, D^2. Write your answer in your table. Now calculate the sum of the squared differences in rank (ΣD^2).

f Use the formula below to calculate the Spearman's rank correlation coefficient.

$$r_s = 1 - \left(\frac{6 \times \Sigma D^2}{n^3 - n}\right)$$

where r_s is Spearman's rank correlation coefficient

ΣD^2 is the sum of the differences between the ranks of the pairs

n is the number of pairs.

If the coefficient is positive, the correlation is positive; if it is negative, the correlation is negative.

g Use the critical values in Table 13.6 to determine whether or not there is a significant correlation between distance of the lamp and length of bubble. Use the terms *probability*, *null hypothesis*, *chance*, *reject* and *significant* in your answer.

n	5	6	7	8	9	10	11	12	14	16
critical value of r_s	1.00	0.89	0.79	0.76	0.68	0.65	0.60	0.54	0.51	0.51

Table 13.6: Critical values.

We can now decide on the probability that the null hypothesis is correct. To determine this, we compare the value of the coefficient with a critical value in a table. In most statistical tests in biology, we use a probability of 0.05 as our baseline.

If our coefficient is larger than the critical value, this means there is only a probability of <0.05 (or 5%) that our null hypothesis is correct. This means that the correlation is significant and we reject the null hypothesis. There is a probability of less than 0.05 that the correlation is due to chance.

If our coefficient is smaller than the critical value, this means there is a probability of >0.05 (or 5%) that our null hypothesis is correct. This means that the correlation is not significant and we do not reject the null hypothesis. There is a probability of greater than 0.05 that the correlation is due to chance.

3 Test the following sets of data to determine if there are significant positive or negative correlations. In each case, formulate a null hypothesis and give a full explanation in terms of probability and chance.

   a The association between the concentrations of chlorophyll *b* in the leaves of a tree with increasing height (Table 13.7).

Distance from base of tree / m	Chlorophyll concentration / $\mu mol\ m^2$ leaf area
0.0	645
0.5	524
1.2	524
2.3	627
3.5	345
4.6	350
4.9	253
5.1	145

Table 13.7: Association between the concentrations of chlorophyll *b* in the leaves of a tree with increasing height.

   b The association between the number of ladybirds per 0.25 m² quadrat in a maize field and the dry mass of maize seeds (Table 13.8).

Quadrat number	Number of ladybirds	dry mass of maize seeds / g
1	8	225
2	4	178
3	5	201
4	2	101
5	7	223
6	9	254
7	3	154
8	7	227
9	6	213
10	8	244

Table 13.8: Association between the number of ladybirds per 0.25 m² quadrat in a maize field and the dry mass of maize seeds.

Exercise 13.4 Free response questions

An important skill is to be able to answer open essay-style questions. You need to read questions carefully, look at the mark allocation and think of the relevant points. Marks are awarded for scientifically accurate, A Level standard answers. In this exercise, you will:

- develop your understanding of how light energy is harvested in photosynthesis
- develop your understanding of how to write extended answers by marking example answers.

Look at three example answers to the question below. Use the mark scheme to decide how many marks each would get. Read the answers carefully and think how easy each was to mark. Remember, even if more points are made, you cannot get more than the maximum number of marks.

1   **Describe how chloroplasts use light energy to manufacture ATP.**   [8]

   a   Plants use chlorophyll to photosynthesise. Chlorophyll is what gives them a green colour and it is found in the chloroplasts. Chloroplasts can absorb plenty of carbon dioxide, water and light energy to photosynthesise. The chloroplasts can also move towards light so that they can maximise the amount of light absorbed. Because plants are green, they reflect green light and cannot use it in photosynthesis.

   The light is used to oxidise chlorophyll molecules and the electron is passed on to electron transport chains where it is used to make ATP and reduced NADP. To replace the chlorophyll's electron, water releases an electron that sticks onto the chlorophyll and can then be used again. This is called photolysis. The waste product of photolysis is oxygen gas, which we can use to breathe. This all takes place on the chloroplast membranes.

   b   Chloroplasts are designed for photosynthesis. There are two main stages of photosynthesis – the light dependent stage and the light independent stage. The light dependent stage is where ATP and reduced NADP are made. The thylakoid membranes, in the grana, are where the light dependent reaction occurs. Light energy is absorbed by collections of primary and accessory pigments that are bound together in two photosystems: photosystem 1 (PS1) and photosystem 2 (PS2). Chlorophyll a is the primary pigment and sits at the reaction centre of the photosystems surrounded by molecules of accessory pigments such as carotene and chlorophyll b. Chlorophyll a absorbs light in the red and blue areas of the spectrum. Chlorophyll b and carotene also absorb extra wavelengths of light and pass the energy on to chlorophyll a. The light energy is used to oxidise the chlorophyll by ejecting an electron. The electron is passed on to electron transport chain proteins which use the energy of the electron to pump H^+ ions into the centre of the thylakoids. The H^+ diffuses back out into the stroma of the chloroplast through H^+ channels attached to ATP synthase enzyme. The movement of the H^+ ions provides energy to phosphorylate ADP + Pi into ATP. The electrons can pass through carriers from PS2 to PS1 through non-cyclic photophosphorylation and from PS1 back to PS1 via electrons through cyclic photophosphorylation.

c Chlorophyll is found in two photosystems – PSI and PSII – that are located in the thylakoid membranes of the chloroplast. The chlorophyll absorbs light from the red and blue end of the spectrum and is oxidised. An electron is passed along a series of electron carriers and uses its energy to make ATP from ADP + Pi. The electron moves from PSII to PSI and then to the enzyme NADP reductase to combine with hydrogen ions to make reduced NADP. The ATP and reduced NADP are used to make sugars in the Calvin cycle, which takes place in the stroma. Photolysis occurs as well, which releases oxygen that can be used in respiration. DCPIP can be used to measure the rate of electron release from chlorophyll. Plants also have other pigments such as carotene, pheophytin and xanthophyll, which can also absorb light.

Mark scheme

1. chlorophyll *a* is a primary pigment ;
2. carotenoids / chlorophyll *b* is an accessory pigment ;
3. in **grana / thylakoids** ;
4. reference PSI and PSII; A P700 and P680 ;
5. primary pigment / chlorophyll *a*, in reaction centre ;
6. chlorophyll *a* / primary pigment is oxidised / loses electron ;
7. chlorophyll *a* absorbs red (ignore orange) and blue (ignore purple) ;
8. accessory pigments (accept named, e.g. carotene) absorb other wavelengths ;
9. electron is transferred to electron transport chain / electron carriers ;
10. H⁺ diffusion into **stroma** ;
11. reference to ATP synthase ;
12. ADP + P_i → ATP / ADP is phosphorylated ; P_i not just P ;

KEY WORDS

grana: (singular: **granum**): stacks of membranes inside a chloroplast

thylakoid: a flattened, membrane-bound, fluid-filled sac which is the site of the light-dependent reactions of photosynthesis in a chloroplast

stroma: the chloroplast's equivalent of cytoplasm, and is the site of enzyme reactions relating to photosynthesis

---

2 Look at the example answers to the question below. Use the mark scheme to decide how many marks each would get. Again, read the answers carefully and think how easy each was to mark. Remember, even if more points are made, you cannot get more than the maximum number of marks.

Describe how products of the Calvin cycle are used to make other groups of molecules. [7]

a The Calvin cycle is the process that fixes carbon. It occurs in the stroma of chloroplasts and is the combination of RuBP and carbon dioxide to make GP. The GP is reduced by NADPH (that is from the light dependent reaction) into TP which is in turn converted into RuP and other products such as carbohydrates. The RuP is phosphorylated by ATP into RuBP. ATP is also used to convert the GP into TP.

b The Calvin cycle makes sugars which are converted into other carbohydrates. Nitrates are used to add to the carbohydrates to make amino acids. The amino acids are combined to make proteins and lipids.

c Calvin cycle intermediates are used to make other compounds. GP is used to make certain amino acids after the addition of nitrates. TP is used to make glucose and fatty acids. Amino acids are combined by condensation reactions to make proteins. Both a and b glucose are made. If a glucose molecules are combined, starch (amylose and amylopectin) is produced. If b glucose molecules are combined, cellulose (for cell walls) is produced.

Mark scheme

1 Glycerate 3-phosphate (GP) ;
2 (GP) is used to make amino acids ;
3 Triose phosphate (TP) ;
4 (TP) is used to make carbohydrates / lipids / amino acids / fatty acids ;
5 nitrates are required for amino acid synthesis ;
6 condensation reactions ;
7 (condensation of) *a* glucose produces, starch / amylose / amylopectin ;
8 (condensation of) *b* glucose produces cellulose ;
9 (condensation of) amino acids produces polypeptides / proteins ;

EXAM-STYLE QUESTIONS

1 Figure 13.5 shows apparatus that can be used to measure the rate of photosynthesis of algae. The algae can be trapped inside a gel bead made of sodium alginate. The beads can be placed into a boiling tube containing a solution of sodium hydrogen carbonate. As the algae photosynthesise, they release oxygen which becomes trapped in the alginate beads and causes them to rise.

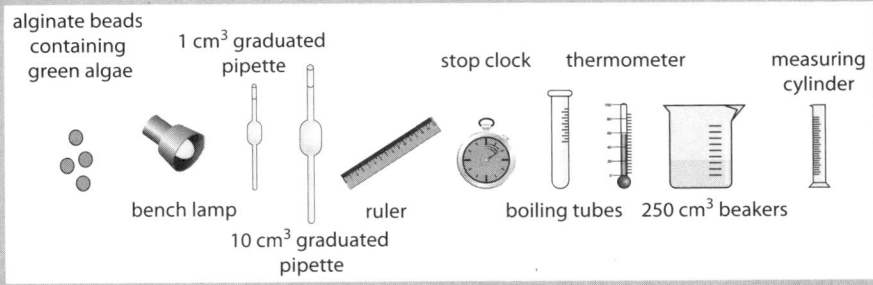

Figure 13.5

CONTINUED

a **i** **Describe** how you could use the apparatus above to investigate the rate of photosynthesis at different concentrations of sodium hydrogen carbonate (a source of carbon dioxide). You will be provided with a stock of 10% sodium hydrogen carbonate solution and distilled water and may assume that you have access to normal school or college laboratory equipment such as boiling tube racks. [7]

ii **State** the dependent variable. [1]

b Alginate beads were used to investigate the effect of three light colours, red, yellow and blue, on the rate of photosynthesis. The results are shown in Table 13.9.

The following hypothesis was made:

Red and blue light will cause faster photosynthesis than yellow light so the alginate beads will rise faster

Light colour	Time taken for beads to rise 10 cm / seconds										
repeat measurements	1	2	3	4	5	6	7	8	9	10	Mean
red	96	126	134	145	156	156	165	166	173	376	146.3
yellow	210	225	243	244	247	259	254	273	281	286	252.2
blue	122	144	157	157	167	168	175	179	181	185	163.5

Table 13.9

i **Explain** how the mean values have been calculated to generate a reliable mean. [1]

ii The standard error for the results was calculated.

The standard error for red is 7.6.

The standard error for yellow is 7.2.

The formula for standard error is:

$$SE = \frac{s}{\sqrt{n}}$$

where SE is the standard error

s is the standard deviation

n is the sample size.

Calculate the standard error for blue. Show all your working. [3]

iii **State** what the standard deviation shows. [2]

COMMAND WORDS

Describe: state the points of a topic / give characteristics and main features.

Explain: set out purposes or reasons / make the relationships between things evident / provide why and/or how and support with relevant evidence.

Calculate: work out from given facts, figures or information.

State: express in clear terms.

CONTINUED

    iv Plot a graph to show the effect of light colour on the mean time taken for the alginate beads to rise 10 cm. Add error bars. [4]

    v Explain whether or not the data support the hypothesis. [2]

    [Total: 20]

2 a Describe the process by which carbon is fixed continuously in the palisade mesophyll cells during photosynthesis. [8]

  b **Predict** and explain the changes in carbon dioxide concentration at different heights in a tropical rain forest over a 24-hour period. [7]

    [Total: 15]

3 Figure 13.6 shows the effect of increasing light intensity on the net rate of release of oxygen of two woodland plants, a 'sun plant' and a 'shade plant'. The 'sun plant' is a tall tree, while the 'shade plant' grows on the woodland floor.

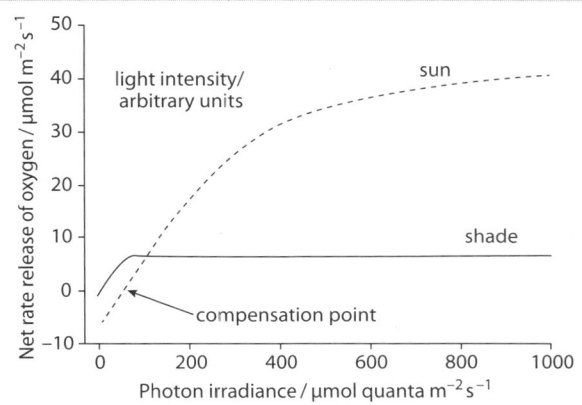

Figure 13.6

  a **Compare** the effect of increasing light intensity on the sun plant and the shade plant. [3]

  b Explain the effect of increasing light intensity on the sun plant. [4]

  c **Suggest** and explain a reason for the difference in the effect of increasing light intensity on the two plants. [2]

    [Total: 9]

4 a **Outline** the roles of chlorophyll *a* and the accessory pigments in photosynthesis. [5]

  b Different types of algae are found at different depths in the ocean. The algae found at different depths have different combinations of accessory pigments. Figure 13.7 shows the ranges of depth at which different algae are found and the approximate depths that different light colours penetrate through the water.

COMMAND WORDS

Predict: suggest what may happen based on available information.

Compare: identify/comment on similarities and/or differences.

Suggest: apply knowledge and understanding to situations where there is a range of valid responses, in order to make proposals / put forward considerations.

Outline: set out the main points.

CONTINUED

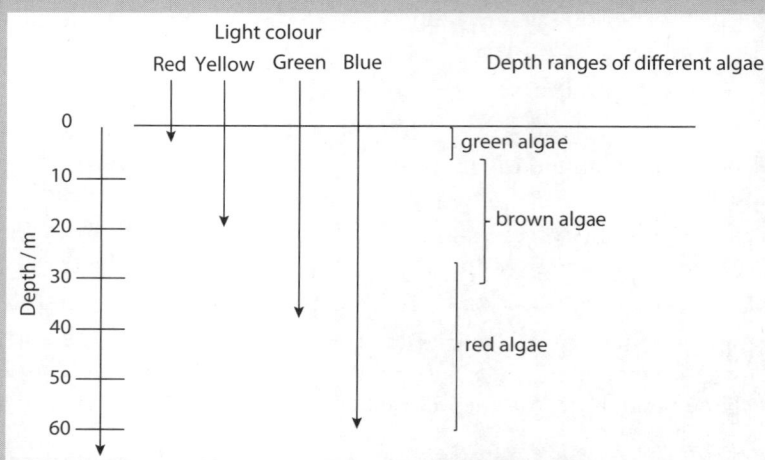

Figure 13.7

Figure 13.8 shows the absorption spectra of different photosynthetic pigments. Table 13.10 shows the different pigments found in the different types of algae.

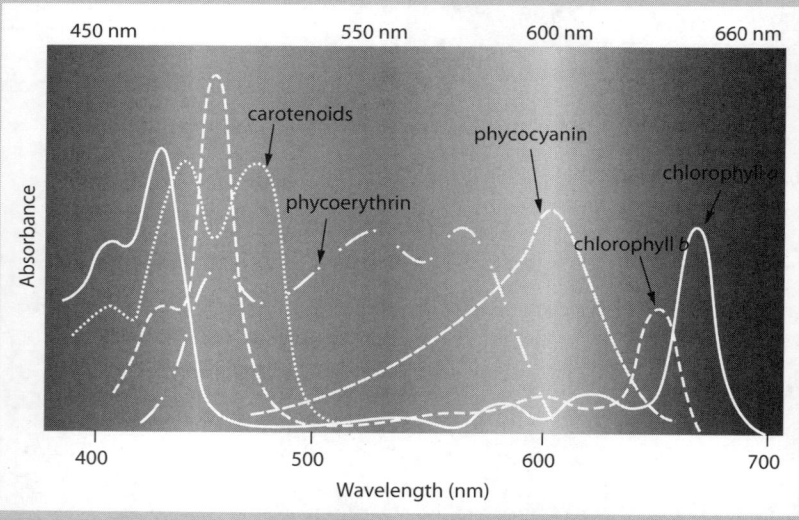

Figure 13.8

Type of algae	Pigments present
green	chlorophyll a, chlorophyll b, carotenoids
brown	chlorophyll a, chlorophyll b, carotenoids, phycocyanin
red	chlorophyll a, phycocyanin, phycoerythrin, carotenoids

Table 13.10

CONTINUED

Use Figures 13.7 and 13.8 and Table 13.10 to explain the depth ranges of the different types of algae. [4]

[Total: 9]

5 Figure 13.9 shows the structure of a chloroplast.

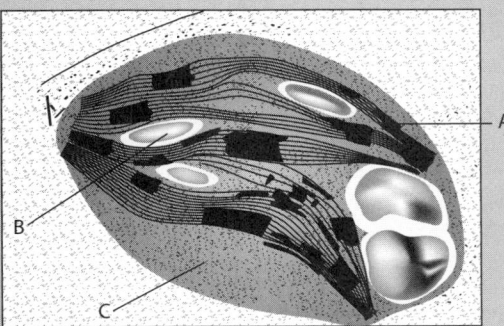

Figure 13.9

a Name the structures labelled **A**, **B** and **C** in Figure 13.9. [3]

b Carbon fixation of isolated chloroplasts was investigated. Chloroplasts were placed into an isotonic buffer solution and illuminated. Radiolabelled $^{14}CO_2$ was added and the relative amounts of radiolabelled GP and RuBP measured every minute for ten minutes. The light was switched off and the relative amounts of GP and RuBP were again measured every minute for ten minutes. The results are shown in Figure 13.10.

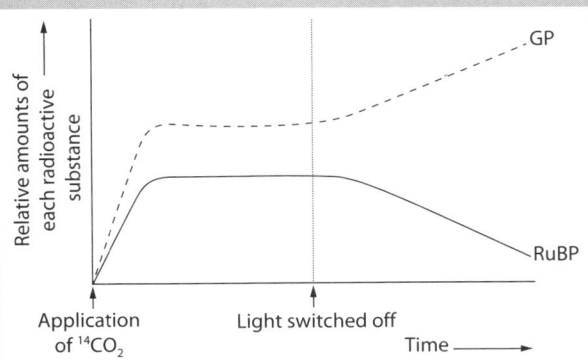

Figure 13.10

Use your knowledge of the Calvin cycle to:

i Explain the changes in the amounts of radiolabelled GP and RuBP in the light. [4]

ii Explain the changes in the amounts of radiolabelled GP and RuBP after the light was switched off. [3]

[Total: 10]

> Chapter 14
Homeostasis

CHAPTER OUTLINE

The questions in this chapter cover the following topics:

- the importance of homeostasis and how the nervous and endocrine systems coordinate homeostasis in mammals
- the structure of the kidneys, and their role in excretion and osmoregulation
- how blood glucose concentration is regulated
- what is meant by negative feedback
- how cell signalling is involved in the response of liver cells to adrenaline and glucagon
- the principles of operation of dip sticks to test for the presence of glucose in urine
- the factors that cause stomata to open and close
- how abscisic acid causes guard cells to close stomata.

Exercise 14.1 The action of the drug metformin in the control of blood glucose concentrations

Type 2 diabetes is caused by the inability of the liver and muscle cells to respond to insulin. It is not treated initially with injections of insulin, since **insulin** levels in people when diagnosed are often normal or higher than normal. The drug metformin is often used to first help to lower blood glucose levels. In this exercise, you will:

- develop your understanding of the regulation of blood glucose concentration and secondary signalling
- develop your analytical skills by examining data for the effect of a drug, metformin, on blood glucose concentration
- develop your understanding of the use of the statistical test, standard error.

1   An experiment was carried out where two groups of 12 rats were injected with either water or the drug metformin which had been dissolved in water. Blood glucose concentration was measured in all the rats at the start, after 30 minutes and after 60 minutes. After 60 minutes, half of the rats from each group were injected with either saline solution or **glucagon** dissolved in saline. Blood glucose concentration was then measured every ten minutes for 30 minutes.

> **KEY WORDS**
>
> **insulin:** a small peptide hormone secreted by the β cells in the islets of Langerhans in the pancreas to bring about a decrease in the concentration of glucose in the blood
>
> **glucagon:** a small peptide hormone secreted by the α cells in the islets of Langerhans in the pancreas to bring about an increase in the concentration of glucose in the blood

a   This is quite a complicated experiment, so it can help to represent it in a flow chart. Copy out Figure 14.1 and fill in the blanks.

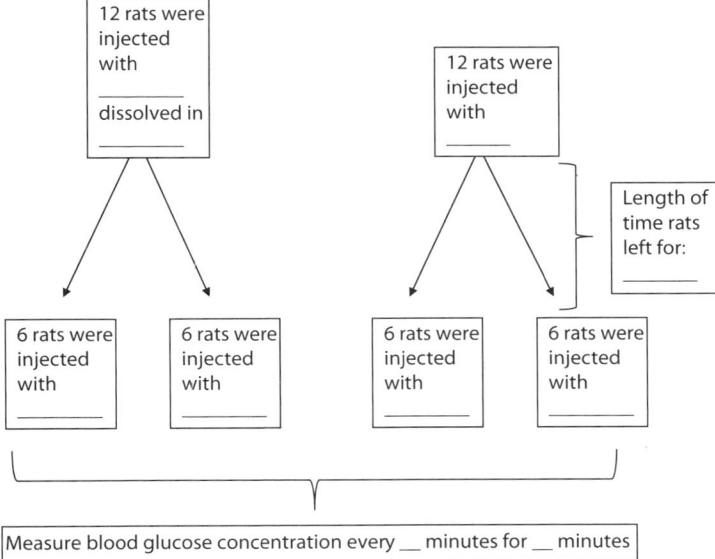

Figure 14.1: Experimental design flowchart.

b   List the independent variables (there are two).
c   Suggest three factors that should be controlled.
d   Explain why one group of rats were also injected with water and saline rather than metformin and glucagon.
e   Explain why patients with type 2 diabetes can be found to have higher than normal concentrations of insulin in their blood.

2   The results of the experiment are shown in Figure 14.2.

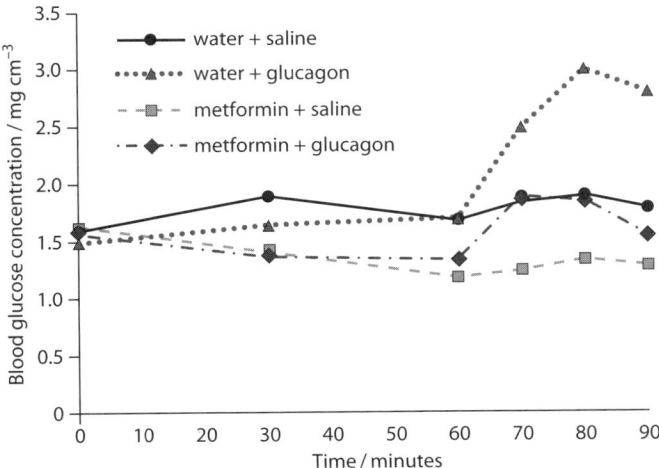

Figure 14.2: Blood glucose concentrations of rats after treatment with metformin and glucagon.

Describing complex data patterns is an important skill and it is important to look for different parts of a graph and be clear what you are comparing. Look carefully at the graph in Figure 14.2 and respond to the following.

a Compare the effects of injection with water and metformin on blood glucose concentration from time 0 minutes to time 60 minutes. Here you are looking for differences between the two treatments with water and the two treatments with metformin.

b The glucagon and saline were injected at 60 minutes. Make the following three comparisons:

- First, compare water + glucagon with water + saline. This shows you the normal effect of glucagon in the absence of metformin. It is a baseline effect of glucagon.

- Second, compare the rats injected with water + glucagon with those injected with metformin + glucagon. This shows you whether metformin affects glucagon action.

- Third, compare the rats with metformin + saline with the rats with metformin and glucagon. This shows you how much blood glucose concentration will rise when adding glucagon with or without metformin.

c Look at your comparisons. Now make a valid conclusion about whether or not metformin affects blood glucose concentration and how it might cause this.

d Suggest how you could make your conclusion more valid and what information you would require to do so.

3 At the end of the experiment, the effects of metformin and glucagon on the production of cyclic AMP (cAMP) in rat liver cells were also investigated. Liver tissue was removed from the same rats and the cAMP levels measured. The results are shown in Table 14.1. Note that cAMP is expressed as its mass as a proportion of total cellular protein mass.

Substances injected	Mass of cAMP / pmol µg^{-1} cellular protein							
	1	2	3	4	5	6	Mean	Standard deviation
water followed by saline	9	10	8	6	7	10	8.3	1.6
water followed by glucagon	65	68	53	54	58	61	59.8	6.0
metformin followed by saline	5	7	12	9	7	6	7.7	2.5
metformin followed by glucagon	24	23	18	16	17	18	19.3	3.3

Table 14.1: Effects of metformin and glucagon on the production of cyclic AMP (cAMP) in rat liver cells.

a Draw a graph to show the effects of the different treatments on the rat liver cells on mean cAMP mass. In Chapter 13 you learned how to calculate and use the standard error of the mean and 95% confidence limits. Calculate the standard error and 95% confidence limits and use them to add error bars to your graph.

14 Homeostasis

   b Explain what the data now show you about the effects of metformin and glucagon on cAMP production. Does it suggest that glucagon utilises a secondary messenger system? Does it suggest that one of the ways in which metformin reduces blood glucose is to reduce the response of liver cells to glucagon?

4 **a** You need to be able to understand the role of cAMP as a secondary messenger. Table 14.2 contains a set of statements about the action of the hormone glucagon. Place the statements in the correct order.

A	G protein is activated and in turn activates an enzyme, adenylate cyclase, to convert ATP into cAMP.
B	**Phosphorylase kinase** is activated by protein kinase.
C	Protein kinase is activated by cAMP.
D	Glucagon binds to glucagon receptor on surface of liver cell membrane.
E	Glucose diffuses out of liver cells through GLUT2 transporter channels.
F	Glycogen phosphorylase is activated by phosphorylase kinase.
G	Glycogen phosphorylase converts glycogen into glucose.

Table 14.2: Statements about the action of the hormone glucagon.

> **KEY WORDS**
>
> **phosphorylase kinase:** an enzyme that is part of the enzyme cascade that acts in response to glucagon; the enzyme activates glycogen phosphorylase by adding a phosphate group

   b Now use the statements to explain why metformin can reduce blood glucose concentration.

Exercise 14.2 Osmoregulation and isotonic drinks

Many sports drinks available are described as isotonic. Many of them suggest that rehydration is far more effective if drinks contain fluid that is isotonic to normal body fluids rather than being pure water. In this exercise, you will:

- develop your understanding of the process of osmoregulation by reviewing an experiment that investigates the effect of different sodium chloride concentrations in drinks on urine output
- develop your understanding of molarity.

1 Imagine that you are going to plan an experiment into the effects of consuming drinks of different sodium chloride concentrations on the reabsorption of water by the kidneys after athletes have performed strenuous exercise. First, you would need to make up a range of drinks with different concentrations of sodium chloride.

   a Take a look back at Chapter 2, Exercise 2.2, to refresh your memory on how to calculate the concentrations of different solutions. The table below shows the different concentrations of the drinks that could be used in this investigation. The concentrations are given as percentage by total volume of solution. Copy out Table 14.3 and write down the mass of sodium chloride required to make up 1 dm³ of each solution.

b Calculate the molarity of each solution and enter it into Table 14.3.

The molecular masses you need are:

Sodium: 23

Chlorine: 35.5

Concentration of sodium chloride solution / % by volume	Mass of sodium chloride dissolved in 1 dm³ / g	Concentration of sodium chloride solution /mol dm^{-3}
0.0		
0.3		
0.6		
0.9		
1.2		
1.5		
1.8		

Table 14.3: Results table.

2 In this experiment, 18 male students aged between 18 and 21, and all of the same approximate fitness, were asked to pedal on a bicycle for one hour at the same power output. After this hour, the students were given a drink of a particular sodium chloride concentration between 0% and 1.2%. The students all had the same fluid intake for four hours before the experiment.

0.9% concentration of sodium chloride is isotonic to normal body fluids.

Five different salinity drinks were used, and three students received each salinity. The volume of drink was the same in all cases = 1000 cm³. Three students received no drink. The total volume of the each student's urine was then measured over the next three hours. The results are shown in Table 14.4.

Concentration of saline / % by volume	Volume of urine produced over 3 hours / cm³			
	Student 1	Student 2	Student 3	Mean
0.0	348	374	325	349.0
0.3	325	359	345	343.0
0.6	254	275	288	272.3
0.9	201	245	242	229.3
1.2	204	195	202	200.3
no drink	143	129	133	235.0

Table 14.4: Results table.

a State the independent and dependent variables.
  b State four of the variables that were controlled.
  c Describe the effect of increasing the concentration of sodium chloride in the drink on the volume of the urine.
  d Suggest an explanation for the pattern you have described. You should think about water potentials, osmoreceptors, salt loss and diffusion of salts into the blood.
  e Drinking solutions with very high concentrations of sodium chloride can lead to a high concentration of sodium chloride in the blood. Predict and explain the effects of drinking hypertonic, 1.8% sodium chloride solution on the body.

3 Place the following statements about osmoregulation in order:

Letter	Statement
A	Osmoreceptors in hypothalamus detect the lower water potential and impulses are sent to the posterior pituitary.
B	Blood water potential becomes lower (more negative).
C	ADH binds to receptors in membranes of collecting duct cells.
D	Phosphorylase enzyme is activated.
E	ADH is released into blood.
F	Water passes through aquaporins into lumen of collecting duct.
G	Vesicles containing aquaporins move to the cell surface membrane and fuse with it.
H	Blood water potential rises and ADH release stops.
I	High rates of sweating lead to evaporation of water.

Table 14.5: Statements about osmoregulation.

Exercise 14.3 Measuring percentage errors and comparing blood glucose monitors

It is important to know the level of accuracy that apparatus can measure to. If we know how much error each measurement may have, we can be more or less confident in our results. It is also important with clinical measuring devices such as glucose monitors that we know how much we can trust a reading. In this exercise, you will:

- develop your understanding of how to measure percentage error
- develop your understanding of how glucose monitors work
- develop your understanding of experimental accuracy.

Some measuring equipment such as balances will often give a measure of error such as 0.01 g. This means that the balance is accurate to ± 0.01 g of the reading given.

For example, a balance with a measuring error of ±0.05 g was used to measure out 1 g of sodium carbonate. The actual mass measured out will lie somewhere between 1.00 g ± 0.05 g (0.9–1.05 g).

If we carry out two measurements in a sequence, the errors will be added together. For example, in our previous example, if we were investigating the change in mass of our sodium carbonate after heating it:

Start mass = 1.00 g ± 0.05 g.

Finish mass = 0.55 g ± 0.05 g.

The change in mass will be 1.00 g – 0.55 g = 0.45 g.

The measuring error could have occurred twice, so will be 0.05 g × 2 = 0.10 g.

So, our final result for the change in mass will be 0.45 g ± 0.10 g (0.35–0.55 g).

When measuring volumes of liquids, unless otherwise stated on the equipment, we generally assume that the error is half of the smallest graduation on the scale.

For example, a 1 cm^3 graduated pipette has increments of 0.1 cm^3. This means that the pipette is accurate to ± (0.1 ÷ 2) = ± 0.05 cm^3.

1   Calculate the ranges that the true values will lie between for the following.

   a   2.500 g of glucose measured using a balance accurate to 0.005 g

   b   25 g of glucose measured in five batches of 5 g using a balance accurate to 0.005 g

   c   the change in mass of 5 g of starch after digestion by amylase left a mass of 2.4 g, using a balance accurate to 0.01 g

   d   a temperature of 33 °C on a thermometer with the smallest graduations on the scale of 1 °C

   e   the change in body temperature of a reptile over a one hour period. Start temperature = 17 °C, finish temperature = 21 °C. Smallest graduations on the thermometer scale are 0.1 °C

   f   the change in volume of oxygen produced by pond weed, using a gas syringe. Start volume = 0.1 cm^3, finish volume = 11.2 cm^3. Smallest graduations on the scale = 0.1 cm^3.

It is also useful to express the amount of error in a measurement as a percentage error. To do this, you use the formula below:

Percentage error (%) = (error ÷ reading) × 100

In other words, for our first example where we measured out 1 g of sodium carbonate, the percentage error would be:

(0.05 ÷ 1.00) × 100 = 5%

Looking at percentage error of equipment should help you to choose the correct size of apparatus. For example, suppose you want to measure out 25 cm^3 of 1% glucose solution and have three measuring cylinders:

- 250 cm^3 cylinder smallest graduations on scale: 5 cm^3

- 50 cm³ cylinder smallest graduations on scale: 2 cm³
- 10 cm³ cylinder smallest graduations on scale: 1 cm³.

The percentage errors when using each of the measuring cylinders would be:

- 250 cm³ cylinder: (2.5 cm³) ÷ 25 cm³ × 100 = 10%
- 50 cm³ cylinder: (1.0 cm³) ÷ 25 cm³ × 100 = 4%
- 10 cm³ cylinder: (1.5 cm³) ÷ 25 cm³ × 100 = 6%.

If we use the 10 cm³ cylinder, we would have to measure out two sets of 10 cm³ and one set of 5 cm³, which would give us a combined error of 0.5 × 3 = 1.5 cm³.

So, for this experiment, we would choose to use the 50 cm³ measuring cylinder as it would give the lowest percentage error.

Generally (although not always), you should choose the smallest piece of measuring equipment that would measure your volume in one batch. The smaller the volume being measured, the higher the percentage error becomes in a particular piece of apparatus.

For example, the percentage error with 250 cm³ of liquid measured in a 250 cm³ measuring cylinder graduated into 5 cm³ is:

$$(2.5 \text{ cm}^3 \div 250 \text{ cm}^3) \times 100 = 1\%$$

The same measuring cylinder used to measure out 1 cm³ of glucose gives a percentage error of:

$$(2.5 \text{ cm}^3 \div 1 \text{ cm}^3) \times 100 = 250\%$$

which is obviously not acceptable.

2 Calculate the percentage errors of the following:

  a 34 cm³ water measured in a 100 cm³ beaker with graduations of 2 cm³
  b the change in distance moved by the fluid in a respirometer: start height 5 cm, finish height 2.5 cm, graduations of 1 mm on the scale
  c 10 cm³ of 2% sucrose solution measured out using a 1 cm³ syringe, scale graduations of 0.1 cm³
  d the change in temperature of calorimeter using a thermometer, scale graduations of 0.5 °C, start temperature 17 °C, finish temperature 64 °C
  e a 27.3 s time, assuming that human reaction time error is 0.1 s.

Percentage error is often used in a different way when placed on clinical equipment. People with diabetes need to constantly monitor blood sugar levels. This can be done using several methods, such as:

- using urine dip sticks (shown in Figure 14.3). These use two different enzymes to produce a brown colour that will change in intensity according to the concentration of glucose.

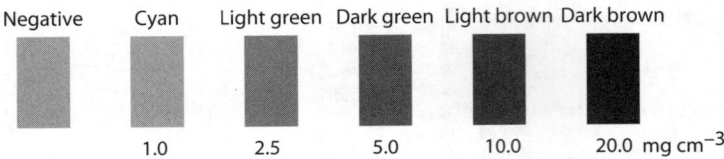

Figure 14.3: Reference chart for urine glucose dip stick.

- using a blood glucose concentration monitor. These use the same enzyme but also generate a measurable electric current.

3   a   Use your knowledge of the enzymes in dip sticks to draw out and complete the flow chart in Figure 14.4 to show how the colour changes in the dip sticks occur.

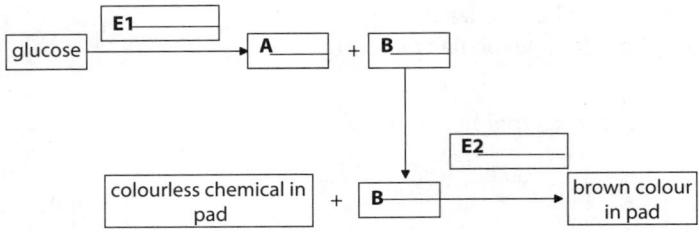

Figure 14.4: Flow chart showing how the enzymes in glucose urine dip sticks produce a brown colour.

   b   Describe whether the dip-stick test for the presence of glucose in urine is a quantitative, semi-quantitative or qualitative test. Explain why it could not be used to predict or detect **hypoglycaemia**.

   Blood glucose monitors offer a more quantitative method of measuring blood glucose concentration. The glucose monitors detect the glucose concentration in the blood at a particular time and can identify low as well as high concentrations. There are many different brands available which have varying levels of accuracy (which depend on how the skill of the user and the sensitivity of the equipment). In 2013, the International Organization for Standardization (ISO) set the following guidelines for the accuracy of blood glucose monitors:

   - accurate to within ±0.15 mg cm^{-3} at concentrations of under 1 mg cm^{-3}
   - accurate to within ±20% at concentrations of 1 mg cm^{-3} or more.

   c   Use the guidelines to calculate the maximum acceptable range and percentage error of blood glucose concentration from a reading of 0.55 mg cm^{-3}.

   d   Use the guidelines to calculate the maximum acceptable range of readings for blood glucose concentrations from a reading of 1.25 mg cm^{-3}.

   e   Suggest why it is important for people with type 1 insulin-dependent diabetes to consider the accuracy of the blood glucose monitors when injecting insulin.

> **KEY WORD**
>
> **hypoglycaemia:** a condition caused by a very low level of blood sugar (glucose)

f   Blood glucose concentrations are often screened to identify those who have diabetes and those who are at risk of developing diabetes. Those who show elevated blood glucose concentrations but are not full diabetics are classed as prediabetics or are said to be suffering from prediabetes. If untreated, prediabetes will often develop into type 2 diabetes. Two measures of blood glucose are carried out.

- The fasting test. Blood glucose concentrations after 24 hours of fasting are measured.
- Oral blood glucose tolerance test. The patient is given a measured dose of glucose and blood glucose concentrations are tested after two hours.

Figure 14.5 shows the recommended guidelines on blood glucose concentrations used to classify a person as having diabetes, prediabetes or neither.

	A1C / %	Fasting plasma glucose / mg cm^{-3}	Oral glucose tolerance test / mg cm^{-3}
Diabetes	6.5 or above	1.26 or above	2.00
Prediabetes	5.7 to 6.4	1.00 to 1.25	1.40 to 1.99
Normal	About 5	0.99 or below	1.39 or below

Figure 14.5: Blood glucose concentrations that enable classification of diabetic, prediabetic or normal patients. The higher the test result, the higher the risk of diabetes.

i   Suggest why both tests are carried out.

ii Evaluate the ISO guidelines on blood glucose monitor accuracy when used to identify diabetic and prediabetic patients.

g   A further test for diabetes is a test called the A1C test. This test is based on the attachment of glucose to haemoglobin. In the body, red blood cells are constantly forming and dying, but typically they live for about three months. The A1C test reflects the average of a person's blood glucose levels over the past three months. The A1C test result is reported as a percentage. The higher the percentage, the higher a person's blood glucose levels have been. A normal A1C level is below 5.7%. Figure 14.5 shows the typical A1C test results for normal, prediabetic and diabetic patients. Figure 14.6 shows the fluctuations in blood glucose concentration of a diabetic patient over one week and the A1C value.

Use the data to evaluate how effective the A1C test would be in identifying patients with prediabetes.

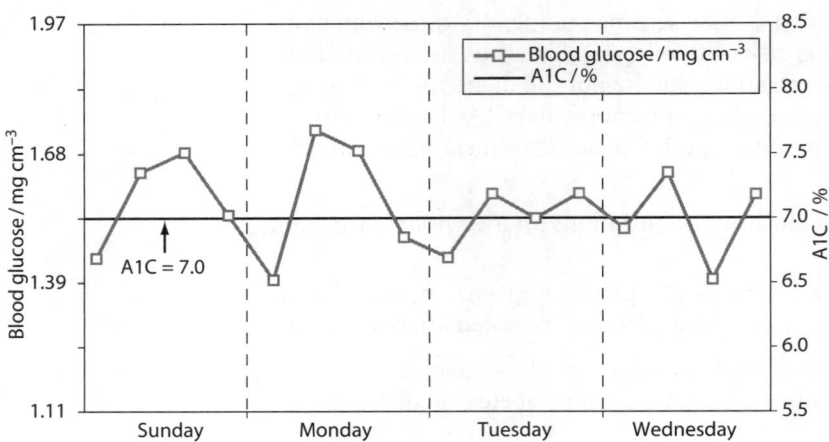

Figure 14.6: Fluctuations in blood glucose concentration of a diabetic patient and A1C value of same patient.

Exercise 14.4 Stomatal opening in plants

Plants open and close their stomata in response to a variety of factors, including light intensity, humidity, water stress and high carbon dioxide. In this exercise, you will:

- develop your understanding of how plants open and close their stomata
- develop your understanding of the difference between the mean, **median** and **mode**
- develop your understanding of how to use the interquartile range.

1 Plants often open and close their stomata in a rhythmic pattern that continues for a while if the light/dark pattern is stopped. An experiment was carried out into the natural rhythm of stomatal opening in *Tradescantia* sp. plants. The plants were illuminated daily from 06:00 hours until 18:00 hours and then placed in darkness from 18:00 hours until 06:00 hours the next day. This cycle was continued for two weeks. Each day at 13:00 hours, over 80% of stomata were found to be open on leaves, and at 00:00 hours fewer than 5% of stomata were found to be open. After continuing the cycle of light and dark for two weeks, the plants were placed into complete darkness and the percentage of stomata open determined at 12:00 hours, 24 hours later, and then again 48 hours later. The results for 50 different leaves are shown below.

> **KEY WORDS**
>
> **median:** the middle value of all the values in the data set
>
> **mode (modal class):** the most common value, or class, in the set of results

14 Homeostasis

Percentage of stomata open at 12:00 hours after 24 hours	Percentage of stomata open at 12:00 hours after 48 hours
54, 56, 58, 59, 58, 62, 62, 61, 63, 64, 65, 67, 68, 68, 67, 69, 67, 68, 68, 70, 73, 72, 71, 72, 71, 71, 73, 74, 73, 75, 74, 73, 74, 75, 78, 76, 78, 77, 78, 79, 78, 79, 78, 81, 84, 82, 83, 83, 87, 89	44, 44, 48, 48, 49, 49, 49, 49, 51, 51, 51, 52, 52, 52, 52, 52, 52, 53, 54, 56, 56, 57, 57, 57, 58, 58, 58, 58, 59, 59, 59, 60, 61, 61, 61, 61, 62, 62, 63, 63, 64, 64, 66, 67, 68, 68, 69, 69, 69, 73, 74

Table 14.6: Natural rhythm of stomatal opening in *Tradescantia* sp. plants.

We can plot frequency distributions to compare the two sets of data.

First, select suitable intervals; for this, intervals of 5% are ideal. The table below is a frequency table for the data. It has been completed for percentage of stomata open after 24 hours.

a Copy and complete Table 14.7 by writing down the frequencies of the different percentages of stomata open after 48 hours.

Range	Number of leaves after 24 hours	Number of leaves after 48 hours
≤45	0	
46–50	0	
51–55	1	
56–60	4	
61–65	6	
66–70	9	
71–75	14	
76–80	9	
81–85	5	
≥86	2	

Table 14.7: Frequencies of the different percentages of stomata open after 48 hours.

To represent these data, we can plot a histogram. A histogram can be plotted when both axes have continuous variables, as opposed to a bar chart which is plotted when one axis has categorical data.

To plot the histogram, write down the intervals on the *x*-axis, as shown in Figure 14.7, and label the *x*-axis 'Percentage of stomata that are open / %'. The *y*-axis is the frequency of each group – label this axis 'Frequency'.

> **TIP**
>
> Remember:
>
> Histograms have bars that touch. Bar charts have gaps between the bars.

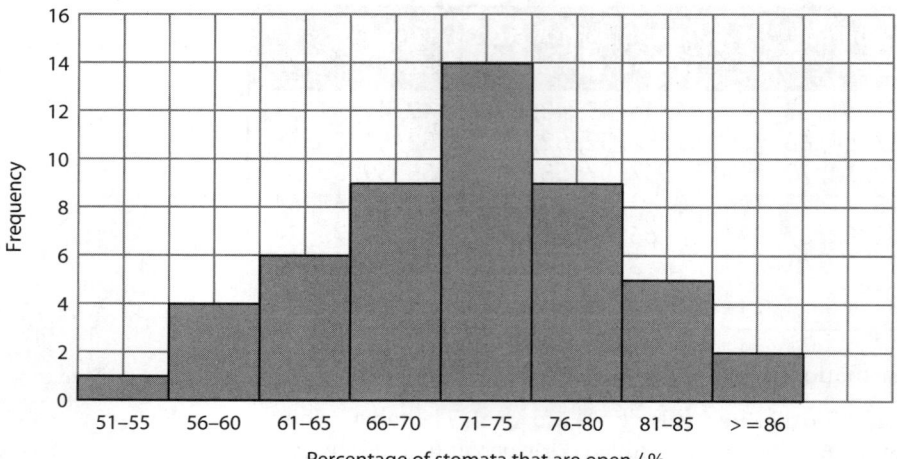

Figure 14.7: Histogram showing the frequency of leaves with different percentages of stomata open.

- **b** Plot a histogram to show the frequency of stomata open after 48 hours.
- **c** Calculate the mean number of stomata open after 24 hours and 48 hours.

    Sometimes, means do not tell the whole story and it is useful to determine the median (middle value) and modal class (most frequent group).

    Follow the following steps to calculate the median value:

    Step 1 Arrange the raw data in order of smallest to largest. For the percentage of stomata open after 24 hours, this is:

    54, 56, 58, 58, 59, 61, 62, 62, 63, 64, 65, 67, 67, 67, 68, 68, 68, 68, 69, 70, 71, 71, 71, 72, 72, 73, 73, 73, 73, 74, 74, 74, 75, 75, 76, 77, 78, 78, 78, 78, 78, 79, 79, 81, 82, 83, 83, 84, 87, 89

    Step 2 If there are an odd number of values, the median is the middle value. If there are an even number of values, the median is the mean of the two middle values.

    In the data for percentage of stomata open after 24 hours, there are 50 values. The median will be the mean of the 25th and 26th values:

    $$(72\% + 73\%) \div 2 = 72.5\%$$

- **d** Now calculate the median of the percentage of stomata open after 48 hours.

    The mode is the most frequent number from the raw data, which for the percentage of stomata open after 24 hours is 78%.

    The true mode is not always as useful as the modal class, which is the range on the histogram that has the highest frequency. Figure 14.7 shows that for the percentage of stomata open after 24 hours, this is 71–75%.

- **e** Calculate the mode and modal class for the percentage of stomata that are open after 48 hours.

    Sometimes it is helpful to look at the interquartile ranges. This is the range of your data into which the middle 50% of your data fall and so ignores the extreme values. This is shown in Figure 14.8.

14 Homeostasis

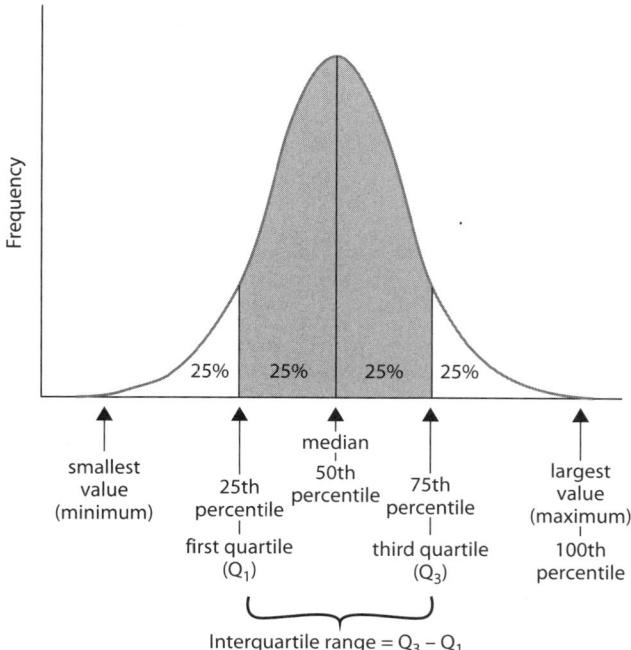

Figure 14.8: Frequency distribution showing the interquartile range.

To calculate the interquartile ranges, we need to calculate the first quartile, Q_1 and the third quartile, Q_3, as shown in Figure 14.8. Do this as follows:

- Determine the median value and n, the number of values there are.
- The first quartile point will be value $(n + 1) \div 4$, the second quartile will be value $(n + 1) \div 2$, the third quartile will be $3(n + 1) \div 4$ and the final 100% quartile will be at n. If the value for the first and third interquartiles are not whole numbers, we again take a mean of the higher and lower number.

For the percentage of stomata open after 24 hours:

Median = mean of the 12th and 13th value; 72.5%

The first quartile value, $Q_1 = (n + 1) \div 4 = (50 + 1) \div 4 = 12.75$. This means that the first quartile, Q_1 will be the mean of value 12 and 13.

$$Q_1 = (67 + 67) \div 2 = 67\%$$

The third quartile value, $Q_3 = 3(n + 1) \div 3 = 3(50 + 1) \div 4 = 38.25$. This means that the third quartile, Q_3 will be the mean of values 38 and 39.

$$Q_3 = (78 + 78) \div 2 = 78\%$$

Our interquartile range is thus 78% − 67% = 11%. This means that 50% of our values for percentage of stomata open after 24 hours are between 67% and 78%.

f Calculate the interquartile range for the number of stomata open after 48 hours.

g Compare the percentage of stomata open at 12:00 hours after 24 hours with the percentage of stomata open after 48 hours. Use the means, medians, modal classes and interquartile ranges in order to suggest what conclusions can be made from these data.

EXAM-STYLE QUESTIONS

1 a **Explain** what is meant by the term *homeostasis* and outline why it is important to living organisms, using specific examples. **[5]**

b A student carried out an investigation into the effect of drinking iced water on the skin and core body temperature. They measured skin and core body temperature, consumed 250 g of iced water and then monitored the temperature continuously using electronic temperature probes for 35 minutes. The results are shown in Figure 14.9.

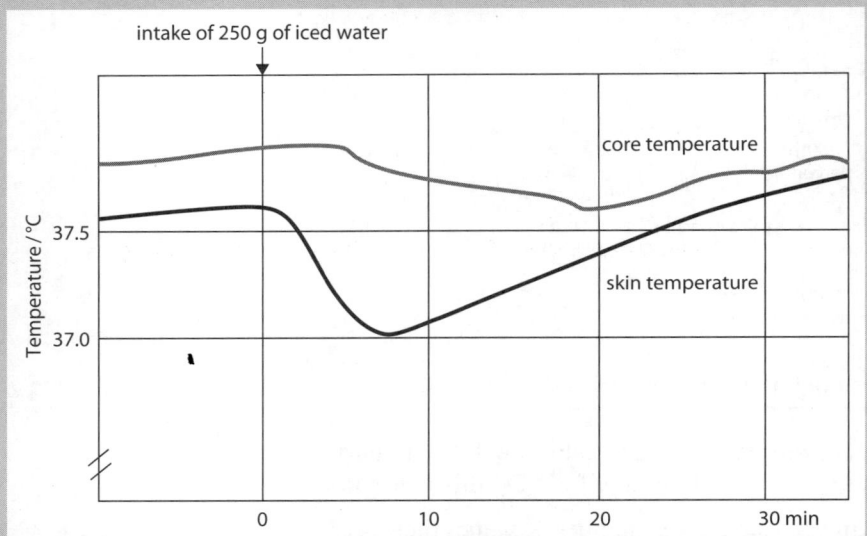

Figure 14.9

i **Compare** the effects of the iced water on the core and skin temperatures. **[3]**

ii Using your knowledge of thermoregulation, explain the changes in core and skin temperatures. **[6]**

[Total: 14]

COMMAND WORDS

Explain: set out purposes or reasons / make the relationships between things evident / provide why and/or how and support with relevant evidence.

Compare: identify / comment on similarities and/or differences.

CONTINUED

2 Figure 14.10 shows the effect of increasing intensity of exercise on the muscle, core body and skin temperature.

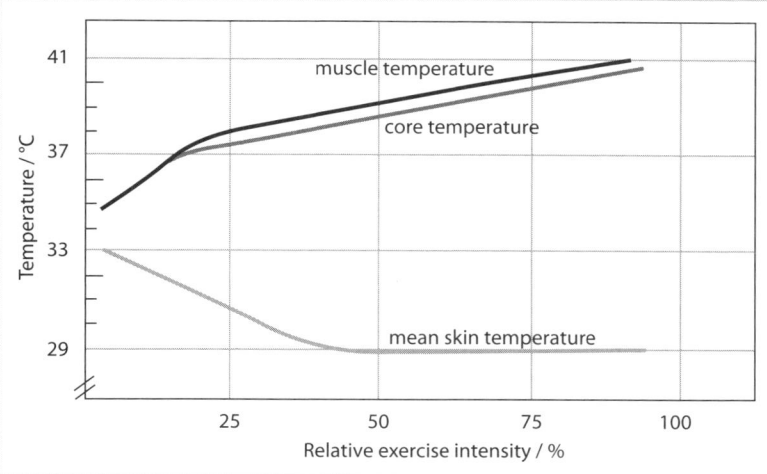

Figure 14.10

a i **Describe** and explain the effect of increasing intensity on the temperature of the muscles and core blood temperature. [4]

  ii **Suggest** and describe an explanation for the effect of increasing exercise on skin temperature. [4]

b After exercise, athletes often take a cold shower to cool down. An experiment was carried out into the effect of water temperature of a shower on core body temperature. Athletes exercised and then took a shower for ten minutes. Three different athletes were exposed to different temperatures of water in the shower. The core body temperature was measured every two minutes during the shower and for ten minutes after. The results are shown in Table 14.8.

Temperature of shower water / °C	Core body temperature / °C										
	0 min	2 min	4 min	6 min	8 min	10 min	12 min	14 min	16 min	18 min	20 min
4	39.5	39.4	38.9	38.5	38.4	38.5	38.7	38.9	38.5	38.3	38.1
12	41.2	40.9	40.4	39.8	39.7	38.7	38.6	38.4	38.3	38.1	37.9
16	42.1	41.1	40.5	39.7	38.5	38.5	38.5	38.2	38.1	37.8	37.6

Table 14.8

  i **State** the independent variable. [1]
  ii Suggest two variables that should have been controlled. [1]

COMMAND WORDS

Describe: state the points of a topic / give characteristics and main features.

Suggest: apply knowledge and understanding to situations where there is a range of valid responses, in order to make proposals – that is, use information provided put forward considerations.

State: express in clear terms.

CONTINUED

  iii Explain whether a reliable conclusion could be drawn from the experiment. [2]

  iv **Compare** the effect of different shower temperatures on the cooling of the athletes' core body temperature. [3]

  v Suggest and explain a reason for the effect of shower temperature of 4 °C on the core body temperature. [3]

  [Total: 18]

> **COMMAND WORD**
>
> **Compare:** identify/comment on similarities and/or differences.

3 Figure 14.11 shows a diagram of the renal capsule.

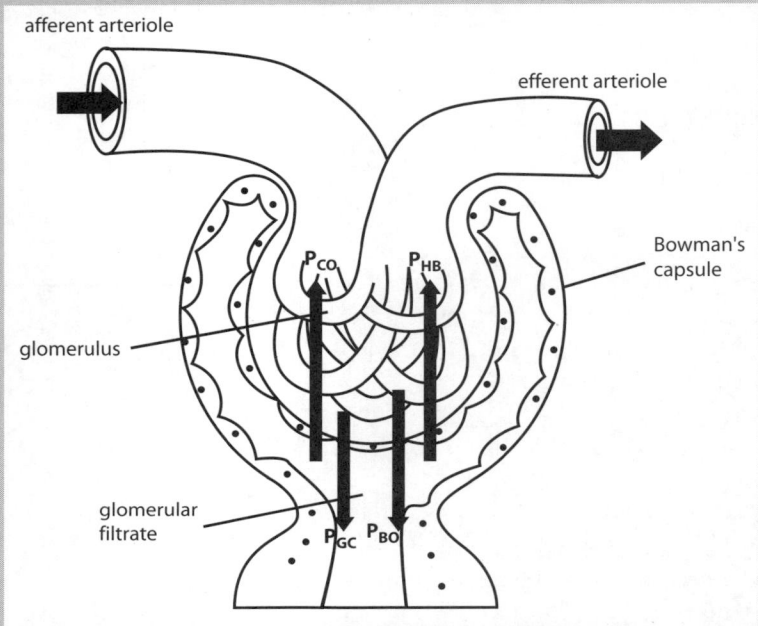

Figure 14.11

Ultrafiltration pressure is the net force that moves the filtrate from the glomerular capillaries into the Bowman's capsule. It is calculated using the following equation:

(P_F) filtration pressure = (hydrostatic glomerular capillary pressure (P_{GC}) + Bowman's capsule oncotic pressure (P_{BO})) − (capillary oncotic pressure (P_{CO}) + hydrostatic Bowman's capsule pressure (P_{HB}))

a i The following pressures were determined for a normally functioning kidney:

$P_{GC} = 8.0$ kPa, $P_{BO} = 0.0$ kPa, $P_{CO} = 3.8$, $P_{HB} = 2.0$ kPa

Use these pressures to determine the filtration pressure, P_F. [1]

CONTINUED

ii Oncotic pressure is the pressure exerted by proteins in the plasma or filtrate attracting water towards them. Use your knowledge of the structure of the renal capsule to explain why the Bowman's capsule oncotic pressure, P_{BO} is 0. [2]

Under normal conditions, 625 cm³ of blood plasma enters the afferent arteriole every minute and 500 cm³ of blood plasma exits the glomerulus via the efferent arteriole.

iii **Calculate** the amount of filtrate produced in one day. [2]

b If hydrostatic blood pressure falls, the kidneys release an enzyme called renin which converts a plasma protein called angiotensinogen into a protein called angiotensin I. Angiotensin I is then converted by another enzyme called ACE into angiotensin II. Angiotensin II stimulates contraction of the circular muscles in the efferent arterioles of the glomerulus. This is shown in Figure 14.12.

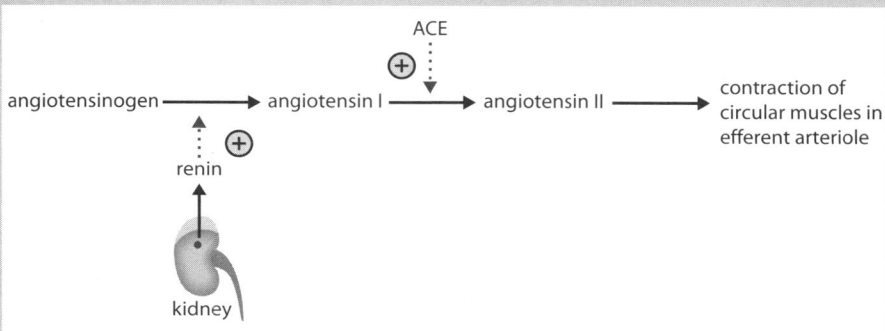

Figure 14.12

i Suggest why a cascade of enzyme activation is used to generate active angiotensin II. [1]

ii ACE inhibitors are drugs used in the treatment of hypertension. Explain how ACE inhibitors result in reduced filtration rates. [3]

[Total: 9]

> **COMMAND WORD**
>
> **Calculate:** work out from given facts, figures or information.

4 A medical student was presented with three urine samples and several simple laboratory tests.

The urine samples came from:

- diabetic patient
- patient identified as having prediabetes; the blood glucose level does not rise as high as a patient with full diabetes
- patient with high blood pressure causing high levels of protein to be present in the urine. The samples were not labelled and the student had to identify each one by carrying out a range of tests.

CONTINUED

The student was also asked to determine the concentration of glucose in the urine of the diabetics.

a Describe how you could identify each urine sample using biochemical tests. Include all practical details.

Table 14.9 shows the relative concentrations of some substances in the regions of the nephron labelled **A**, **B**, **C** and **D** on Figure 14.13. [6]

Area of nephron	Relative concentration of substance compared to concentrations at position A			
	Protein	Glucose	Urea	Salt
A	1	1	1	1
B	0	0	50	45
C	0	0	175	125
D	0	0	500	295

Table 14.9

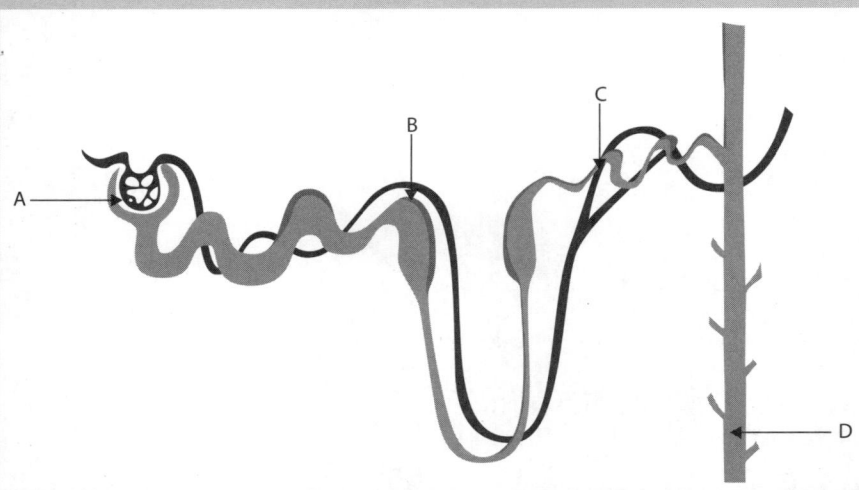

Figure 14.13

b Explain the changes in relative concentration of the substances as they pass through the nephron:
  i protein [3]
  ii glucose [3]
  iii urea and salt. [4]

c Explain in detail why the concentration of urea at **D** will be higher after exercising without drinking. [5]

[Total: 21]

CONTINUED

5 a Describe how mammals break down excess amino acids. [4]

b The loop of Henle is described as a counter-current multiplier. Explain the term *counter-current multiplier* and describe how the loop of Henle maximises the reabsorption of water. [7]

c Explain why sodium and potassium ions are required for the reabsorption of glucose by the cells of the proximal convoluted tubule. [4]

[Total: 15]

6 Figure 14.14 shows the fluctuations in blood plasma insulin, glucagon and glucose concentrations over a 24-hour period in a human. The times that meals were taken are shown on the graph.

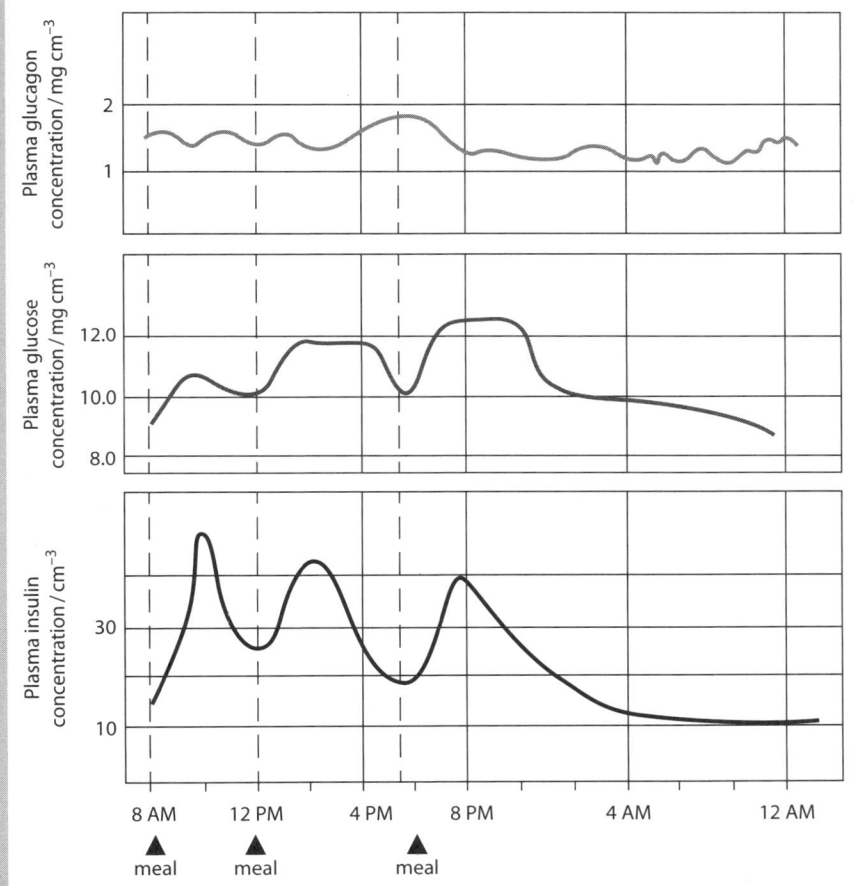

Figure 14.14

a Explain the changes in concentrations of blood glucose plasma insulin and plasma glucagon over the 24-hour period. [7]

CONTINUED

b Figure 14.15 shows the changes in blood glucose concentration over a 24-hour period for three patients, **A**, **B** and **C**. One patient is normal, one suffers from diabetes and the third from undiagnosed prediabetes. Prediabetes is diagnosed where a patient is less able to control their blood glucose concentration than normal. It will often develop into full type 2 diabetes if left untreated.

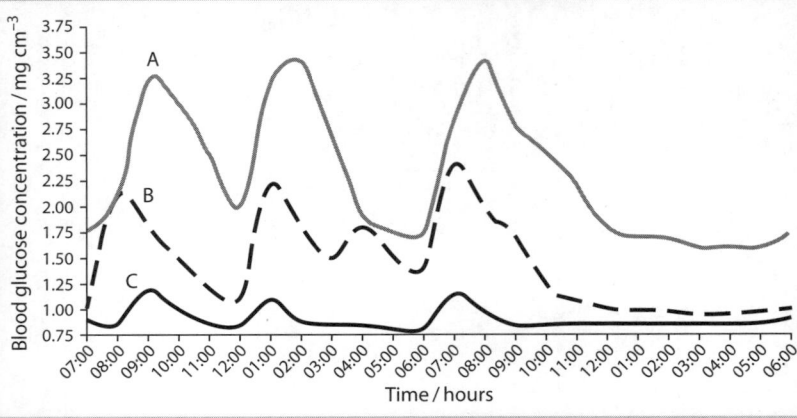

Figure 14.15

  i State which of the patients is most likely to be suffering from diabetes. Explain your answer. [1]

  ii The patients were told to eat three meals during the 24-hour period. One patient also had a snack as well as the three meals. State which patient had an additional snack and state the approximate time that the snack was taken. [1]

Glucose undergoes selective reabsorption in the kidney. Figure 14.16 shows the effect of increasing blood glucose concentrations on the reabsorption and excretion of glucose into the urine.

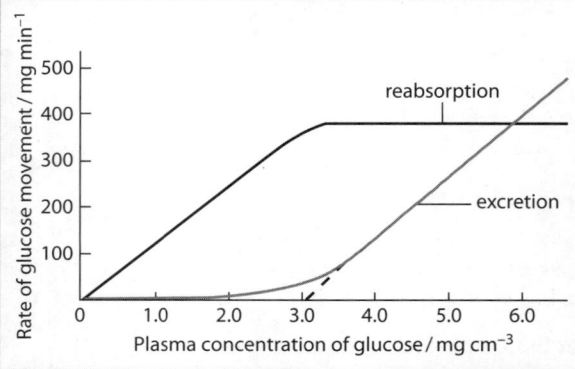

Figure 14.16

14 Homeostasis

CONTINUED

iii Describe and explain the effect of increasing blood glucose concentration on glucose reabsorption and excretion. **[4]**

iv The patient suffering from diabetes and the patient with undiagnosed prediabetes were given sets of glucose dip sticks that identify the presence of glucose in urine.

The dipsticks confirmed the diagnosis of diabetes in one patient but did not indicate any blood glucose regulation problems in the prediabetic patient. Use Figures 14.15 and 14.16 to explain these results. **[3]**

[Total: 16]

7 Cobalt chloride paper is coloured blue when dry and turns pink when wet. It can be placed on the lower surface of a leaf and secured between two microscope slides as shown in Figure 14.17.

a i Explain why the cobalt chloride paper will turn from blue to pink on the lower surface of the leaf. **[1]**

ii **Predict** and explain the effect of increasing light intensity on the time taken for the cobalt chloride paper to turn from blue to pink. **[2]**

iii Figure 14.18 shows some laboratory apparatus. Plan a controlled investigation using a plant, cobalt chloride paper as shown in Figure 14.17 and the apparatus shown in Figure 14.18 to investigate the effect of light intensity on stomatal opening. **[7]**

> **COMMAND WORD**
>
> **Predict:** suggest what may happen based on available information.

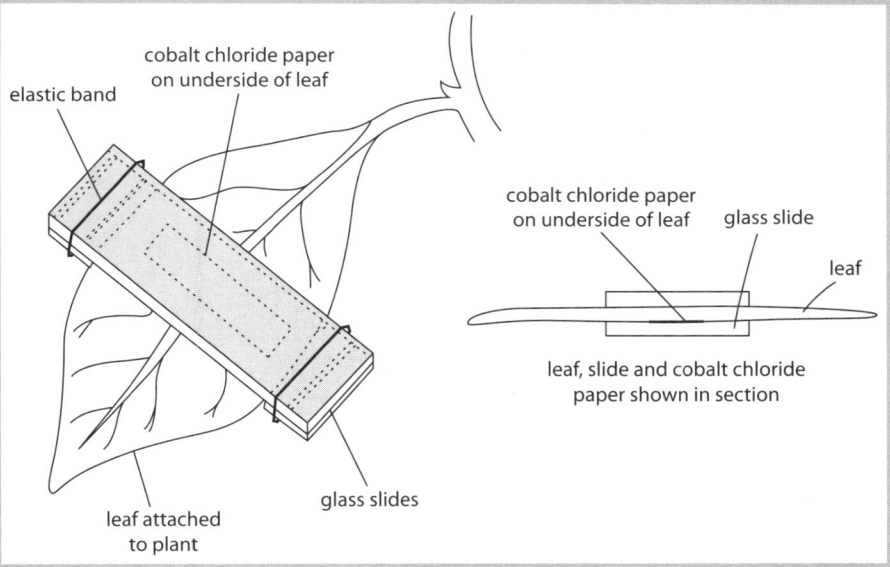

Figure 14.17

CONTINUED

Figure 14.18

b Stomatal conductance is the amount of gas that can pass through the stomata and is a measure of stomatal opening or closing. Low conductance is the result of few stomata opening or narrow stomatal apertures.

Abscisic acid (ABA) is produced by plants that undergo water stress due to dehydration. Scientists carried out an experiment to investigate whether there is a correlation between the mass of ABA found in leaves and stomatal conductance.

The results from the leaves from ten plants are shown in Table 14.10.

Leaf number	ABA mass / µg g^{-1} dry leaf	Rank$_1$, r_1	Stomatal conductance / arbitrary units	Rank$_2$, r_2
1	6		17	8
2	6		26	6
3	2		65	1
4	9		24	7
5	12	2	4	9
6	14	1	2	10
7	8		27	5
8	8		43	4
9	4		56	3
10	5		58	2

Table 14.10

CONTINUED

i Formulate a null hypothesis for the relationship between mass of ABA and stomatal conductance. [1]

ii Copy out the table and use it to calculate the Spearman's rank correlation coefficient for the results. [3]

Use Table 14.11 to determine whether or not there is a significant correlation between mass of ABA in the leaves and stomatal conductance.

n	5	6	7	8	9	10	11	12	14	16
Critical value of r_s	1.00	0.89	0.79	0.76	0.68	0.65	0.60	0.54	0.51	0.51

Table 14.11

iii Explain your answer using the terms *probability* and *chance*. [3]

[Total: 17]

Chapter 15
Control and coordination

CHAPTER OUTLINE

The questions in this chapter cover the following topics:
- the ways in which mammals coordinate responses to internal and external stimuli
- the structure and function of neurones, including their roles in simple reflexes
- the roles of sensory receptor cells
- the transmission of nerve impulses
- the structure, function and roles of synapses
- the ultrastructure of striated muscle and how muscles contract in response to impulses from motor neurones
- how responses in plants, including the rapid response of the Venus fly trap, are controlled.

Exercise 15.1 The nerve impulse

You need to understand the nature of the nerve impulse in detail. This includes how the resting potential is maintained, and how depolarisation and repolarisation occur. In this exercise, you will:

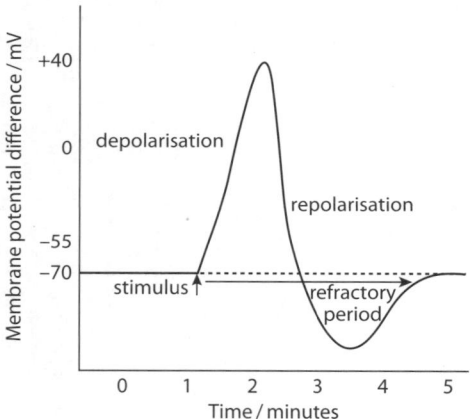

Figure 15.1: An action potential.

- develop your understanding of the roles of the different neurone membrane proteins and how drugs and toxins interfere with them
- develop your understanding of the nature of an action potential
- develop your analytical skills by investigating the effects of toxins on neurones.

Figure 15.1 shows the potential difference across the membrane of a **neurone**, showing the **resting potential** and an **action potential**. There are several different types of membrane protein that generate both the resting potential and action potential.

KEY WORDS

neurone: a nerve cell; a cell which is specialised for the conduction of nerve impulses

resting potential: the difference in electrical potential that is maintained across the cell surface membrane of a neurone when it is not transmitting an action potential; it is normally about −70 mV inside and is partly maintained by sodium–potassium pumps

action potential: a brief change in the potential difference from −70 mV to +30 mV across the cell surface membranes of neurones and muscle cells caused by the inward movement of sodium ions

1 a Copy and complete Table 15.1 below to show which membrane channels and proteins are active during the resting potential, **depolarisation** and **repolarisation** phases, and explain the direction of ion movement for each.

Membrane protein	Phase					
	Resting		Depolarisation		Repolarisation	
	Open / closed / active	direction of ion movement	Open / closed / active	Direction of ion movement	Open / closed / active	direction of ion movement
sodium channel	open	sodium ions enter neurone	open	sodium ions enter neurone	open	sodium ions enter neurone
potassium channel						
voltage-gated sodium channel						
voltage-gated potassium channel						
sodium–potassium exchange pump						

Table 15.1: Membrane channels and proteins that are active during the resting potential, depolarisation and repolarisation phases.

> **KEY WORDS**
>
> **depolarisation:** the reversal of the resting potential across the cell surface membrane of a neurone or muscle cell, so that the inside becomes positively charged compared with the outside.
>
> **repolarisation:** returning the potential difference across the cell surface membrane of a neurone or muscle cell to normal following the depolarisation of an action potential.

b There are many different drugs and neurotoxins that affect resting potentials and action potential generation. Ouabain is an inhibitor of the sodium–potassium (Na^+–K^+) exchange pump. It is a naturally occurring toxin made by *Acokanthera schimperi* plants in East Africa and was used to make poison darts. The African crested rat deliberately chews the roots and bark of the tree and smears its saliva onto especially absorbent fur on its back as a defence against predators.

Figure 15.2 shows a resting potential across a neurone membrane. Copy and complete the graph to predict the effect of the application of ouabain on the resting potential. Explain the graph that you have sketched.

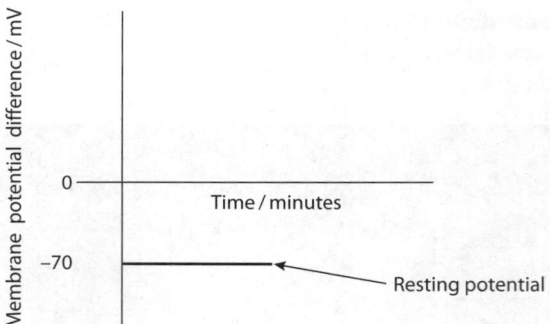

Figure 15.2: Resting potential in a neurone prior to the application of ouabain.

c   Tetrodotoxin (TTX) is a neurotoxin that is produced by a wide variety of species including the pufferfish, *Takifugu* sp. It binds to voltage gated sodium channels, blocking their activity. The effects of poisoning due to consumption of the fish include loss of sensation in the mouth, difficulty in breathing and muscle paralysis.

Use your knowledge of action potentials to explain the effect of TTX. You need to understand the sequence of events that occur at **synapses**.

2   a   Table 15.2 shows the sequence of events beginning when an action potential reaches the presynaptic terminal of a neuromuscular junction. Place the events in the correct order.

A	Action potential arrives at presynaptic terminal.
B	**Acetylcholinesterase** enzyme digests acetylcholine in synaptic cleft and presynaptic membrane absorbs the products of the digestion.
C	Postsynaptic receptor / sodium channels open causing Na$^+$ to diffuse into postsynaptic neurone.
D	Vesicles move towards the presynaptic membrane and fuse with it.
E	Ca^{2+} diffuses into the presynaptic terminal.
F	**Acetylcholine (ACh)** diffuses across the synaptic cleft and binds to postsynaptic receptors proteins in postsynaptic membrane.
G	Acetylcholine is released into the synaptic cleft.
H	If threshold voltage is reached, voltage gated sodium channels open resulting in action potential in postsynaptic neurone.
I	Voltage gated Ca^{2+} channels open.

Table 15.2: Sequence of events beginning when an action potential reaches the presynaptic terminal of a neuromuscular junction.

Many questions will give you examples of unfamiliar neurotoxins and drugs that act at synapses and ask you to work out the effects of them. You are not expected to learn the effects of every drug or toxin – you need to apply your knowledge of synapses to each situation.

> **KEY WORDS**
>
> **synapses:** the junctions between neurones or between neurones and their target tissues such as skeletal muscle
>
> **acetylcholinesterase:** an enzyme in the synaptic cleft and on the post-synaptic membrane that hydrolyses acetylcholine to acetate and choline
>
> **acetylcholine (ACh):** a molecule made up of coenzyme A and a 2C acetyl group, important in the link reaction; a type of neurotransmitter released by cholinergic synapses

b Curare is a toxin found in several South American plants including *Chondrodendron tomentosum*. It was used to make poison darts and arrow heads. α-Bungarotoxin is a toxin found in snake venom from the Taiwanese banded krait. Both toxins bind to and block the postsynaptic ACh receptors. This will prevent muscles from contracting, causing paralysis. Figure 15.3 shows the sites of action of several toxins, drugs and conditions on a neuromuscular junction.

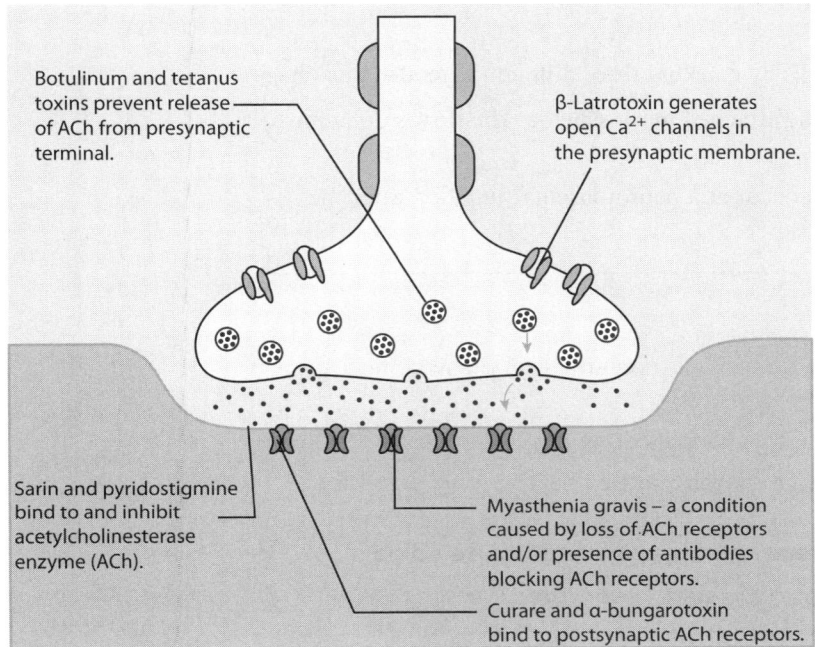

Figure 15.3: Sites of toxin action at neuromuscular junction.

Read the text below and predict and explain the effects of these toxins:

   i α-Latrotoxin is the venom found in the bite of the black widow spider. It generates a large number of open calcium channels in the presynaptic membrane. Predict and explain its effects.

   ii Botulinum and tetanus toxins prevent fusion of vesicles with the presynaptic membrane. Predict and explain their effects.

   iii The nerve gas agent sarin and the drug pyridostigmine inhibit the action of the acetylcholinesterase enzyme (ACh) in the synaptic cleft. Predict and explain their effects.

   iv Myasthenia gravis is a condition that causes muscle weakness. In some cases, there are fewer ACh receptors found on the postsynaptic membrane and in others, autoimmune antibodies bind to the ACh receptors. Explain why pyridostigmine is often used to help treat the condition.

   v Explain why pyridostigmine is often used to treat cases of curare and α–bungarotoxin poisoning but is not useful in cases of tetanus or botulinum poisoning.

Exercise 15.2 The use of technical language

Remember when answering questions to always use precise and correct language; you need to be as clear as possible and provide a level of detail that is appropriate for the question and reflects the number of marks that will be awarded.

In this exercise, you will:

- develop your extended writing skills by marking three different extended answers.

1 Mark the following essays using the attached mark scheme. The first is done for you as an example with comments.

Question: Describe the events that occur at a neuromuscular junction after the stimulation of a motor neurone. [9]

> **Mark scheme**
>
> 1 action potential / impulse / depolarisation arrives at presynaptic terminal / button / knob ;
> 2 (voltage-gated) calcium / Ca^{2+} channels open ;
> 3 calcium / Ca^{2+} diffuses into the **presynaptic** neurone (only penalise once for Ca^+) ;
> 4 vesicles move towards **presynaptic** membrane and **fuse** / **bind** ;
> 5 acetylcholine is released into the **cleft** / **exocytosis** ;
> 6 (acetylcholine) diffuses across the cleft ;
> 7 (acetylcholine) binds with **receptors** on **postsynaptic membrane** ; (must be membrane not just neurone) (only penalise once for wrong neurotransmitter / use of term neurotransmitter rather than acetylcholine)
> 8 sodium channels open in postsynaptic membrane / postsynaptic neurone ;
> 9 sodium diffuses into postsynaptic neurone / equivalent ;
> 10 if **threshold potential** / equivalent is reached, actions potential occurs / impulse is generated ;
> 11 **acetylcholinesterase** hydrolyses acetylcholine / ACh ;
> 12 **choline** is reabsorbed by **presynaptic** neurone / equivalent ;

KEY WORDS

threshold potential: the critical potential difference across the cell surface membrane of a sensory receptor or neurone which must be reached before an action potential is initiated

15 Control and coordination

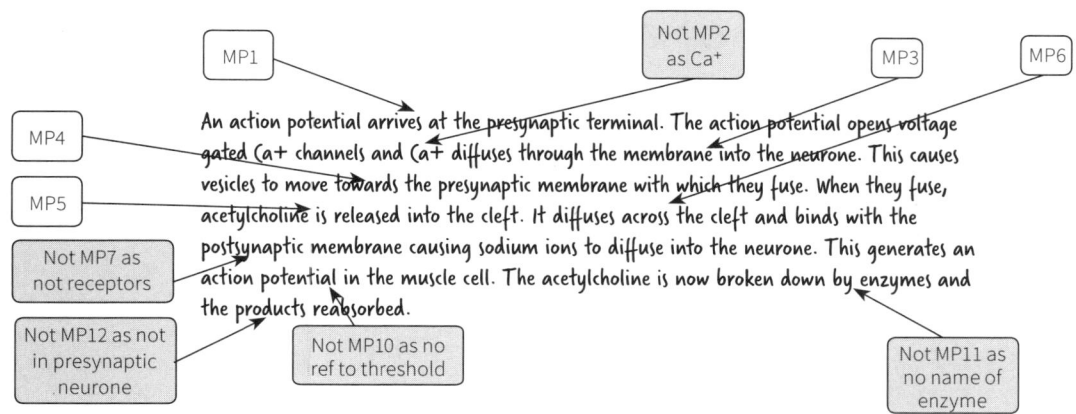

This is a good answer, but could be improved. It does not include sufficient detail and precision.

Now, try marking these answers.

a An impulse reaches the presynaptic terminal. This causes calcium channels to open and calcium ions move inside the presynaptic neurone. Calcium then enters the vesicles, which then move towards the end of the presynaptic neurone. The vesicles release the calcium into the cleft and it then diffuses into the postsynaptic membrane, generating an action potential.

b An action potential reaches the end (terminal) of the presynaptic neurone. It opens voltage gated calcium channels so that calcium ions diffuse into the presynaptic neurone. Vesicles move towards the presynaptic membrane and release acetylcholine into the cleft. The acetylcholine then binds to ACh receptors on the postsynaptic neurone. Sodium channels in the postsynaptic membrane open and sodium ions floods in. If a threshold voltage is reached, voltage gated channels then open so that an action potential is generated in the postsynaptic neurone. Acetylcholine is then reabsorbed by the presynaptic membrane to stop it binding to the postsynaptic receptors.

c An action potential arrives at the presynaptic terminal. This opens calcium channels, and calcium ions enter the presynaptic neurone. The calcium stimulates the vesicles to move towards the presynaptic membrane, fuse with it and release the neurotransmitter (acetylcholine or dopamine) into the cleft. The neurotransmitter binds with receptors in the postsynaptic membrane, which causes sodium channels to open leading to an inflow of sodium ions down their gradient. This inflow of sodium ions into the postsynaptic neurone generates an action potential. The enzyme acetylcholinesterase digests up the neurotransmitters and the choline is reabsorbed.

Exercise 15.3 The effect of auxin on root and shoot growth

Auxin hormones can have a range of effects on plant growth. In addition, different concentrations of auxins can have different growth effects on the same tissues. In this exercise, you will:

- develop your understanding of how auxins can inhibit and stimulate root and shoot growth at different concentrations
- develop your understanding of how to make up solutions of different concentrations
- develop your understanding of how to use logarithmic scales when there are variables that have very high ranges.

1 An experiment was carried out to investigate the concentration of auxin that generated the largest increase in root and shoot growth of tomato seedlings.

Tomato seedlings were grown in nutrient agar for five days. Different concentrations of auxin were added either to the nutrient agar, for root stimulation, or to the stem and leaves as spray. Growth of roots and shoots was measured after five days.

The results of the effects of the different concentrations of auxin on root and shoot growth are shown in Tables 15.3 and 15.4.

Concentration of auxin applied / mg dm^{-3}	Log$_{10}$ concentration / mg dm^{-3}	Initial length of root / mm	Final length of root / mm	Ratio of final length / initial length
0.000001	−6	12	24	
0.000010		14	32	
0.000100		12	30	
0.001000		15	34	
0.010000		14	25	
0.100000		12	19	
1.000000		11	13	
10.000000		16	16	
100.000000		14	14	
1000.000000		12	12	

Table 15.3: Effects of the different concentrations of auxin on root growth.

Concentration of auxin applied / mg dm⁻³	Log$_{10}$ concentration / mg dm⁻³	Initial length of shoot / mm	Final length of shoot /mm	Ratio of final length / initial length
0.000001	−6	17	17	
0.000010		16	16	
0.000100		17	17	
0.001000		14	15	
0.010000		18	27	
0.100000		21	82	
1.000000		23	90	
10.000000		17	87	
100.000000		18	32	
1000.000000		19	21	

Table 15.4: Effects of the different concentrations of auxin on shoot growth.

The auxin concentrations used range from 0.000001 mg dm⁻³ to 1000 mg dm⁻³. This is obviously a very large range, and if you wanted to plot a graph of this it would require a very long axis or some of the smallest increments would be cramped up together. The solution is to plot a graph using a logarithmic axis. This can be done by taking logarithms to base 10 of the concentrations.

Logarithms are calculated as follows: $\log_{10}$ of a number is the power to which 10 is raised to obtain that number. For example:

The concentration 0.00001 is the same as 10^{-6}, so $\log_{10}$ 0.00001 is −6.

You should be able to use your calculator to determine the $\log_{10}$ values of numbers. How to do this will vary according to the make and model of calculator, but most will have a button labelled 'log'.

Use your calculator to determine the $\log_{10}$ values of:

a   0.00006

b   12 000 000

c   12

d   0.0000325

e   0.0025.

f   Now, copy Tables 15.3 and 15.4 and write down the $\log_{10}$ values for all the concentrations of auxin.

The mean growth of the roots and shoots is expressed as a ratio of the final length to initial length. A ratio of 1.0 would mean that the two values were the same, in other words the root or shoot had not grown at all.

For the root in auxin at a concentration of 0.000010 mg dm⁻³, the ratio of final length to initial length is:

32 mm ÷ 14 mm = 2.3

> **TIP**
>
> You should have found that the $\log_{10}$ values are numbers that are within a similar magnitude of each other and so could be plotted on a graph.

g Calculate the final length to initial length ratios for all the concentrations of auxin for the roots and shoots and enter them in your tables.

h Now we have logarithms of the concentrations, we can plot a graph to visualise the results. Follow the following steps to draw the graph:

Step 1 Label the *x*-axis: $\log_{10}$ concentration of auxin / mg dm^{-3}. Add a scale with even increments for each $\log_{10}$ value of concentration, starting at -6.

Step 2 Label the *y*-axis ratio of final length / start length and add a continuous scale.

Step 3 Plot your points for both root and shoot data, join the points for each and add a key.

i Compare the effect of increasing concentration of auxin on the roots and shoots.

To refer to concentrations of auxin used you will have to reconvert the $\log_{10}$ number back to a concentration. This is called the 'antilog' or 'inverse log' and should be easy to do on a calculator. For example, the antilog of -5 is 0.00001 (or 10^{-5}).

2 In a further experiment designed to identify the optimal concentration of auxin for root and shoot elongation, different ranges of concentrations of auxins would be used. Explain what range of concentrations would be chosen for the root and shoot elongation experiments.

EXAM-STYLE QUESTIONS

1 The sea hare, *Aplysia*, has been used to investigate simple animal behaviours. It is a marine mollusc that has a protruding siphon and gill system for gas exchange, as shown in Figure 15.4. The siphon pumps water over the gills for gas exchange. Touching the siphon initiates a reflex arc that brings about the withdrawal of the gills.

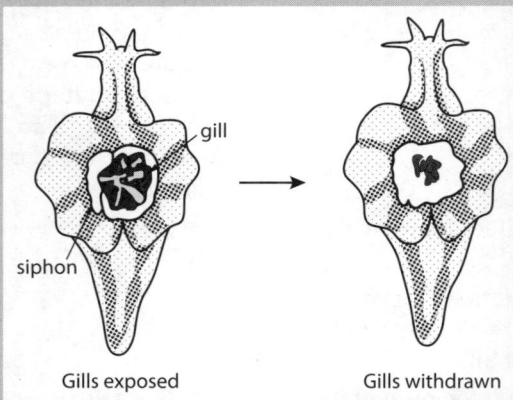

Figure 15.4

CONTINUED

An investigation was carried out to examine the time for which the gills were withdrawn after repeatedly touching the siphon with a paint brush. The siphon was touched with a paint brush and the time taken for the gills to fully open recorded. When the gills were fully open, the siphon was touched again. This was repeated for a total of 10 times. The results are shown in Table 15.5.

Touch number	Rank, r_1	Time gill withdrawn /s	Rank, r_2	D	D^2
1	10	25	1		
2	9	15	2		
3	8	10			
4	7	8			
5	6	6			
6	5	4			
7	4	3			
8	3	2			
9	2	1			
10	1	3			

Table 15.5

a i **State** the dependent variable. [1]
   ii **Give** *two* variables that would need to be standardised. [2]
b It was suggested that there was a correlation between the number of touches and the length of time that the siphon was withdrawn.
   i State a null hypothesis for the association between number of touches of the siphon and gill withdrawal time. [1]
   ii Copy and complete Table 15.5. Use the equation below to **calculate** the Spearman's rank correlation coefficient, r_s: [3]

$$r_s = 1 - \frac{6\Sigma D^2}{n^3 - n}$$

where r_s is Spearman's rank correlation coefficient
ΣD^2 is the sum of the differences between the ranks of the pairs
n is the number of pairs of items in the sample.

> **COMMAND WORDS**
>
> **State:** express in clear terms.
>
> **Give:** produce an answer from a given source or recall/memory.
>
> **Calculate:** work out from given facts, figures or information.

CONTINUED

iii Use your calculated value for r_s and the critical values for r_s given in Table 15.6 to evaluate your null hypothesis. Use the terms *probability* and *chance* in your answer. [3]

n	5	6	7	8	9	10	11	12	14	16
Critical value of r_s	1.00	0.89	0.79	0.76	0.68	0.65	0.60	0.54	0.51	0.51

Table 15.6

c Plot a graph to show the effect of the number of trials on the withdrawal time of the gills. [3]

d *Aplysia* lives in areas of kelp forest that are inhabited by predators. Suggest reasons for the effect of repeatedly touching the siphon on gill withdrawal time. [2]

[Total: 15]

2 Figure 15.5 shows the change in potential difference across the membrane of a neurone.

a **Explain** how the membrane maintains the resting potential, labelled **A** in Figure 15.5. [3]

b Explain how the membrane depolarises and repolarises over section **B** in Figure 15.5. [4]

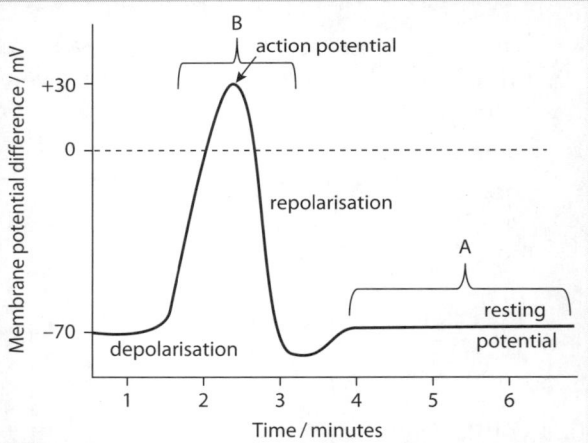

Figure 15.5

c **Suggest** and explain the effect of the application of a respiratory inhibitor on the resting potential. [2]

[Total: 9]

> **COMMAND WORDS**
>
> **Explain:** set out purposes or reasons / make the relationships between things evident / provide why and/or how and support with relevant evidence.
>
> **Suggest:** apply knowledge and understanding to situations where there is a range of valid responses, in order to make proposals / put forward considerations.

CONTINUED

3 Serotonin is a neurotransmitter that has several functions in the body. One of its functions is thought to be as a mood enhancer. Figure 15.6 shows a synapse in the CNS that uses serotonin. Prolonged stimulation of the postsynaptic neurone sends impulses to 'pleasure centres' in the brain.

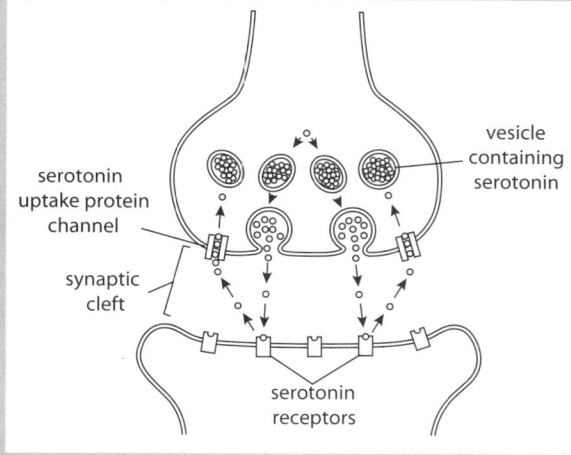

Figure 15.6

a **Describe** how an impulse would be transferred from the presynaptic neurone to the postsynaptic neurone in Figure 15.6. [5]

b Some antidepressant drugs have a similar shape to serotonin but do not directly stimulate an action potential in the postsynaptic neurone. Use Figure 15.6 to suggest how they may work to improve mood. [3]

[Total: 8]

4 a Figure 15.7 shows an electron micrograph of a muscle sarcomere.

  i Calculate the length of the sarcomere labelled **A** in Figure 15.7. [2]

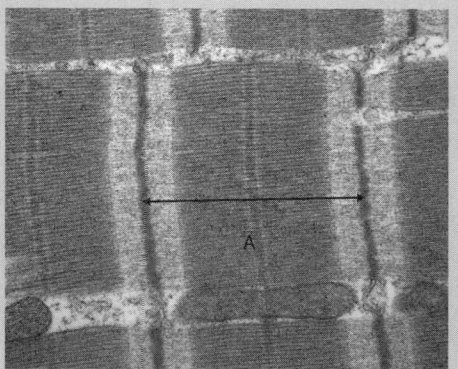

Figure 15.7

> **COMMAND WORD**
>
> **Describe:** state the points of a topic / give characteristics and main features.

CONTINUED

ii Use your answer to part **a i** to **calculate** the number of sarcomeres present if the muscle it was removed from had a length of 20 cm. [1]

iii Figure 15.8 shows a sarcomere from a relaxed muscle. **Sketch** the sarcomere as it would appear in a fully contracted muscle, clearly labelling the **actin** and myosin filaments. [2]

Figure 15.8

b Figure 15.9 shows the maximum forces that a muscle is able to generate when contractions are initiated with sarcomeres of different starting lengths.

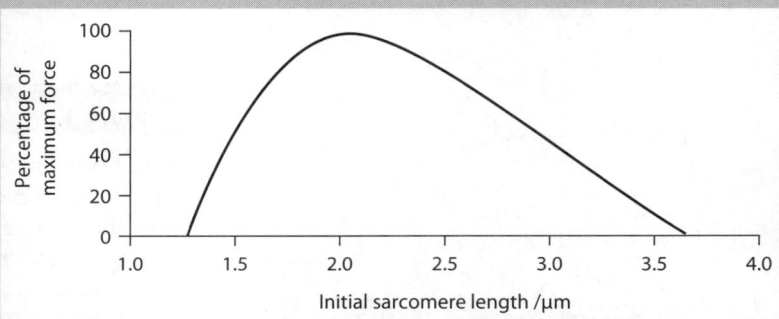

Figure 15.9

i Describe the effect of changing length of sarcomere on the force generated by the muscle. [2]

ii Explain the effect of changing the initial length of the sarcomere on the maximum force generated by the muscle. [3]

[Total: 10]

COMMAND WORDS

Calculate: work out from given facts, figures or information.

Sketch: make a simple drawing showing the key features.

KEY WORD

actin: the protein that makes up the thin filaments in striated muscle

CONTINUED

5 There are several distinct types of muscle fibre found within skeletal muscle, including type I, type IIa and type IIx fibres.

A scientist made a hypothesis that type IIx fibres are mainly used for sprinting at high speed. The scientist investigated this by examining the percentage of skeletal muscles in different animals that were type IIx fibres and comparing it with the maximum sprint speed the animal is capable of. The results are shown in Figure 15.10.

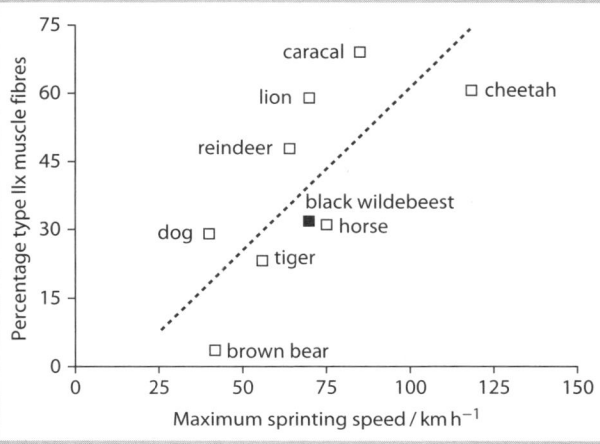

Figure 15.10

a Explain whether the results in Figure 15.10 support the scientist's hypothesis. [3]

b Table 15.7 shows some of the properties of the different muscle fibre types.

	Muscle fibre type		
	Type I fibres	Type IIa fibres	Type IIx fibres
contraction time	slow	moderate	fast
resistance to fatigue	high	moderate	low
power produced	low	medium	high
myoglobin presence	high	medium / low	very low
capillary density	high	medium	low
mitochondria density	very high	high	low
major storage molecules	triglycerides	glucose, creatine phosphate	ATP, creatine phosphate

Table 15.7

CAMBRIDGE INTERNATIONAL AS & A LEVEL BIOLOGY: WORKBOOK

CONTINUED

Use the information to explain why animals adapted for endurance running have a different muscle composition than animals adapted for sprinting. [5]

[Total: 8]

6 A scientist carried out an experiment to investigate the roles of the different parts of a barley seed in germination. The scientist made three sterile starch agar plates and cut out two wells from each of them. The embryos were separated from the remainder of the seeds and both parts were sterilised. The following plates were set up:

- Plate A: embryo and remainder of seed
- Plate B: remainder of seed alone
- Plate C: embryo alone.

The plates were incubated at 37 °C for 24 hours, after which they were flooded with iodine solution. A clear, unstained area was seen around the remainder of the seed in plate A.

The results are shown in Figure 15.11.

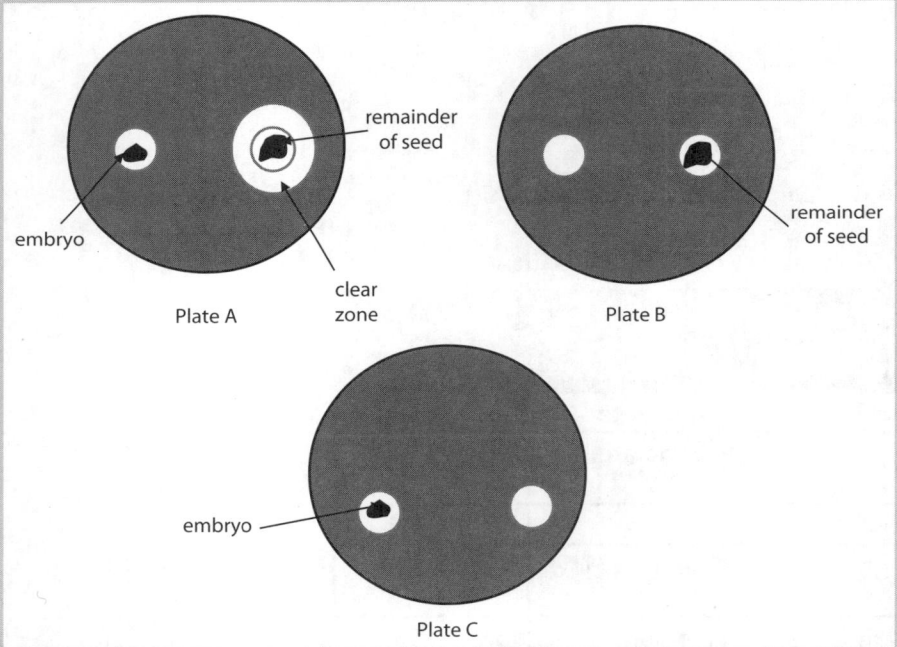

Figure 15.11

a i Explain why the starch agar plates and seeds were sterilised. [2]
   ii Explain why the plates were flooded with iodine. [1]

CONTINUED

    b Use your knowledge of seed germination to explain the results shown in Figure 15.12. [5]

        [Total: 8]

7 a Explain how the design of myofibril structure enables muscle contraction. [9]

  b Describe the events that occur between the arrival of an impulse along a motor neurone and the initiation of muscle contraction. [6]

        [Total: 15]

8 a Describe how auxins bring about changes in plant growth. [9]

  b Describe how a Venus fly trap is able to trap and digest insects. [6]

        [Total: 15]

Chapter 16
Inheritance

CHAPTER OUTLINE

The questions in this chapter cover the following topics:
- the process of meiosis, and its importance in life cycles and in generating genetic variation
- solving genetics problems
- the chi-squared test
- how gene expression is controlled.

Exercise 16.1 Describing meiosis

This exercise tests your knowledge and understanding of the sequence of events in meiosis, and also how it compares with mitosis. You will need to think very carefully about some of the statements in part 4.

1. The second column in Table 16.1 lists the different stages in meiosis. The statements in the third column describe some of the events that take place at different stages during meiosis.

    Copy the columns. Write the numbers 1 to 8 in the first column, to show the sequence in which the stages take place.

Sequence	Name of stage	Events during this stage
	anaphase 1	Centromeres split and chromatids are pulled to the poles by the spindle fibres.
	metaphase 2	Chromatids arrive at the poles and nuclear membranes form around them.
	prophase 1	Pairs of homologous chromosomes line up on the equator of the spindle.
	telophase 1	Individual chromosomes, each made of a pair of chromatids joined at a centromere, line up on the equator of the spindle.
	prophase 2	Homologous chromosomes separate and are pulled to the poles by the spindle fibres.
	telophase 2	Chromosomes form groups at either end of the cell; a nuclear membrane may form.
	metaphase 1	Chromosomes condense and become visible; the homologous chromosomes associate to form bivalents.
	anaphase 2	The nuclear membrane breaks down, and the individual chromosomes are visible.

Table 16.1: Different stages in meiosis.

2 Draw a line from the name of each stage to the description of the events that take place during it.

3 In the 'Name of stage' column, add an asterisk (*) to any stages where crossing over can occur.

4 Copy Table 16.2. For each statement, use a tick if the statement is true for a type of cell division, and a cross if it is not true.

Statement	Mitosis	Meiosis
can produce haploid cells from haploid cells		
can produce haploid cells from diploid cells		
can produce diploid cells from diploid cells		
can only take place in a cell with an even number of chromosomes		
produces genetically identical daughter cells		
involves independent assortment of chromosomes		
involves crossing over between chromatids of homologous chromosomes		
takes place in the formation of gametes		
takes place in gametes		
takes place in a zygote		

Table 16.2: Comparison of mitosis with meiosis.

Exercise 16.2 Terms in genetics

Genetics has a very large number of specialist terms. You need to know and be able to use these correctly – and it helps if you can spell them correctly, too.

Each sentence below requires one or more genetic terms to complete it. Choose the terms from the list below, and then write out the complete sentence, highlighting the term or terms you have chosen. You may need to use some of the terms more than once. In some cases, you will need to use the plural form of the term.

allele   autosomal linkage   codominant   dominant   epistasis   F1
F2   gene   genotype   heterozygous   homozygous   locus
multiple alleles   phenotype   recessive   sex linkage   test cross   X   Y

1 A chromosome is a length of DNA, containing a large number of .... , each of which codes for a particular protein. The position of each one on the chromosome is known as its ....

2 Homologous chromosomes have the same genes at the same .... However, the .... of the genes may not be the same.

3 Some genes are found on the part of the .... chromosome that does not have a matching length on the .... chromosome. Recessive alleles of these genes are more likely to show in the phenotype of male organisms than females, because there is only one copy of the allele present, and this is known as ....

4   In a diploid cell, there are two copies of each ...., whereas a haploid cell has only one copy.

5   The term .... is used to describe the characteristics of an organism, which are partly determined by its ....

6   A .... allele is one that only affects the phenotype if a .... allele of the same gene is not present.

7   When two parents are both .... but for different alleles, their resulting offspring are known as the .... generation. They all have a .... genotype. When these offspring are crossed, they produce the .... generation.

8   The term .... is used to describe two .... that both have an effect on the phenotype when they are present in a .... organism.

9   Some genes have three or more different forms, which are known as ....

10 When two genes are found close together on the same chromosome, other than a sex chromosome, they tend to be inherited together. This is known as ....

11 It is not possible to tell the genotype of an organism that has the phenotype produced by a dominant .... You can use a .... to determine its genotype.

12 Two genes at different loci may both affect the same phenotypic characteristic. For example, one gene may code for an enzyme that acts as the substrate of an enzyme coded for by a different gene. This is known as ....

Exercise 16.3 Answering questions involving a dihybrid cross

It is important to have a reliable strategy for tackling a problem, and also to be careful in the way that you set out your answer.

Here's a genetics problem, involving a dihybrid cross.

The type of hair on a guinea pig is determined by two different genes, which are found on two different chromosomes. One gene, **A/a**, determines the length of hair, with the dominant allele **A** giving short hair and allele **a** giving long hair. The second gene, **B/b**, determines whether the hair is smooth or produces rough 'rosettes'. The dominant allele **B** produces rough hair, whereas **b** produces smooth hair.

16 Inheritance

1. Copy and complete Table 16.3:

genotype	phenotype
AABB	short hair, rough
AaBB	
aaBB	
AABb	
AaBb	
aaBb	
AAbb	
Aabb	
aabb	

Table 16.3: Genotype and phenotype.

- If you are given particular symbols to use to represent alleles in a genetics question, you must use those symbols and not invent ones of your own.
- If you do need to invent your own symbols, you should choose one letter for each gene, and then use an upper case (capital) letter for the dominant allele and a lower case (small) letter for the recessive allele. Here, it would not be sensible to use S/s for the hair length gene, because it will be very difficult to tell the upper case and lower case letters apart when you are writing them.
- It's almost always a good idea to construct a table showing all the possible genotypes and phenotypes. Do this before you start answering the question itself. You'll find it really helpful to have this to refer to as you work through the rest of the question.
- When you write down a dihybrid genotype, write the two alleles of one gene, followed by the two alleles of the second gene.

> **TIP**
> Here are some tips to help you when you are working through genetics problems.

2. Construct a genetic diagram to predict the genotypes and phenotypes of the offspring produced by a cross between a short-haired, smooth-coated guinea pig that is heterozygous for the hair length gene, and a short-haired, rough-coated guinea pig that is heterozygous at both gene loci.

Try doing this yourself, then check your answer against the example below in Figure 16.1.

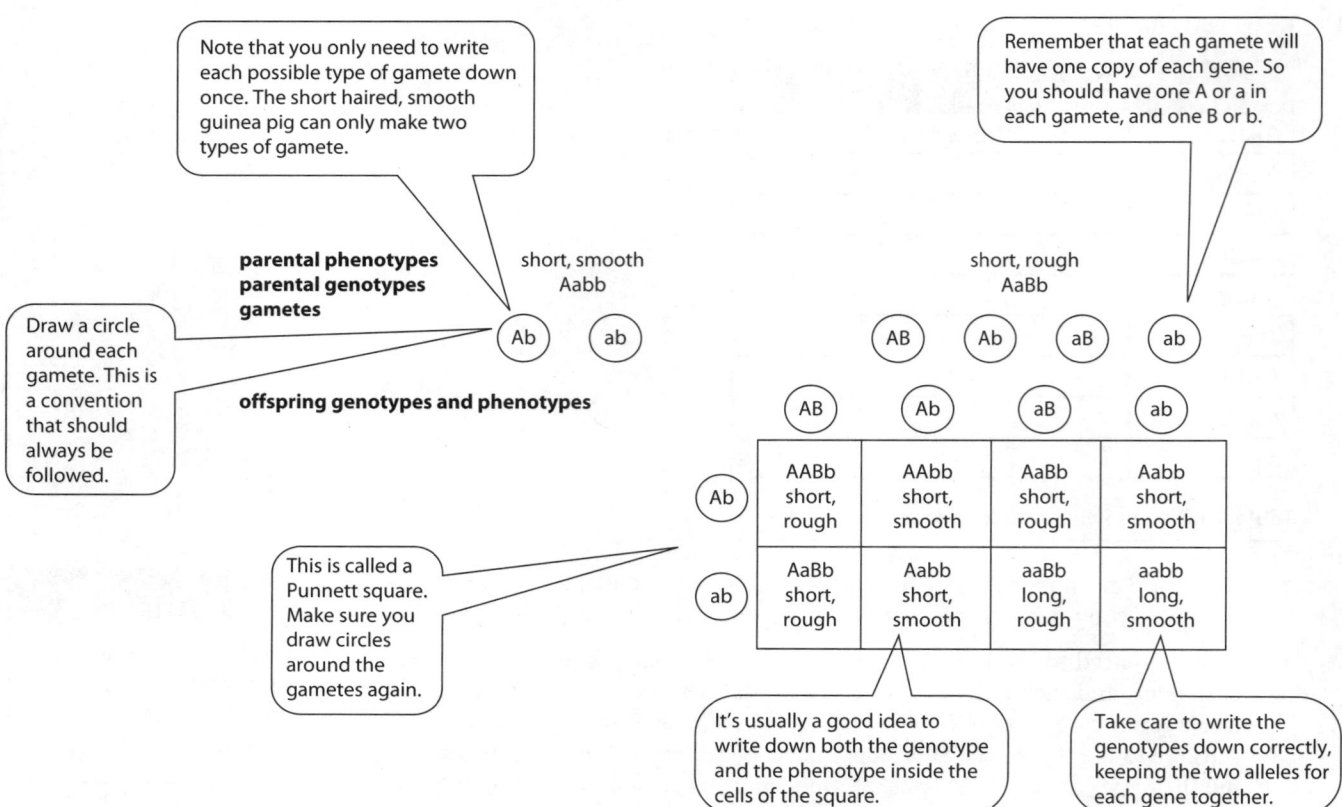

Figure 16.1: How to set out a genetic diagram.

> **TIP**
>
> Remember that what you have shown in the Punnett square is not the number of different offspring that will be produced. It is the relative chances of getting each of the possible genotypes.

> **TIP**
>
> There is an accepted format for setting out a genetic diagram, which is shown in Figure 16.1. Do not take short cuts. It may seem to take a long time to do it properly, but this is the only way to ensure that you will get full marks.

3 What are the expected ratios of the different phenotypes in the offspring of this cross?

4 Eleven baby guinea pigs were produced from this cross. How many would you expect to have long, smooth hair?

5 Imagine that you have a guinea pig with short, rough hair. You want to find out its genotype, which you can do by crossing it with another guinea pig and looking at the phenotypes of their offspring. This is called a test cross.

    a What type of guinea pig would you cross your short-haired, rough-haired guinea pig with? Explain your choice.

    b Explain how you would use the offspring phenotypes to work out the genotype of the short, rough-haired parent. You might want to use a genetic diagram to help your explanation.

Exercise 16.4 Another dihybrid cross question

This cross involves one pair of alleles that are codominant, as well as another where the alleles are dominant and recessive. For codominant alleles, make sure that you use the same upper case letter for both of them, and then use superscripts to represent the different alleles. This problem requires quite a bit of detective work.

A bull without horns and with red hair was bred on several occasions with a cow with horns and white hair. Six calves were born. All the calves were roan (with a mixture of red hairs and white hairs). Two of the offspring had horns, and four did not.

The same bull was also bred many times with a different cow that did not have horns and had white hair. Seven calves were born. All of them had roan hair. Five did not have horns, and two had horns.

1   What were the genotypes of the bull and the two cows? Use genetic diagrams to explain your answers.

Here's a possible strategy for solving this problem.

- Separate the two features – horns and hair colour – in your mind, and think about each one in turn.

- For horns, there are only two phenotypes (horns or no horns), which suggests we have a simple dominant/recessive situation.

- What clues do you have about whether the allele for having horns is dominant or recessive? (The cross with the second cow is the most helpful for working this out.) Now choose suitable symbols for the allele for horns and the allele for no horns.

- You may find it helpful to draw little sketches of the crosses. So, for the first cross, draw a hornless bull and a hornless cow and their six calves. Sometimes, that can get you started on seeing how the feature is inherited.

- For hair colour, there are three phenotypes, which suggests that we have codominance. Choose a 'base' letter for the hair colour gene (e.g. **C**) and then choose two different letters to use as superscripts. For example, you could use C^R for the allele for red hair, and C^W for the allele for white hair.

- At the start of your answer, write down clearly what each of your symbols represents, for example: C^R is the allele for red hair.

- Now construct a table, like the one in Exercise 16.3.1, listing the nine possible genotypes and their phenotypes.

- Now go back to what you were told about the bull, the cows and their offspring. Can you work out any of their genotypes? Start with coat colour – it's always easy to work out genotypes where there is codominance. And where there is a dominant allele and a recessive allele, remember that you always know the genotype of an organism that shows the recessive characteristic.

- Sketch out genetic diagrams to see if your ideas work out. Lastly, draw fully complete genetic diagrams to answer the question.

Exercise 16.5 Autosomal linkage

When two genes are found on the same chromosome, they do not show independent assortment. Instead, the two genes are inherited together and are said to be linked. This is called autosomal linkage, to distinguish it from sex linkage (when a gene is found on a part of the X chromosome that does not have a homologous region on the Y chromosome).

1   A plant with red flowers and serrated leaves, which was heterozygous at both loci, was crossed with a plant with yellow flowers and smooth leaves.

   a   By means of a genetic diagram, predict the ratios of phenotypes in the offspring from these two plants, assuming that the two genes are on different chromosomes. Use the symbols **R/r** for the flower colour gene, and **T/t** for the leaf shape gene.

   b   Now, using the same symbols, construct a genetic diagram predicting the ratios of phenotypes if the two genes are close together on the same chromosome.

> **TIP**
>
> Remember:
>
> - Always start by constructing a table showing all the possible genotypes and phenotypes.
> - Then construct a complete genetic diagram, remembering to draw circles around the gametes.
> - Finally, write a sentence summarising the expected ratios of the different phenotypes.

> **TIP**
>
> Remember:
>
> - If two genes are on the same chromosome, they are linked. They are inherited together.
> - We can tell which alleles are linked together by looking at the organism with the double recessive genotype. It has the genotype **rrtt**, so we know that r and t are found together.
> - The heterozygous parent has the genotype **RrTt**. The alleles **r** and **t** are linked together on one chromosome, and R and T are linked together on the other. So we should write this genotype: **(RT)(rt)**. Instead of the four kinds of gamete you should have shown in **a**, if the genes are linked like this they can make only gametes with the genotypes **RT** and **rt**.

Exercise 16.6 Linkage and crossing over

It's very rare for two genes to be completely linked. Almost always, crossing over between chromatids of the two homologous chromosomes happens during meiosis, so that some gametes are made with different combinations of alleles. This can give us unusual ratios of phenotypes.

1   In a species of insect, the allele for green body is dominant to the allele for brown body. The allele for long wings is dominant to the allele for short wings.

   An insect with green body and long wings was crossed with an insect with brown body and short wings.

The offspring obtained were:

green body, long wings	132
green body, short wings	31
brown body, long wings	29
brown body, short wings	135

Construct a genetic diagram to explain the results of this cross.

> **TIP**
>
> - As we have offspring showing the features resulting from the recessive alleles, the green, long-winged parent must be heterozygous at both loci.
>
> - You should recognise that this looks roughly like a 1 : 1 ratio of the two parental phenotypes (green, long; brown, short), with a small number of 'different' phenotypes. The 'different' ones are called 'recombinants', because their combinations of features differ from those of their parents.
>
> - When you construct the genetic diagram, you should show all four possible gametes from the heterozygous parent, but indicate that two of them will be relatively rare (because they will only occur if crossing over takes place between the two gene loci).

Exercise 16.7 The chi-squared test

Genetics is all about chance, so we almost never get exactly the ratio of phenotypes that we predict. The chi-squared test allows us to determine whether the results we have obtained are a good fit for our ideas about the type of inheritance we are dealing with, or whether the results are so different that we need to come up with a different explanation for them.

Let's imagine we cross two plants, both with green stems and curly petals. The offspring we obtain are:

green stems, curly petals	121
green stems, straight petals	22
red stems, curly petals	23
red stems, straight petals	26

1 a Using the symbols **G/g** for stem colour, and **P/p** for petal shape, construct a genetic diagram to show that you would expect a 9 : 3 : 3 : 1 ratio of phenotypes in the offspring if these two genes are not linked.

   b i Calculate the total number of offspring.
      ii Assuming a 9 : 3 : 3 : 1 phenotypic ratio, calculate the numbers of these offspring you would expect to show each phenotype.

   We can see that the numbers of offspring with each phenotype don't quite match the results we would expect if the ratio is 9 : 3 : 3 : 1. But are they different enough to suggest that the genes are linked?

Let's assume that the genes are not linked. We can write our null hypothesis:

The genes for stem colour and petal shape are not linked.

The chi-squared test will tell us how likely it is that the difference between our observed results and the results we would expect if our hypothesis is correct is just due to chance – or whether the difference is so great that it is unlikely to be due to chance, and our hypothesis is wrong.

The formula for finding chi-squared is:

$$\chi^2 = \sum \frac{(O-E)^2}{E}$$

c Copy Table 16.4:

Phenotypes	green, curly	green, straight	red, curly	red, straight
observed number (O)	121			
expected ratio	9 :	3 :	3 :	1
expected number (E)	108			
O – E	13			
(O – E)²	169			
(O – E)² ÷ E	1.40			

Table 16.4: Calculating Chi-squared.

i The cells for the green curly phenotype have already been completed. Complete the cells for the other three phenotypes.

ii Use your answer to **i** to show that the value for chi-squared is 25.20.

Now we have to use this value to work out the probability that the difference between our observed results and the expected results is due to chance. This requires us to look at a table of chi-squared values, like Table 16.5:

Degrees of freedom	Probability that the differences between observed and expected results are due to chance			
	0.1	0.05	0.01	0.001
1	2.71	3.84	6.64	10.83
2	4.60	5.99	9.21	13.82
3	6.25	7.82	11.34	16.27
4	7.78	9.49	13.28	18.46

Table 16.5: Chi-squared values.

First of all, decide on the number of degrees of freedom in your data. You can calculate this using the formula:
$$v = c - 1$$
where:

v is the number of degrees of freedom

c is the number of classes.

   iii In this question, the 'classes' are the different phenotypes. Calculate the number of degrees of freedom.

   iv Looking along the row for this number of degrees of freedom, find the chi-squared value that most closely matches your calculated value.

You should find that there is no number anywhere near as large as yours. Our value of chi-squared is way off to the right of the table somewhere. This shows that the probability that the differences between the observed and expected values differ just by chance is much lower than 0.001.

In this instance, we can take a probability of 0.05 to be our cut-off point. If the probability is greater than 0.05, then we accept that the differences between our observed and expected results could be due to chance. In other words, our hypothesis could be correct.

If the probability is less than 0.05, then it is extremely unlikely that the differences are due only to chance.

   v On this basis, what can you say about the null hypothesis that we used to calculate the expected results? Justify your answer.

Now work out this chi-squared problem for yourself.

2 A homozygous fish with a black stripe and a forked tail was crossed with a homozygous fish with a blue stripe and a smooth tail. All the offspring (the F_1 generation) had black stripes and forked tails.

   a Construct a genetic diagram to predict the results of crossing two of these offspring (the F_2 generation). Remember to explain the symbols that you are using.

   b The phenotypes in the F_2 generation were as follows:

black stripe, forked tail	189
black stripe, smooth tail	52
blue stripe, forked tail	53
blue stripe, smooth tail	63

Suggest an explanation for these results.

   c Construct a null hypothesis, and then use the chi-squared test to determine whether the results differ significantly from those you would expect from your hypothesis. Show all of your working.

Exercise 16.8 Interacting genes

Quite frequently, one gene affects another. For example, one gene might encode an enzyme that is needed in order for a protein encoded by a second gene to be converted to a product. Without both of these genes being present, no product can be produced. This type of gene interaction is called epistasis.

1 In horses, a gene **E/e** controls the production of the black pigment, eumelanin. Allele **E** causes the pigment to be produced, while allele **e** does not produce eumelanin.

A second gene, **A/a**, determines the distribution of black pigment in the coat. Allele **A** restricts the black colour to the mane, tail, ears and lower legs, with the rest of the body being brown. Allele **a** causes black pigment to be produced all over the body.

A horse that is brown with black mane, tail, ears and lower legs is said to be bay (see Figure 16.2). A horse that is brown all over is chestnut.

Figure 16.2: A bay horse.

a Copy and complete Table 16.6.

genotype	phenotype
EEAA	bay
EEAa	
EEaa	
EeAA	
EeAa	
Eeaa	
eeAA	
eeAa	chestnut
eeaa	

Table 16.6: Genotypes and phenotypes.

b A bay mare was crossed with a black stallion. The foal was chestnut.

  i   What is the genotype of the black stallion? Explain your answer.

  ii What are the two possible genotypes of the bay mare and the two possible genotypes of the chestnut foal? Explain your answer.

TIP

In part **b**, start off by writing down what you can work out about the genotype of the chestnut foal. Then think about how he got these alleles. Then work out the genotype of the black stallion.

Exercise 16.9 Control of gene expression

The *lac* operon in *Escherichia coli* is one of the best-studied examples of how the expression of a gene is controlled. In this exercise, you will practise using the correct names for the various components of the operon and neighbouring DNA, and describe how the process works.

1   The boxes below represent the various components that control the translation of the genes for β-galactosidase in the bacterium *Escherichia coli*.

   a   Copy the boxes, but arrange them as they are organised in the bacterial chromosome.

   | promoter for regulatory gene | β-galactosidase gene | operator |

   | promoter for structural gene | regulatory gene |

   b   Identify the component of the *lac* operon and neighbouring DNA with each of the following roles. Some components may fit more than one description:

   i   codes for mRNA that is used in the production of β-galactosidase

   ii   codes for mRNA that is used in the production of the *lac* repressor protein

   iii   is a binding site for the *lac* repressor protein

   iv   is a binding site for mRNA polymerase, which can then begin to transcribe the structural gene

   v   is a structural gene

   vi   is a binding site for mRNA polymerase, which can then begin to transcribe the regulatory gene.

   c   Explain what happens to cause the production of the enzyme β-galactosidase when the bacterium takes up lactose.

EXAM-STYLE QUESTIONS

1 a The micrograph in Figure 16.3 shows a plant cell in anaphase 1 of meiosis.

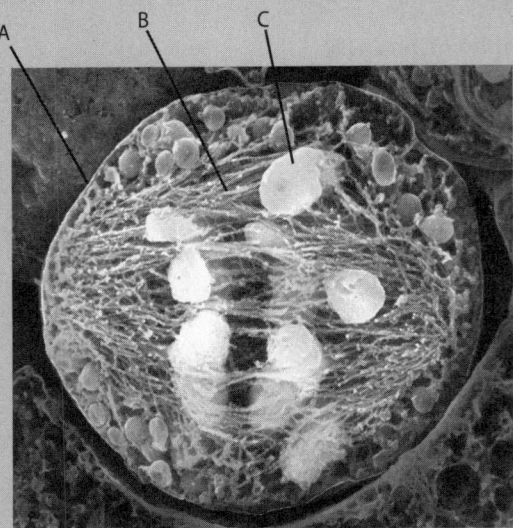

Figure 16.3

   i Name the type of microscope that was used to make this image. **Explain** your answer. [2]

   ii **Identify** the parts of the cell labelled **A** and **B**. [2]

   iii The structure labelled **C** is a chromosome. **Describe** what is happening to the chromosomes at this stage of meiosis. [2]

b Explain how independent assortment of chromosomes during meiosis results in genetic variation in the daughter cells. [4]

[Total: 10]

> **TIP**
> This question tests your knowledge and understanding of meiosis.

> **COMMAND WORDS**
>
> **Explain:** set out purposes or reasons / make the relationships between things evident / provide why and/or how and support with relevant evidence.
>
> **Identify:** name / select /recognise
>
> **Describe:** state the points of a topic / give characteristics and main features.

246

CONTINUED

2 Figure 16.4 shows the sex chromosomes of a male cat and a female cat. The alleles of a gene at a locus on the X chromosome are also shown.

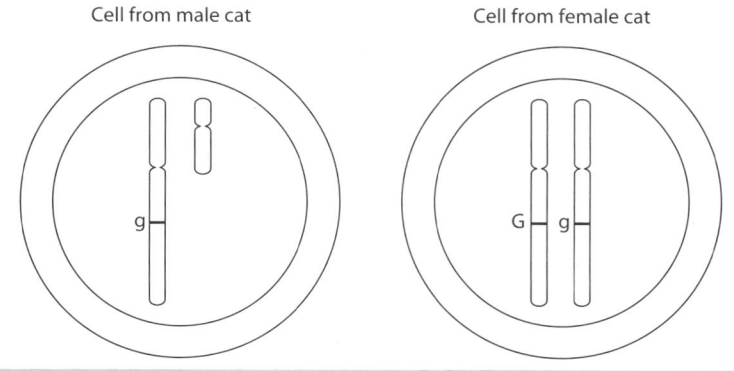

Figure 16.4

This gene affects fur colour. Allele **G** produces ginger (reddish orange) fur, while allele **g** produces black fur.

In female cats, only one of the X chromosomes is functional in each cell. This is determined randomly in the early embryo, so if the cat is heterozygous at this gene locus, one of the X chromosomes is expressed in some groups of cells and the other X chromosome in others. This results in a cat with black and orange patches, a colour known as tortoiseshell.

  a **i** **State** the phenotypes of the cats whose cells are shown in Figure 16.4. [1]

     ii Use a genetic diagram to show the expected ratio of colours in the offspring of a cross between the male cat and the female cat. [5]

  b Another gene, found on one of the autosomes, also affects fur colour. The dominant allele **W** reduces the formation of cells that produce the pigments that give the fur its colour, so the cat is white.

     A male cat with genotype X^gYWW is crossed with a female cat with genotype X^GX^gww.

     State the colour or colours of their offspring. Explain your answer. [2]

  c A third gene, with alleles **T/t**, codes for the production of tyrosinase.

     Explain why a cat with the genotype **tt** is albino, no matter what genotype it has for other genes that affect coat colour. [4]

[Total: 12]

TIP

This question begins with an example of sex linkage, involving just one gene locus – a monohybrid cross. Part **b** then moves on to consider how this gene interacts with another. There is a very quick and simple way to answer this part, which you might be able to think of. The final part asks you to remember what you have learnt about the enzyme tyrosinase.

COMMAND WORD

State: express in clear terms.

CONTINUED

3 Male pattern baldness is a condition in which hair is lost from the top of the head. This condition is at least partly caused by elevated levels of an enzyme called 5α-reductase, which converts testosterone to dihydrotestosterone (DHT).

DHT is a steroid that is transported in the blood to all cells of the body. Its receptor, the androgen receptor, is found in the cytoplasm of cells at the base of hair follicles.

The androgen receptor is normally bound with proteins called heat shock proteins. When DHT is present, it binds with the androgen receptor, freeing this receptor from the heat shock proteins. The androgen receptor then moves to the nucleus and binds with specific DNA sequences that regulate gene expression. This in turn affects the production of proteins that are involved in the loss of hair, although the exact mechanism by which this happens is not yet understood.

a Name the substance described in the passage above that acts as a transcription factor.

Explain your answer. [2]

b DHT is a steroid. **Suggest** why the androgen receptor is found in the cytoplasm and not in the cell surface membrane. [2]

c Finasteride is a drug whose molecules have a shape that is similar to the normal substrate of 5α-reductase.

  i Explain how finasteride could inhibit the production of DHT. [3]

  ii Explain how this drug could prevent male pattern baldness. [3]

d Different forms of the gene that controls the production of the androgen receptor appear to be associated with the likelihood of developing male pattern baldness. A study was carried out in which the presence of an allele *Stu*I was determined in a sample of 552 men. Each of the men was also assessed for baldness. The results are shown in the 'observed' row in Table 16.7.

The researchers used the chi-squared test to determine whether or not there was an association between baldness and the *Stu*I allele.

	bald, *Stu*I present	bald, *Stu*I absent	not bald, *Stu*I present	not bald, *Stu*I absent
observed	415	31	82	25
expected	138	138	138	138

Table 16.7

  i State the null hypothesis that was used to calculate the expected values. [1]

  ii State the number of degrees of freedom in these data. [1]

  iii The calculated value of chi-squared was 755.01.

  Use Table 16.8 to explain how this value can be used to interpret the results obtained by the researchers.

TIP

Despite all that has been learnt in recent years about the human genome, there is still a huge amount to be discovered about how our genes affect our phenotypes. Many bald men would love to be able to regrow their hair, and research is taking place to find drugs that might be able to achieve this. This question ranges across several different areas of the syllabus, including some topics that you learnt about in the AS year.

COMMAND WORD

Suggest: apply knowledge and understanding to situations where there is a range of valid responses, in order to make proposals / put forward considerations.

CONTINUED

Degrees of freedom	Probability that the differences between observed and expected results are due to chance			
	0.1	0.05	0.01	0.001
1	2.71	3.84	6.64	10.83
2	4.60	5.99	9.21	13.82
3	6.25	7.82	11.34	16.27
4	7.78	9.49	13.28	18.46

Table 16.8

[4]

[Total: 16]

4 Figure 16.5 shows a length of DNA in a bacterium that controls the production of the amino acid tryptophan. The structural genes code for enzymes that catalyse the synthesis of tryptophan.

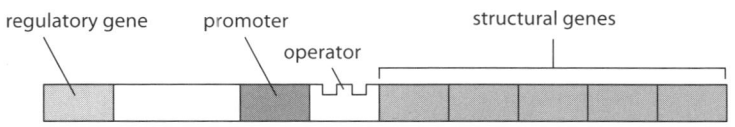

Figure 16.5

When tryptophan is present, it binds to a repressor protein. This protein then allows the repressor protein to bind with the operator.

a **Give** the correct biological term for the length of DNA shown in Figure 16.5. [1]

b State the function of the regulatory gene in Figure 16.5. [1]

c With reference to Figure 16.5, explain how the presence of a large quantity of tryptophan in a cell will affect the production of more tryptophan. [3]

d Suggest the advantage of this control mechanism to the bacterium. [2]

e State whether the enzymes for the production of tryptophan are repressible or inducible enzymes. Explain your answer. [2]

[Total: 9]

5 In humans, blood group is controlled by multiple alleles, for which the symbols I^A, I^B and I^o are used. Albinism is controlled by a recessive allele of a gene that determines the production of melanin.

a Suggest suitable symbols for the alleles of the gene that controls the production of melanin. [1]

b A woman with blood group A and normal skin pigmentation, and a man with blood group B and normal skin pigmentation, had a child with blood group O and albinism.
Construct a genetic diagram to explain how this could happen. [5]

c What is the chance that the couple's next child will also have blood group O and albinism? [1]

[Total: 7]

TIP

You have learnt about how the synthesis of enzymes that catalyse the metabolism of lactose is controlled in bacteria. In this question, you are asked to apply what you have learnt in an unfamiliar context.

COMMAND WORD

Give: produce an answer from a given source or recall/memory.

TIP

In this question, you are asked to construct a genetic diagram for a dihybrid cross involving multiple alleles at one of the gene loci.

Chapter 17
Selection and evolution

> **CHAPTER OUTLINE**
>
> The questions in this chapter cover the following topics:
> - the causes of variation, and why variation is important in selection
> - using the *t*-test
> - natural selection and selective breeding
> - using the Hardy–Weinberg equations
> - evolution and speciation.

Exercise 17.1 Using the *t*-test to analyse variation

In Chapter 5, Exercise 5.3, we looked at using tally counts to draw a bar chart. You will use that skill again here, to analyse two sets of measurements from two populations of plants. You will then go on to use a statistical test to determine whether there is a significant difference between the measurements for the two populations.

1 The leaf lengths of 30 plants belonging to the same species, growing in two different areas, were measured. These were the results. The lengths are in mm.

Area A

19, 12, 13, 13, 16, 18, 17, 16, 19, 17, 16, 18, 15, 17, 19, 18, 18, 18, 15, 17, 16, 14, 16, 18, 13, 17, 15, 21, 14, 14

Area B

29, 27, 24, 26, 30, 27, 23, 19, 26, 28, 25, 26, 30, 29, 29, 24, 25, 29, 12, 31, 26, 28, 22, 29, 17, 28, 23, 29, 30, 31

First, you are going to construct histograms to get a visual representation of the variation within each of these two populations.

a Begin by looking at the complete range of lengths in each population – the spread between the smallest and largest lengths. Use this to decide how to group the lengths into around 10 different classes.

Next, make two tally charts to show the numbers of plants in each area with leaf lengths in each of your classes. Look back to Chapter 5, Exercise 5.3, if you are not sure how to do this.

b Use your tally charts to construct a pair of histograms to display these results.

2 Use the raw results to calculate the mean leaf length for each population.

It looks as though these populations really are different from one another. However, when we look at the two histograms, we can see that there is overlap between the range of leaf lengths, which might make us question whether the difference between them really is significant. We can use the *t*-test to decide whether the difference between the means is significant or not. As for the chi-squared test, we need to construct a null hypothesis. This can be:

There is no difference between the mean leaf lengths of these two populations.

3 a Step 1 Calculate the standard deviation for the data sets for each area.

Step 2 Substitute into the *t*-test formula:

$$t = \frac{|\bar{x}_1 - \bar{x}_2|}{\sqrt{\left(\frac{s_1^2}{n_1} + \frac{s_2^2}{n_2}\right)}}$$

where

t is the value of *t* that you will calculate

x_1 is the mean for area A

x_2 is the mean for area B

s_1 is the standard deviation for area A

s_2 is the standard deviation for area B

n_1 is the number of individual measurements in area A

n_2 is the number of individual measurements in area B.

Step 3 Next, decide on the number of degrees of freedom in your data, using the formula:

$$v = n_1 + n_2 - 2$$

where *v* is the number of degrees of freedom.

Step 4 Now, look at the table of probabilities (below) for difference values of *t*. The numbers in the table cells are values of *t*.

You may remember, from the chi-squared test, that:

- Table 17.1 gives you the probability that the difference between the results is due to chance rather than being a significant difference.
- We need to select a critical value – in this example, we will take a probability of 0.05 as our cut-off value.

If we choose 0.05 as the critical value, this means that when a value of *t* gives us a probability of greater than 0.05, we can say that the difference between the means is just due to chance. This leads us to the conclusion that the differences are not significant. In other words, our results support the null hypothesis.

However, if the value of *t* shows a probability of less than 0.05 that the difference between the means is due to chance, then we can say that there is a significant difference, and our two populations really are significantly different from one another. In other words, our results do not support the null hypothesis.

TIP

Remember:

your answer should be to the same number of significant figures as, or one more than, the smallest number of significant figures in the raw data.

TIP

Look at Exercise 12.1 in Chapter 12 to remind yourself how to calculate standard deviation. If you have set up a spreadsheet, you could use that. If not, you could use a website to do the calculation quickly for you, or use a function on your calculator.

Degrees of freedom	Probability that the difference between the means is due to chance			
	0.10	0.05	0.01	0.001
15	1.75	2.13	2.95	4.07
20	1.73	2.09	2.85	3.85
30	1.70	2.04	2.75	3.65
>30	1.64	1.96	2.58	3.29

Table 17.1: Probability that the difference between the means is due to chance for different degrees of freedom.

b Look up your calculated value of t in the probability values table. What can you say about the significance of the difference between the means for the populations in area A and area B?

TIP

Remember to choose the row that is for the correct number of degrees of freedom.

Exercise 17.2 Answering questions about selection

It's very important to make absolutely clear that mutations happen by chance, (not 'on purpose' or in response to the selection pressure) and that selection acts on variation that is already in place, rather than causing new variants to emerge. You also need to take care over using the terms gene and allele – remember that an allele is a particular form of a gene. Finally, don't be too general – remember to make each of your points about the example in the question.

1 Here are three example answers to a question. Start by reading the question carefully – you might also like to try to answer it yourself. Then read each of the three answers, and comment on what has been done well and what is not so good about each one.

Question: Explain the role of natural selection in the evolution of antibiotic resistance in bacteria. **[5]**

Example X

If a bacterium is exposed to an antibiotic, susceptible ones will die while those that can resist will survive. Mutation may arise, therefore the frequency of alleles may change. Some will become resistant due to mutation and some will not. This is due to selective advantages.

Example Y

Natural selection acts by exerting selection pressures on bacteria, which includes antibiotics. Increased selection pressures increase the chance that one or more bacteria might have genes that confer resistance to the antibiotics. These bacteria can reproduce with members of the same species. Therefore, more antibiotic resistant strains are created.

Example Z

In a bacterial population, some individual bacteria will, by chance, have alleles that give them resistance against an antibiotic. These alleles will have arisen by mutation. Normally, these individuals have no better chance of survival than all the others, but when they are exposed to an antibiotic, the ones with the alleles for resistance have a better chance of survival. They are more likely to reproduce, while the ones without these alleles will die. So, the next generation of bacteria will inherit the allele for resistance and, after a while, all the bacteria in the population will have this resistance allele.

Exercise 17.3 Using the Hardy–Weinberg equations

The Hardy–Weinberg equations enable us to work out the frequency of an allele in a population. Students most often have problems with this if they are not clear in their minds what the symbols used in the equations, p and q, stand for. And, of course, you also need to be able to substitute numbers in the two equations correctly, and manipulate the equations to help you to find out what you need to know.

The Hardy–Weinberg equations are two equations that we can use when a trait in a population is controlled by a gene with two alleles, where one allele is dominant and the other is recessive.

Imagine there is a population of animals where the length of ears is controlled by a gene with two alleles, **E** and **e**. Allele **E** is the dominant allele and gives long ears. Allele **e** gives short ears.

We can use the symbol p to represent the frequency of the dominant allele, in this case **E**. The 'frequency of E' means the proportion of all the alleles of the gene that are allele **E**.

Similarly, we can use the symbol q to represent the frequency of the recessive allele, in this case **e**.

p is the frequency of the dominant allele.

q is the frequency of the recessive allele.

The frequencies are expressed as decimal values:

- If every allele in the population is **E**, then p is 1.
- If no alleles in the population are **E**, then p is 0.
- If 50% of the alleles in the population are **E**, then p is 0.5.

Because there are only the two alleles, the frequencies of the two alleles together always add up to 1.

So:

- If 50% of the alleles in the population are **E**, then the other 50% of the alleles must be **e**. So p is 0.5 and q is 0.5
- If 70% of the alleles in the population are **E**, then the other 30% must be **e**. So p is 0.7 and q is 0.3.

This is shown by the first of the two Hardy–Weinberg equations:

$$p + q = 1$$

Now let's think about the genotypes of the organisms in the population. There are three possible genotypes, **EE**, **Ee** or **ee**.

There are four possible ways in which an organism can inherit these alleles:

- The chance of an animal inheriting allele **E** from each parent is $p \times p$ or p^2.
- The chance of an animal inheriting allele **e** from each parent is $q \times q$ or q^2.
- The chance of an animal inheriting allele **E** from its father and allele **e** from its mother is $p \times q$, or pq.
- The chance of an animal inheriting allele **E** from its mother and allele **e** from its father is $p \times q$, or pq.

Overall, the frequency of all of these chances must add up to 1.

So we can write the second Hardy−Weinberg equation:

$$p^2 + 2pq + q^2 = 1$$

where

p^2 is the frequency of homozygous dominant genotypes

$2pq$ is the frequency of heterozygous genotypes

q^2 is the frequency of homozygous recessive genotypes.

Now let's look at how these two equations can be used. Here's a question to be answered:

In a population of animals, 18 animals out of every 100 have short ears. The other 82 have long ears. What is the frequency of heterozygous organisms in the population?

Because the allele for long ears is dominant, heterozygous organisms with the genotype **Ee**, and those with the genotype **EE**, all have long ears – we can't tell which is which. But we do know that the genotype of each of the short-eared animals is **ee**.

The frequency of the homozygous recessive genotypes is represented by q^2. So we can say that q^2 is 0.18.

From this, we can now work out the frequency of the recessive allele, **e**, which is represented by q:

$$q = \sqrt{0.18}$$
$$= 0.42$$

To answer the question, we need to know the frequency of the dominant allele, **E**. This is represented by p, and we know that $p + q = 1$.

$$p = 1 - q$$
$$= 1 - 0.42$$
$$= 0.58$$

We want to know the frequency of the heterozygous organisms in the population. This is represented by $2pq$:

$$2pq = 2 \times 0.42 \times 0.58 = 0.49$$

In other words, about 49% of the organisms in the population are heterozygous, with the genotype **Ee**.

Here's a question for you to try:

1 In a population of 350 woodlice, 22 animals are red-brown, while the rest are grey.

Assuming that the allele for the production of grey pigment is dominant, and the allele for red-brown pigment is recessive, calculate:

a the frequency of the allele for red-brown pigment in the population

b the frequency of animals that are homozygous for the allele for grey pigment in the population.

Exercise 17.4 Assumptions when using the Hardy–Weinberg equations

The Hardy–Weinberg equations assume that the chances of inheriting the two different alleles are affected only by their relative frequencies in the population. In this exercise, we will use this principle to determine whether selection is acting against a particular genotype or allele.

The Hardy–Weinberg equations assume that the frequencies of alleles do not change from generation to generation. This will only happen if:

- there is no significant migration into the population, or emigration from it
- every organism in the population is equally likely to breed
- mating is random – that is, organisms do not 'choose' a mate by their phenotype, so organisms of either phenotype are equally likely to mate with each other
- new mutations do not occur
- the population is large.

Drongos are small, black birds that live in many tropical regions of the world. In a population of 1270 drongos, 15 albino birds were counted. It is known that the allele for albino colouration is recessive.

1 Use the Hardy–Weinberg equations to calculate the frequencies of the two alleles for colouration in this population.

2 a Two years later, a second survey of the same drongo population found that, out of 2654 individuals, 17 albino birds were counted.

Use your answers from **1** to calculate the number of albino birds that you would expect in this population.

b Name a statistical test that you could use to determine whether the observed number of albino birds in the population differs significantly from the expected number.

3 What do your answers to **1** and **2 a** suggest about selection pressures on the two phenotypes in the drongo population?

EXAM-STYLE QUESTIONS

1 Spruce trees, *Picea* sp., are grown for timber in North America. They are frequently damaged by a pest called the spruce budworm, which is a caterpillar that feeds on the needles (leaves) of the trees and can kill even mature trees. However, some spruce trees have varying degrees of natural resistance to the budworm.

   a **Outline** how a selective breeding programme could be carried out to produce populations of spruce trees that are resistant to the spruce budworm. [5]

   b Researchers have discovered a gene in white spruce, *Picea glauca*, whose expression is related to the degree of resistance to the budworm. The gene is present in all white spruce trees, but is expressed to different degrees. The degree of resistance to the budworm is passed on from parents to offspring.

   Suggest why the gene is expressed to a greater extent in some individual spruce trees than others. [3]

   c The gene codes for the enzyme β-glucosidase, which catalyses the cleavage of molecules of acetophenone sugars to produce compounds called piceol and pungenol, which are active against spruce budworm.

   Figure 17.1 shows the results of tests to determine the concentration of these compounds in resistant and non-resistant spruce trees in June and August in 2010.

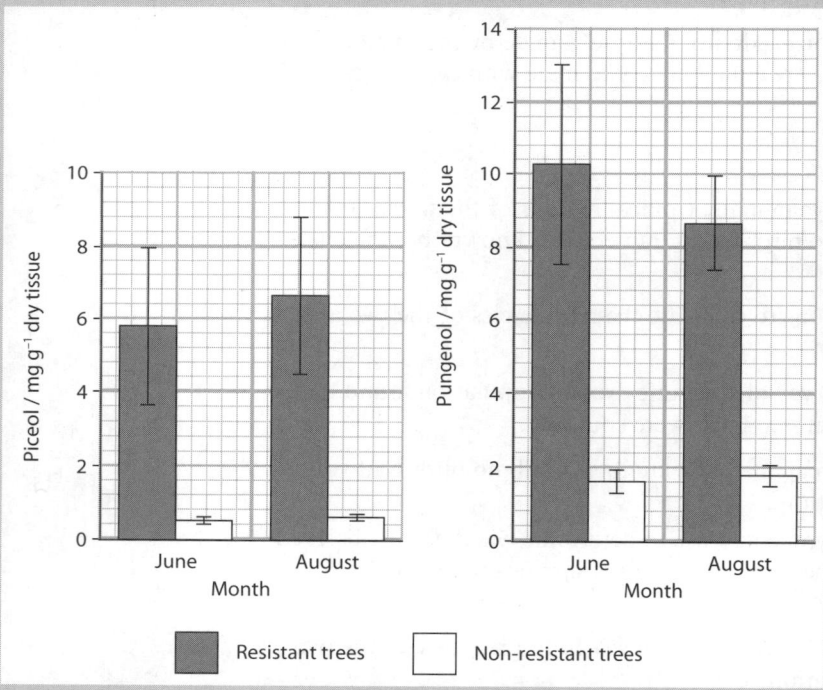

Figure 17.1

TIP

We generally think of selective breeding in relation to crops that are grown in fields, and farm animals, but selective breeding can also be used for trees to be harvested for timber. It is a very slow process for trees, as breeders may have to wait many years for the young trees to grow large enough to reproduce. In this question, you will think about how the discovery of the involvement of a gene for a desired characteristic could speed up the process of selective breeding.

COMMAND WORD

Outline: set out the main points.

Suggest: apply knowledge and understanding to situations where there is a range of valid responses, in order to make proposals / put forward considerations.

17 Selection and evolution

CONTINUED

i **Compare** the results for piceol and pungenol. [4]

ii Error bars are shown on the bar charts. **Explain** how the lengths of the error bars are determined, and what they indicate. [3]

iii Suggest how the discovery of this gene could enable faster development of strains of spruce tree that are resistant to spruce budworm. [2]

[Total: 17]

2 *Mimulus cardinalis* and *Mimulus lewisii* are two species of monkeyflower that grow in moist, streamside habitats in North America. *M. cardinalis* grows at altitudes between 0 and about 2000 m above sea level, and has red flowers that are pollinated by hummingbirds. *M. lewisii* grows at altitudes of about 1600 m to 3000 m, and has pink flowers that are pollinated by bumblebees. Figure 17.2 shows these two species of monkeyflowers.

Figure 17.2

DNA analysis suggests that these two species are very closely related. They can be crossed artificially to produce fertile hybrids, but hybrids are only very rarely found in the wild.

a Use the information above to suggest why hybrids of these two monkeyflowers are only rarely found in the wild. [3]

COMMAND WORDS

Compare: identify/comment on similarities and/or differences.

Explain: set out purposes or reasons / make the relationships between things evident / provide why and/or how and support with relevant evidence.

TIP

Speciation occurs when two populations are genetically isolated from one another, and then evolve along different pathways, becoming so different that they are no longer able to interbreed successfully. Once they have become two separate species, they are prevented from interbreeding by various types of reproductive barriers. This question asks you to describe and explain data about the possible mechanisms that keep two closely related species of flowering plant separate from one another.

CONTINUED

b The fruits of these monkeyflowers contain large numbers of tiny seeds. Researchers grew both species in controlled conditions, and then pollinated the flowers either with pollen from their own species, or with pollen from the other species. The results are shown in Table 17.2.

Experiment number	Species from which pollen was taken	Species onto whose stigma the pollen was placed	Mean number of seeds contained in resulting fruits
1	M. cardinalis	M. cardinalis	2700
2	M. lewisii	M. cardinalis	1200
3	M. cardinalis	M. lewisii	800
4	M. lewisii	M. lewisii	1400

Table 17.2

  i Use a suitable calculation to compare the number of seeds produced by *M. cardinalis* when pollinated with pollen from its own species and with pollen from *M. lewisii*. [2]

  ii With reference to the data, suggest explanations for the difference between the number of seeds produced in experiments 2 and 3. [3]

c The researchers grew plants from the seeds obtained from experiments 1, 2, 3 and 4. They allowed the plants to grow to adulthood and flower. They then collected pollen from each of the plants and tested its fertility by determining whether it could germinate and produce pollen tubes.

The results are shown in Figure 17.3.

Suggest explanations for the results shown in Figure 17.3. [3]

d With reference to all of the information provided in this question, list *three* mechanisms that prevent interbreeding between *M. cardinalis* and *M. lewisi*. [3]

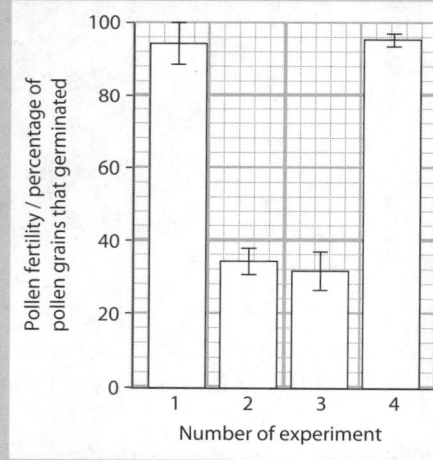

Figure 17.3

[Total: 14]

CONTINUED

3 Cone snails are predatory marine molluscs. Cone snails produce venom that paralyses their prey, making it easier for the snail to feed on them. In most species of cone snails, these venoms contain various neurotoxins. One of these neurotoxins is a protein that inhibits voltage-dependent sodium ion channels.

  a Suggest how this neurotoxin could paralyse the cone snail's prey. [4]

  b In some species of cone snail, the venom also contains insulin.

  All molluscs, including cone snails, produce insulin that has a different structure from vertebrate insulin. The molluscan-type insulin produced by cone snails is called con-ins G2.

  Cone snails also produce a different type of insulin called con-ins G1.

  Below are the amino acid sequences from the polypeptide chain A in con-ins G2, con-ins G1 and insulin from a zebra fish. Each letter stands for a different amino acid.

  con-ins G2 T G Y K G I A C E C C Q H Y C T D Q E F I N Y C P P V T E S S S S S S S A
  con-ins G1 G V V E H C C H R P C S N A E F K K Y C
  zebra fish G I V E Q C C H K P C S I F E L Q N Y C N

   i   Compare the amino acid sequence for con-ins G1 with those for con-ins G2 and zebra fish insulin. [3]

   ii The letter C stands for the amino acid cysteine. Disulfide bonds form between cysteine molecules.

        Explain what the information in the table implies about the tertiary and quaternary structure of con-ins G1 and insulin from zebra fish. [3]

   iii **Discuss** what the information in the amino acid sequences suggests about the closeness of the evolutionary relationship between molluscs and fish. [3]

  c In 2015, researchers discovered that the cone snail *Conus geographicus* releases con-ins G1 into the water when it is hunting fish. This causes hypoglycaemia in fish, so that they stop swimming and can more easily be captured by the cone snail. In contrast, other species of cone snails, which hunt worms and do not hunt fish, only produce con-ins G2, and do not produce con-ins G1.

  d Suggest how *C. geographicus* may have evolved from worm-hunting cone snails. [6]

  [Total: 19]

TIP

This question requires you to use knowledge and understanding of facts and concepts from several different areas of the syllabus, including protein structure, the transmission of nerve impulses, the action of insulin and evolution by natural selection.

COMMAND WORD

Discuss: write about issue(s) or topic(s) in a structured way.

CONTINUED

4 Rice is a cereal crop that is grown for food in many countries. In the 1970s, the discovery of a mutant rice plant with short stems, broad dark green leaves and short, round grains led to the development of a dwarf variety called Daikoku. It was found that Daikoku was dwarf because it was completely insensitive to gibberellin. Gibberellin normally causes the expression of genes that are involved with stem elongation.

a Outline the advantages of growing dwarf varieties of cereals such as rice. [2]

b In the early 21st century, it was discovered that Daikoku contains a recessive allele of a gene, *gid*1, that codes for a protein that normally functions as a gibberellin receptor. This receptor, known as GID1, interacts with the rice DELLA protein. The protein coded for by *gid*1 has one amino acid different from the protein coded for by the normal, dominant allele *GID1*.

 i Gibberellin cannot bind with the receptor coded for by *gid*1. Explain why this is so. [2]

 ii Explain why a plant with two copies of *gid*1 remains dwarf, even if treated with gibberellin. [5]

c Rice blast is a serious fungal disease that greatly reduces yields of rice. Some varieties of rice have natural resistance to rice blast.

A breeding programme could be used to produce a variety of rice that has the dwarf phenotype of Daikoku, and is also resistant to rice blast.

 i **State** the parents that would be used in this breeding programme. [1]

 ii Outline the steps that would be taken in the breeding programme, after the two parents have been crossed. [5]

[Total: 15]

TIP

This question relates to the incorporation of mutant alleles into cereal crops to produce dwarf varieties. These alleles often determine gibberellin synthesis, but in some cases they affect the gibberellin receptors rather than the production of gibberellin itself. Part **b** requires you to remember what you have learnt about gibberellin and DELLA proteins.

COMMAND WORD

State: express in clear terms.

Chapter 18
Classification, biodiversity and conservation

CHAPTER OUTLINE

The questions in this chapter cover the following topics:
- the concept of biodiversity
- Simpson's index of diversity and Pearson's linear correlation
- carrying out ecological investigations
- classifying organisms and viruses
- ways in which biodiversity can be conserved.

Exercise 18.1 Collecting and analysing data on species richness and species diversity

Various techniques can be used to assess the number of different species in an area, and their relative abundance. This exercise will help you to decide on appropriate techniques to use in different contexts, and to analyse the data that are obtained.

1 One aspect of biodiversity is the number of species in an area (**species richness**) and their relative abundance (**species diversity**). Table 18.1 summarises some of the techniques that can be used to collect data on the number of species and their abundance in a defined area.

KEY WORDS

species richness: the number of different species living in an area. It does not measure abundance.

species diversity: a measure of the number and evenness of species living in an area. It indicates which species are present and how abundant each one is.

Technique	How it is used	Type of data collected	Notes
Frame quadrat	Quadrats are placed randomly within a defined area. The species within the quadrat are identified and their relative abundance is assessed.	For slow-moving animals and large plants including trees (which can be counted as individuals): numbers of each species in the quadrats; for small plants (which often cannot be counted individually): percentage cover or abundance measured on an abundance scale (e.g. the Braun–Blanquet scale).	Appropriate sizes of quadrats need to be used, e.g. a quadrat with 10 m sides might be appropriate to assess trees in a forest, while a quadrat with 0.25m sides would be appropriate on a rocky shore or in a grassy field.
Line transect	A tape is placed on the ground to make a 'line', and the species touching the tape are recorded.	Species present at different distances along the line.	This is useful when you want to know how the species change as environmental factors change (e.g. from sunlight into shade).
Belt transect	A tape is placed along the ground as for a line transect. Frame quadrats are then placed along the line, either continuously or at intervals (interrupted belt transect). Abundance is assessed as for frame quadrats.	Abundance of each species at different distances along the line.	As for line transect, but more data are collected and assessment of relative abundance along the line is more reliable.

Table 18.1: Techniques used to collect data on the number of species and their abundance in a defined area.

Decide which technique or techniques you could use in each of the following situations, and outline what you would do.

a   You want to know whether the species of seaweed that grow on a rocky shore change as you go up the shore from mean sea level.

b   You want to compare the species that grow around the edge of a rice field with those that grow around the edge of a wheat field.

2   Imagine two communities both containing ten different species. In one **community**, nearly all the organisms are of just two of these species, while the other eight are rare. In the other community, numbers of all the different species are quite similar. This second community has a higher species diversity than the first.

Table 18.2 shows data that a student collected from two different areas of grassland. The two areas were the same size, and the student used the same sized quadrats, and the same number of quadrats in each area. The quadrats were placed randomly.

KEY WORDS

quadrat: a square frame which is used to mark out an area for sampling of organisms

transect: a line along which samples are taken, either by noting the species at equal distances (line transect) or placing quadrats at regular intervals (belt transect)

community: all of the different species living in an area at the same time

Species	Number in area P, n	Number in area Q, n
A	282	310
B	141	103
C	376	292
D	0	49
E	132	2
F	21	0
G	210	14
H	63	1

Table 18.2: Data collected from two different areas of grassland.

> **TIP**
> Note that you do not actually need to identify the species by name to be able to calculate species diversity.

You are going to calculate Simpson's index of diversity for area **P**, and then for area **Q**.

The formula for calculating Simpson's index is:

$$D = 1 - \left(\Sigma \left(\frac{n}{N} \right)^2 \right)$$

where

D = Simpson's index of diversity

n = the number of each species in the sample

N = the total number of organisms, of all species.

a Calculate the total number of organisms, N, found in area **P**.
b Calculate $\frac{n}{N}$ for each of the species in area P.
c Square each of your values of $\frac{n}{N}$.
d Add up all these squared values.
e Subtract your answer from 1.
f Repeat the calculation for area **Q**.
g Which area has the greater species diversity?

> **TIP**
>
> You may find it helpful to do steps **a** to **d** in this calculation in a chart or spreadsheet like this:
>
> | Species | n | n ÷ N | (n ÷ N)² |
> |---|---|---|---|
> | A | | | |
> | B, etc. | | | |
> | Total | N = | $\Sigma \left(\frac{n}{N} \right)^2 =$ | |
>
> **Table 18.3:** Sample results table.

3   We can use the **mark–release–recapture** technique to estimate the population of a mobile animal.

   To estimate the size of a woodlouse population, you could:

   - Capture a fairly large number of woodlice, say 50.
   - Mark each one with a small spot of non-toxic paint. Record the number of woodlice you have marked.
   - Then release the woodlice where you found them, and leave them for long enough to mix back into the rest of the population – perhaps two or three days.
   - Catch a second sample, and count the number of marked animals and the total number of animals caught.
   - Calculate an estimate of the size of the whole population using the formula:

   $$\text{Estimated size of population} = \frac{\text{number caught and marked in first sample} \times \text{total number in second sample}}{\text{number of marked animals in second sample}}$$

   a   A student caught and marked 62 woodlice. She released them, then caught a second sample two days later. There were 73 woodlice in the second sample, of which 9 were marked. Calculate the estimated number in the population.

   b   Suggest why this method gives us only an estimated number, not a true value for the population size.

> **KEY WORD**
>
> **mark–release–recapture:** a method of estimating the numbers of individuals in a population of mobile animals

> **TIP**
>
> This is known as the Lincoln index. An easy way to remember the formula is that you multiply the two largest numbers together and divide by the smallest one.

Exercise 18.2 Using Pearson's linear correlation

In ecology, we often want to know if the distribution and abundance of a particular species correlates with that of another species, or with a particular biological factor such as light intensity. Pearson's linear correlation tells us if there is a linear correlation between the two variables.

In Chapter 13, you practised using Spearman's rank correlation to determine whether or not there was a correlation between two continuous variables. This is also very useful in ecology. However, there is also another correlation test that we can use, if we think that there could be a linear correlation between the two variables.

When you are wondering about a possible correlation between two variables, it is a good idea to begin by drawing a scatter graph. The pattern of results on the graph can give you a good idea of whether or not the two variables may have a relationship with one another. Here are three examples.

> **TIP**
>
> Note that scatter graphs tend to be used when there is *no* independent variable. You do not have control over or choose values for either of the variables. You simply measure their values. It therefore does not matter which variable you put on which axis.

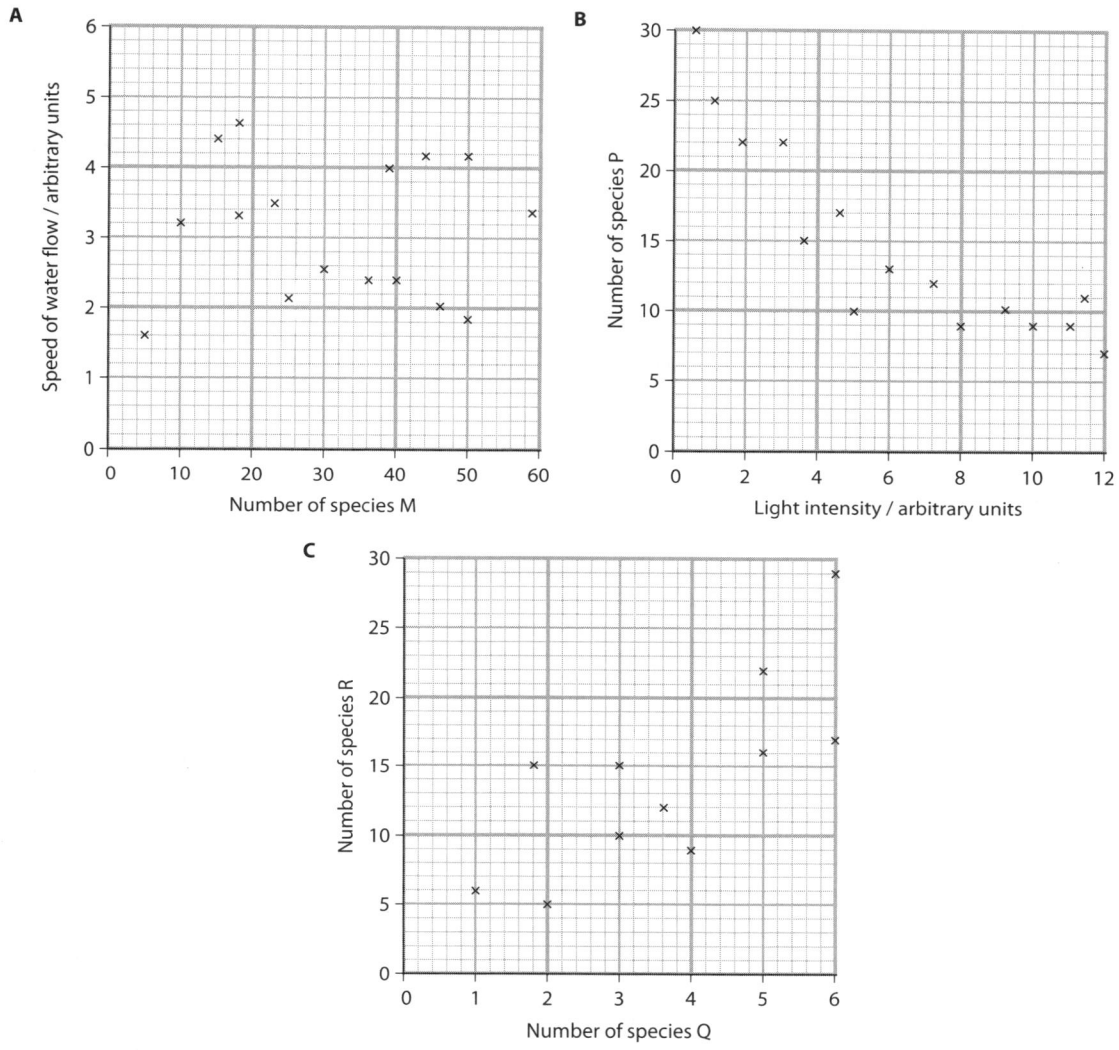

Figure 18.1: Scatter graphs of pairs of variables.

In graph **A**, it doesn't look as though there is any relationship between the two variables. It is not worth doing any statistical test to look for a correlation.

In graph **B**, it looks as though the numbers of species P might be lower in high light intensities. There seems to be a **negative correlation**. If we drew a best fit line through the points, however, it probably would not be a straight line, but would be a curve. The relationship is not linear. We could use Spearman's rank correlation to test this relationship.

In graph **C**, it looks as though the numbers of species R and the numbers of species Q are related. If large numbers of species R are found, there are also large numbers of species Q. We could possibly draw a straight best fit line through these points. The relationship looks as though it might be a positive **linear correlation**. We could use Pearson's linear correlation test to decide whether this relationship exists.

> **KEY WORDS**
>
> **negative correlation:** a relationship between two variables such as when one increases, the other decreases
>
> **linear correlation:** a relationship between two variables such that a straight best-fit line can be drawn when one is plotted against the other

1   A student wanted to know if the number of galls on a leaf was correlated with the surface area of the leaf.

The student collected 20 leaves from a lime tree, *Tilia* sp. She measured the surface area of one side of each leaf, and counted the number of galls on the leaf.

a   Suggest the variables that the student should control when collecting the leaves for her study.

b   Suggest how the student could calculate the area of each leaf.

c   Table 18.4 shows the student's results.

Figure 18.2: Galls on a leaf of *Tilia* sp. Each gall is caused by an insect that lays an egg in the leaf. Substances in the egg affect cell division and differentiation in the leaf, causing it to produce these growths in which the insect egg develops.

Leaf	Surface area / cm²	Number of galls
1	101	180
2	143	139
3	53	119
4	61	98
5	111	120
6	92	131
7	149	151
8	123	105
9	47	88
10	63	107
11	120	149
12	62	134
13	141	156
14	83	101
15	76	137
16	142	123
17	21	101
18	104	142
19	65	98
20	137	114

Table 18.4: Student's result table recording leaf area and number of galls per leaf.

i   Plot a scatter graph to display these results.

ii Describe the relationship that there appears to be between these two variables.

18 Classification, biodiversity and conservation

The formula for Pearson's linear correlation is:

$$r = \frac{\Sigma xy - n\bar{x}\bar{y}}{(n-1)s_x s_y}$$

where

r = Pearson's correlation coefficient
x is the value of one variable (in this case, the surface area of the leaf)
y is the value of the other variable (in this case, the number of galls)
n is the number of readings (in this case, 20)
$\bar{x}$ is the mean value of x
$\bar{y}$ is the mean value of y
s_x is the standard deviation for the surface area of the leaf
s_y is the standard deviation for the number of galls.

2 a Construct a table (see Table 18.5) and complete it by filling in the values of x and y, and calculating xy for each pair of values. The first row has been done for you.

Leaf	Surface area / cm² x	Number of galls y	xy
1	101	180	18 180
2			
3			
4			
5			

Table 18.5: Results table.

i Calculate Σxy by adding up all the values in the last column.
ii Calculate $\bar{x}$ and $\bar{y}$.
iii Calculate $n\bar{x}\bar{y}$.
iv Calculate s_x and s_y. (Use a calculator that can do this quickly for you, or input the numbers to a website or spreadsheet that will do the calculation.)
v Substitute your values into the formula, and calculate r.
vi Does your value of r suggest that there is a positive linear correlation between the surface area of the leaf and the number of galls?

> **TIP**
>
> A value of r between 0.5 and 1.0 indicates a high degree of correlation. A value of r between 0.3 and 0.5 indicates a medium degree of correlation. A value of r of 0 indicates no correlation. If your value for r does not lie between 0 and 1, you've gone wrong somewhere!

Exercise 18.3 Answering structured questions about conservation

It's important to ensure that you write the best answers that you can, using the knowledge that you have. This exercise asks you to identify strong and weak points in three example answers and then write an answer yourself.

1   Here is part of a question, worth four marks, about conserving gorillas. Read the question, and then each of the answers. Rank the answers according to which one you think is best and which is least good, and justify your ranking. Finally, try writing an answer yourself.

Question: The western lowland gorilla, *Gorilla gorilla,* lives in west central Africa. It is largely herbivorous, feeding on foliage and fruit that is rarely in short supply. This species is currently listed as endangered.

Explain how captive breeding programmes in zoos could help to protect the western lowland gorilla. [4]

Answer X

Breeding programmes in zoos help an endangered species by maintaining the animal alive so they can reproduce and produce offspring. They then take care of the offspring until they reproduce. Keeping them captive means they will not be harmed by outside enemies. They can be kept safe in a healthy environment, so that they can breed.

Answer Y

Captive breeding ensures the gorillas can continue to reproduce even when threatened in the wild. Zoos can ensure that the animals are healthy, which improves the chances of successful breeding. They could coordinate with one another to ensure that only unrelated gorillas are bred together to maximise genetic diversity. They could use reproductive techniques such as IVF, or use frozen embryo transfer, to increase the number of young gorillas produced. Eventually, the zoos would hope to be able to transfer some gorillas back to the wild to increase their numbers there, if their habitat can be made safe.

Answer Z

They can help as the gorillas are protected from any harm and competition with other animals. Competition for food will be non-existent as they can be given plenty of food and kept healthy. Reproduction can take place without danger for the young.

> **TIP**
>
> Questions on conservation can be challenging for students to answer to their full potential. Often the problem is that students answer using their general knowledge rather than using specific subject knowledge that they have gained through their A level course. Remember to use correct terminology, phrases and explanations that you have learnt during your course and reflect your biological understanding.

EXAM-STYLE QUESTIONS

1 An investigation was carried out to test the hypothesis that a higher level of diversity in a community results in greater plant biomass.

The study was done in a grassy field that had been mowed regularly but never ploughed. Three plant species dominated the community:

- Kentucky blue grass, *Poa pratensis*
- wild strawberry, *Fragaria virginiana*
- dandelion, *Taraxacum officinale*.

The researchers cleared all the vegetation from 45 plots of 40 cm × 40 cm. They then planted a total of 14 plants from these three species in each of these plots, to achieve three levels of evenness in the species in the plot. Fifteen plots were planted, with the ratios of the three different species as follows:

1.5 : 1 : 1   5 : 1 : 1   12 : 1 : 1

a The researchers used a diversity index to calculate the species evenness in each of the three different ratios of species planted in the plots. The diversity indices were 0.45, 0.61 and 0.96.

Match these three diversity indices to the three different treatments. [1]

b Within each of the evenness treatments, five plots were planted with *Poa* as the dominant species, five with *Fragaria* as the dominant species and five with *Taraxacum* as the dominant species.

  i How many plants of the dominant species would be planted in a plot in which the ratio was 5 : 1 : 1? [1]

  ii **Suggest** *two* variables that the researchers should have tried to keep the same in all of the plots. [2]

c The above- and below-ground plant material was harvested from all 45 plots at the end of the growing season, and its biomass determined. Table 18.6 shows the results.

Diversity index (species evenness)	Total biomass / g per plot					
	Poa dominant		*Fragaria* dominant		*Taraxacum* dominant	
	Mean	SE	Mean	SE	Mean	SE
0.45	319.9	54.4	301.2	27.3	246.5	57.5
0.61	336.9	26.1	410.5	30.3	205.5	33.8
0.96	348.7	41.0	381.5	35.3	358.3	27.5

Table 18.6

TIP

It is not easy to do controlled experiments in the field to investigate the effects of biodiversity. This question describes one such experiment that was carried out in the year 2000. You may need to read the information at the start of the question more than once in order to get a clear picture of exactly what the researchers did. You may find it useful to make some sketches to help you to understand the investigator's method.

COMMAND WORD

Suggest: apply knowledge and understanding to situations where there is a range of valid responses, in order to make proposals / put forward considerations.

CAMBRIDGE INTERNATIONAL AS & A LEVEL BIOLOGY: WORKBOOK

CONTINUED

  i On graph paper, plot the results for the *Fragaria* dominant plots as a bar chart. [3]

  ii Use the values for standard error to show error bars on your bar chart. [2]

  iii **Discuss** the extent to which these results support or disprove the hypothesis that was being tested. [3]

  [Total: 12]

TIP

Zoos carefully design captive breeding programmes for mammals using knowledge of the genetic diversity and physiology of the animals concerned. This question asks you to think about these issues in the context of an endangered small cat species.

COMMAND WORDS

Discuss: write about issue(s) or topic(s) in a structured way.

Explain: set out purposes or reasons / make the relationships between things evident / provide why and/or how and support with relevant evidence.

2 The fishing cat, *Prionailurus viverrinus*, is a small cat that lives in wetland habitats in South and South-East Asia. Fishing cats are largely nocturnal and feed mainly on fish, which they catch by wading or swimming into water. They are classified by the IUCN, the International Union for Conservation of Nature, as an endangered species.

Figure 18.3 shows a fishing cat.

  a i Suggest *two* reasons why fishing cats are endangered. [2]

   ii Fishing cats are listed in Appendix 1 of CITES, the Convention on International Trade in Endangered Species.

      Explain how this can help in their conservation. [2]

  b Several zoos in North America keep fishing cats, and a captive breeding programme has begun. In 2006, there were a total of 71 fishing cats in these zoos, all bred from 8 founder animals imported from Asia. The genetic diversity of these 71 animals was 0.85, meaning that 85% of gene loci had more than one allele present in the population.

  A computer model was used to predict how the genetic diversity of this population would change in the next 50 years, if breeding was restricted to these 71 animals and their offspring, or if new animals were imported from Asia. The model assumed that the maximum number of fishing cats that could be kept in zoos in North America would be 100.

Figure 18.3

CONTINUED

Figure 18.4 shows some of these predictions.

A If no new cats are introduced to the population

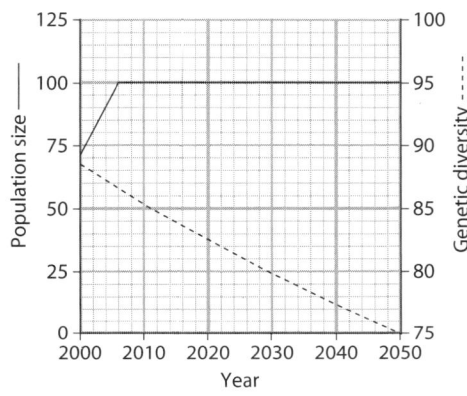

B If one new cat is introduced to the population every five years

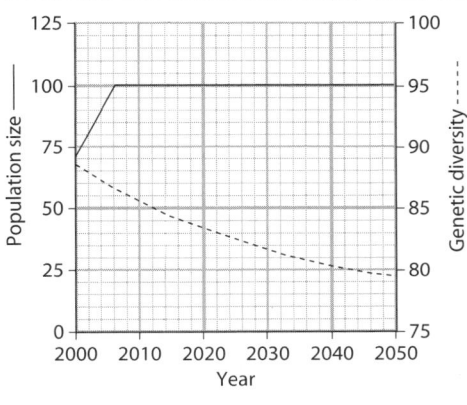

C If one new cat is introduced to the population every year

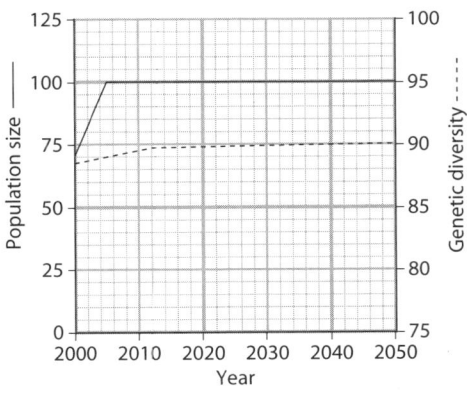

Figure 18.4

CONTINUED

   i With reference to the data in Figure 18.4, explain why it is important to continue to import fishing cats from elsewhere to add to the population in North American zoos. [5]

   ii Suggest *one* reason why the importation of new animals to North American zoos might not be desirable. [1]

   iii There is concern that genetic drift might lead to loss of genetic diversity in the remaining wild populations of fishing cats. Explain how this could occur. [3]

  c Many species of small cats, including fishing cats, breed only once a year, and often have only small numbers of young. Assisted reproduction techniques such as embryo transfer could potentially increase the number of young cats that are born, and also help to maintain or increase genetic diversity.

   i **Outline** how embryo transfer could potentially be used in a captive breeding programme for fishing cats. [5]

   ii At present, scientists do not have detailed knowledge of the hormonal control of reproductive cycles in most small cat species. Suggest how this limits the use of assisted reproduction techniques for these animals. [3]

   [Total: 21]

3 A student investigated whether the abundance of a plant species growing in a shady habitat was significantly different from that in a sunny habitat.

   The student measured the area covered by the plant species, in cm^2, in each of ten areas of $1\ m^2$ in the shady habitat, and another ten areas of $1\ m^2$ in the sunny habitat.

   a i What was the independent variable in this investigation? [1]
      ii What was the dependent variable? [1]
   b **State** a suitable null hypothesis for the student's investigation. [1]
   c The student used random sampling in his survey. Explain what is meant by random sampling, and why it was important to use this method. [2]
   d The results were as follows:

   Area of ground (in cm^2) covered by the plant in a $1\ m^2$ area:

   shady habitat: 43, 52, 31, 61, 56, 47, 22, 42, 33, 40

   sunny habitat: 54, 56, 32, 65, 23, 43, 59, 41, 45, 31

   Calculate the mean area covered per m^2 in each habitat. [2]

   e The student calculated the standard deviation, *s*, for the two samples:

   s for the shady area = 11.945

   s for the sunny area = 13.609

COMMAND WORDS

Outline: set out the main points.

State: express in clear terms.

Calculate: work out from given facts, figures or information.

TIP

In Chapter 17, you practised using the *t*-test to compare the variation in two different populations, by testing for the significance of difference between their mean values. In this example, you will use the *t*-test again, but this time in an ecological context.

18 Classification, biodiversity and conservation

CONTINUED

He used a *t*-test to determine whether the difference between the means of the two samples was significantly different.

i   Explain why the *t*-test is a suitable statistical test to use for these data. [3]

ii The formula for the *t*-test is:

$$t = \frac{|\bar{x}_1 - \bar{x}_2|}{\sqrt{\left(\frac{s_1^2}{n_1} - \frac{s_2^2}{n_2}\right)}}$$

where

t is the value of *t* that you will calculate
$\bar{x}_1$ is the mean for one sample
$\bar{x}_2$ is the mean for the other sample
s_1 is the standard deviation for one sample
s_2 is the standard deviation for the other sample
n_1 is the number of individual measurements in one sample
n_2 is the number of individual measurements in the other sample.
Calculate the value of *t*. Show your working. [3]

iii Table 18.7 shows the probability that chance could have produced a particular value of *t* for a range of degrees of freedom.

degrees of freedom, $v = (n_1 - 1) + (n_2 - 1)$

Use your value of *t*, and the table below, to decide whether the null hypothesis that you gave as your answer to **b** is supported or not. Explain your answer. [4]

Degrees of freedom	Probability that the difference between the means is due to chance			
	0.10	0.05	0.01	0.001
15	1.75	2.13	2.95	4.07
18	1.73	2.10	2.88	3.92
20	1.73	2.09	2.85	3.85
22	1.72	2.07	2.82	3.79
30	1.70	2.04	2.75	3.65
>30	1.64	1.96	2.58	3.29

Table 18.7

[Total: 17]

CONTINUED

4 a Explain what is meant by a taxonomic hierarchy, with reference to the classification of organisms in the Eukarya domain. **[5]**

Organisms in domain Archaea and domain Bacteria are prokaryotes.

b List *three* differences between Archaea and Bacteria. **[3]**

c Explain the difference between the terms *biological species concept* and *morphological species concept*. **[4]**

d Viruses are not classified within any of the three domains, and they are not grouped into species.

  i Explain why viruses are not classified in the three domains. **[2]**

  ii List *two* features of viruses that are used in their classification. **[2]**

[Total: 16]

5 a Explain the meaning of the following terms:

  i ecosystem **[2]**

  ii niche **[1]**

  iii biodiversity **[2]**

b A researcher carried out an investigation to test the hypothesis that the abundance of limpets on a rocky shore shows a negative correlation with the abundance of seaweeds.

Before planning the investigation, the researcher decided on the statistical test that he would use to determine whether any correlation indicated by the results was significant.

  i Suggest an appropriate statistical test that could be used. **[1]**

  ii Outline how the researcher could collect data, to test this hypothesis. **[5]**

  iii Explain how the researcher could display the data, in order to look for a correlation between limpet and seaweed abundance. **[3]**

[Total: 14]

> **TIP**
>
> This question is about classification. It tests your recall and understanding of several learning objectives in Section 18.1 of the syllabus.

> **TIP**
>
> The first part of this question requires some precise statements about three technical terms. In the second part, you may be able to use your own experience of carrying out a similar investigation. In part **b ii**, be sure to give clear descriptions of exactly what the researcher should do, but remember that the command word is 'outline', so do *not* go into too much detail.

> Chapter 19
Genetic technology

CHAPTER OUTLINE

The questions in this chapter cover the following topics:
- the principles of genetic technology
- using databases
- screening for and treating genetic diseases
- GM crops
- ethical and social issues in genetic technology.

Exercise 19.1 Choosing the right tools to analyse data

During your A level course, you have come across a number of different statistical tests and calculations that you can use to help you to make sense of data that have been collected. This exercise will help you to think about when you should use some of these calculations and tests.

When you plan an experiment, it is important to consider how you will analyse your results. You can then plan what data to collect, and how you will collect them, in order to allow you to use your chosen statistical test to interpret what the data mean.

In your examination, however, you may be given information about an investigation, and asked to suggest how you would analyse the results.

Table 19.1 summarises some of the data analysis calculations that you have been using in your A level course.

Calculation	Criteria for using it	How to interpret the value you calculate	Notes
Lincoln index (mark–release–recapture)	You want to estimate the size of a population of mobile animals. You can mark an animal without affecting its chances of survival. Marked animals mix randomly and fully into the rest of the population after release.	The value gives you an estimate of the size of a population.	The value is only an estimate; the only way to find the absolute number of organisms in a population is to count every one of them.
Simpson's index of diversity, D	You have collected quantitative data about the numbers of individuals belonging to the different species in an area.	The greater the value, the greater the species diversity (that is, the higher the number of species and the greater the evenness of their population sizes).	There are several different versions of the formula for Simpson's index of diversity; make sure you are familiar with the one in Chapter 18.
Hardy–Weinberg equations	In a population, a gene has two alleles. There is random mating between organisms with different phenotypes. There is no significant immigration or emigration from the population. Each phenotype has an equal chance of survival (i.e. there is no selection against any of the phenotypes).	If you use the equations to calculate the frequency of a recessive allele in a population, you can then predict the proportion of people in the population who are likely to be carriers of that allele.	
Standard deviation, s	You have data that show a normal distribution. You want to know how much these data are spread on either side of the mean value.	The greater the standard deviation, the more widely the data are spread on either side of the mean.	

Calculation	Criteria for using it	How to interpret the value you calculate	Notes
Standard error, SE	You have data that show a normal distribution. You want to know how close the mean value is likely to be to the true mean – in other words, how likely it is that if you collected another set of data, the mean would be the same.	The greater the standard error, the less chance that a second set of data would give you exactly the same mean value. You can be 95% certain that the true mean lies with 2 × the standard error either above or below your calculated mean. If this shows overlap between the possible true mean values of two sets of data, you cannot be sure that the true means are different.	Standard error can be used to draw error bars on graphs. These are usually drawn as vertical lines extending to 2 × SE above and below the data plot.
t-test	You have two sets of continuous, quantitative data. You have more than 10 but fewer than 30 readings in each set. Both sets of data come from populations with normal distributions. The standard deviations for the two data sets are similar.	Construct a null hypothesis, predicting that there is no difference between the two data sets. Determine the number of degrees of freedom in your data. Look up your calculated value of t in a table of probabilities. If the probability table tells you that the probability of your null hypothesis being correct (that is, the difference between the two populations being due to chance) is greater than the critical value (e.g. 0.05) then the null hypothesis is supported.	In general, the greater the value of t, the smaller the probability that the null hypothesis is correct. The value of 0.05 means that there is a 5% probablility that the differences between the two data sets are due to chance. This is the critical value that is generally used in biology. If the value is less than 0.05, then we can reasonably assume that the differences are real, and not just due to chance.

Calculation	Criteria for using it	How to interpret the value you calculate	Notes
chi-squared test, χ^2	You have two or more sets of quantitative data, which belong to two or more discontinuous categories.	Construct a null hypothesis predicting that there is no difference between the observed and expected values. Determine the number of degrees of freedom in your data. Look up your calculated value of x^2 in a table of probabilities. If the probability table tells you that the probability of your null hypothesis being correct (that is, the difference between the two populations being due to chance) is greater than the critical value (e.g. 0.05), then the null hypothesis is supported.	See notes on the *t*-test.
Spearman's rank correlation, r_s	You have collected data for two variables, both of which can be ranked. The samples for both sets of data have been collected randomly. You have at least 5 pairs of data, but preferably between 10 and 30. You have plotted a scatter graph, and it looks as though there may be a relationship between the two variables.	Values of r_s can lie anywhere between −1 and +1. A value of −1 indicates a perfect negative correlation, and +1 a perfect positive correlation. A value of 0 indicates no correlation. Use a correlation coefficient table to look up your value of r_s. If your value of r_s is greater than the value with a probability of 0.05, then you can say there is a significant correlation between the two variables.	Remember that correlation does not prove causation.

Calculation	Criteria for using it	How to interpret the value you calculate	Notes
Pearson's linear correlation, r	You have two sets of interval data. You have at least 5 pairs of data, but preferably 10 or more. Both sets of data have an approximately normal distribution. You have plotted a scatter graph, and it looks as though there might be a linear correlation between the two variables.	Values of r can lie anywhere between -1 and $+1$. A value of -1 indicates a perfect negative correlation, and $+1$ a perfect positive correlation. A value of 0 indicates no correlation.	Remember that correlation does not prove causation.

Table 19.1: Data analysis calculations used in A level course.

1   For each of these situations, decide which of the calculations in the table you might use. In some cases, you could use more than one. Justify your choice if you think this is helpful.

   a   You want to know if the yield per plant from a variety of rape genetically modified for herbicide resistance is significantly different from the yield of a non-GM variety.

   b   You have counted the numbers of animals of every visible species in 10 quadrats placed randomly on a rocky shore.

   c   You know the percentage of children born each year with a genetic disease caused by a recessive allele. You want to know the frequency of that allele in the population.

   d   You want to know if the numbers of offspring obtained in a genetic cross are significantly different from the values you would expect from your predictions.

   e   You are using the mark–release–recapture technique for estimating the number of water boatmen in a pond.

   f   You have measured the test scores of all the members of the biology group who also study physics and also the test scores of all the members of the biology group who do not study physics. You are interested in finding out whether the true mean values for the scores of the physicists and non-physicists are likely to be different.

> **TIP**
>
> You can find guidance on how to do the calculations in earlier chapters in this book.

Exercise 19.2 Gene technology terminology

1 Choose the term from the list that matches each description. If you find that you are not absolutely sure about any of them (even if you can guess the answer), this signals that you need to do a little more work on that topic.

DNA ligase
gene therapy
polymerase chain reaction (PCR)
promoter
restriction endonuclease
Taq **polymerase**

gel electrophoresis
microarray
primer
recombinant
reverse transcriptase

> **TIP**
>
> There are many technical terms associated with this topic. This exercise will help you to use each term appropriately.

a   An enzyme used to cut DNA at a specific base sequence.

b   An enzyme derived from bacteria that are adapted to live in hot water springs; the enzyme has a very high optimum temperature, and is used to synthesise new strands of DNA during PCR.

c   An enzyme that catalyses the formation of bonds between phosphate groups and deoxyribose during the synthesis of DNA.

d   An enzyme that produces complementary DNA strands using an RNA template.

e   A length of DNA that controls the expression of a gene.

f   A short length of DNA with a base sequence complementary to the part of a DNA strand that is to be copied during the PCR.

g   An adjective used to describe DNA, or an organism, in which the DNA has been modified so that it contains base sequences from more than one organism.

h   A small piece of glass or plastic on which thousands of tiny pieces of single-stranded DNA are attached in a regular arrangement; the DNA pieces act as probes and allow rapid analysis of DNA of unknown composition.

i   A method of separating lengths of DNA or proteins according to their relative mass or charge.

j   Inserting DNA into the cells of an organism to correct a harmful genetic condition.

k   A (usually fully automated) method of rapidly producing many copies of a small DNA sample.

EXAM-STYLE QUESTIONS

1 a Plants contain an enzyme called glutamine synthase, which catalyses the reaction of glutamate and ammonia to produce glutamine.

Some types of herbicide contain glufosinate. This compound can bind to the glutamate binding site of glutamine synthase. This results in a build-up of ammonia in the plant, and kills it.

Explain how glufosinate results in a build-up of ammonia. [3]

b Several varieties of rape, *Brassica napus*, have been genetically modified for resistance to herbicides containing glufosinate.

Herbicide-resistant rape varieties contain the *bar* gene, which codes for an enzyme called phosphinothricin acetyl transferase (PAT). PAT changes glufosinate into an inactive compound.

Explain how growing rape plants containing the *bar* gene can result in higher yields of rape seed. [3]

c A company that produces and sells GM herbicide-resistant rape seed has developed an F1 hybrid variety, produced by crossing two herbicide-resistant pure-breeding varieties called MS8 and RF3.

i Explain what is meant by the term F1 hybrid. [1]

ii **Suggest** *one* advantage to a farmer of growing a hybrid variety of oil-seed rape, rather than a pure-breeding variety. [1]

iii MS8 (male-sterile) contains a gene called *barnase*, which encodes for a ribonuclease that is expressed only in the anthers of the flowers. This affects RNA synthesis and stops anther development.

RF3 (restoration of fertility) contains a gene called *barstar*, which codes for a ribonuclease inhibitor that is only expressed in the anthers. This can inactivate the ribonuclease encoded by barnase.

Explain how the company can use these two varieties for large-scale production of seed that is 100% hybrid. [3]

iv **Predict** whether plants grown from the hybrid seed would be fertile. Explain your answer. [2]

[Total: 13]

2 a Figure 19.1 shows the positions on the human X chromosome of two genes that are associated with hereditary conditions.

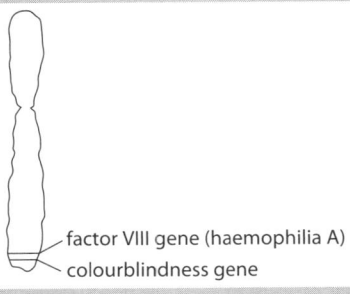

Figure 19.1

TIP

Questions about genetic technology will often introduce unfamiliar contexts and data, asking you to apply your knowledge and understanding. They may also cross over into other topics. Here, the question begins by asking about a topic relating to enzymes, and finishes by asking you to consider the use of a novel technique for producing hybrid herbicide-resistant plants.

COMMAND WORDS

Explain: set out purposes or reasons/make the relationships between things evident/provide why and/or how and support with relevant evidence.

Suggest: apply knowledge and understanding to situations where there is a range of valid responses, in order to make proposals / put forward considerations.

Predict: suggest what may happen based on available information.

> **CONTINUED**

   i **State** the term used to describe the position of a gene on a chromosome. **[1]**
   ii These two genes are sex-linked. Explain what is meant by this term. **[2]**

b The gene that is associated with haemophilia A codes for a protein called factor VIII. A man with haemophilia A has one of several different recessive alleles of this gene, which do not produce functioning factor VIII. Haemophilia A can be treated by injecting the person with factor VIII, and a man with haemophilia who has such treatment whenever required will have a life expectancy similar to that of a man without the condition.

   i It is now possible to screen embryos for the presence of a haemophilia A allele. **Outline** the advantages of this screening. **[3]**
   ii Suggest the ethical considerations that should be taken into account when screening embryos for the presence of a haemophilia allele. **[2]**

c Haemophilia A is treated with factor VIII extracted from donated human blood plasma, or with recombinant factor VIII made by genetically engineered mammalian cells grown in tissue culture.

Doctors and patients may be able to choose which of these types of factor VIII are used. Both are equally good at controlling the condition. However, other issues to be considered are cost (recombinant factor VIII is more expensive) and the likelihood of development of immune reactions to the products. Some patients produce antibodies to the injected factor VIII, preventing it from working, and it is possible that the source of the factor VIII could affect the likelihood of antibodies being produced.

   i Many studies have been carried out to try to determine which type of factor VIII is more likely to cause patients to develop antibodies. Table 19.2 shows the results from 12 of these studies.

Study	Type of factor VIII used	Number of people in the study	Percentage developing antibodies
1	plasma derived	48	6.25
2	plasma derived	16	0.00
3	plasma derived	37	2.70
4	plasma derived	89	28.09
5	plasma derived	19	5.26
6	plasma derived	135	21.48
7	recombinant	59	23.73
8	recombinant	50	18.00
9	recombinant	181	29.28
10	recombinant	35	5.71
11	recombinant	47	36.17
12	recombinant	101	31.68

Table 19.2

Suggest reasons for the wide variation in the results shown in the table. **[4]**

> **TIP**
>
> The first part of this question involves genetics, but then moves on to topics relating to evaluating data and discussing ethical issues. Ethical issues and social issues overlap; in general, ethical issues are ones that relate to values of right and wrong, and social issues relate to effects on situations and conflicts in society. Remember that the command word 'discuss' means that you should address two different viewpoints of an issue.

> **COMMAND WORDS**
>
> **State:** express in clear terms.
>
> **Outline:** set out the main points.

CONTINUED

ii A further independent trial is planned. In this trial:

- The parents of boys who have just been diagnosed with haemophilia will be asked to participate in a randomised trial.
- The parents will be given full information about the known safety and effectiveness of both types of treatment.
- If the parents agree, the boys will be randomly allocated to a group being given plasma-derived factor VIII, or a group being given recombinant factor VIII.
- The boys will be tested at frequent intervals for the production of antibodies, and measures will be taken to deal with this if it occurs.

Suggest why only boys who have just been diagnosed with haemophilia will be included in the study. [2]

iii **Discuss** the ethical and social issues that are raised by the implementation of this study. [4]

[Total: 18]

3 Maize is a cereal crop grown in many parts of the world. In many countries, more than 90% of the maize plants that are grown are F1 hybrids. Hybrid maize has significantly higher yields of grain than pure-breeding maize.

a Explain what is meant by:
i an F1 hybrid [1]
ii pure-breeding. [1]

b The greater vigour and yield of hybrids compared with pure-breeding parents is termed hybrid vigour. Biologists still do not fully understand what causes hybrid vigour. One theory is that harmful recessive alleles that are present in the parents are masked by dominant alleles in the hybrid. Another is that the level of expression of a particular allele might be greater or lower in the hybrid than in either of the parents.

An investigation was undertaken to determine the level of expression of 13 999 genes in two pure-breeding maize varieties, B73 and Mo17, and a hybrid between them.

- Large numbers of seedlings of B73, Mo17 and the hybrid Mo17 × B23 genotypes were grown in controlled conditions.
- Ten replicates of each genotype were grown.
- After 14 days, above-ground tissue was harvested from all three genotypes.
- RNA was extracted from the seedlings.
- The RNA was used to make single-stranded cDNA with a complementary base sequence.
- The cDNA was labelled with fluorescent dyes.
- A microarray with 13 999 spots containing single-stranded DNA with specific sequences (each known to be part of the maize genome) was used to determine the quantity of the different types of cDNA produced from each of the three maize genotypes.

> **COMMAND WORD**
>
> **Discuss:** write about issue(s) or topic(s) in a structured way.

> **TIP**
>
> This question links material relating to Chapter 17, (inbreeding and hybridisation in maize) and Chapter 19 (genetic technology, including the use of microarrays). Some of the questions are quite straightforward, but the question ends with a very difficult three-mark section. Even if you are very unsure about this, it is worth having a try – just think about what you know about the control of gene expression and make a suggestion or two.

CONTINUED

    i   Suggest *two* conditions that would be standardised when growing the seedlings. [2]

    ii Explain why the researchers extracted RNA from the seedlings and not DNA. [2]

    iii Name the enzyme that would be used to make single-stranded DNA from the extracted RNA. [1]

    iv Outline how the microarray would be used, and how it would enable the researchers to determine the level of expression of the genes in the maize seedlings. [5]

  c The researchers found that, of the 13 999 genes tested, 1367 were expressed to different levels in the hybrid than in either of the parents.

    Of these 1367 genes:

    - 1062 were expressed to a degree half-way between that of the two parents
    - 34 were expressed to a higher degree in the hybrid than in either of the parents
    - 10 were expressed to a lower degree in the hybrid than in either of the parents.

    Suggest reasons why a gene in the hybrid could be expressed to different levels than in either of the parents. [3]

    [Total: 15]

4 Genotyping of 104 000 individuals in Iceland has been carried out, and the complete sequences of the genomes of 2636 people have been determined. Databases constructed from this information are providing new insights into genetic variation in humans and its consequences.

  a It has been found that 93% of differences between the base sequences of the 2636 people were caused by a change in a single base, and the remaining 7% were insertions or deletions.

    One group of researchers was interested in loss-of-function mutations, which prevent the gene coding for a useful protein.

    In the 2636 genomes analysed, 6795 loss-of-function mutations in 4924 genes were found. Of these, 3979 mutations were changes in a single base, and 2816 were insertions or deletions.

    i Calculate the percentage of loss-of-function mutations that were insertions or deletions. [1]

    ii Explain the reasons for the difference between the value you have calculated in **i**, and the value of 7% given in the question for part **a**. [3]

TIP

In 2015, details of the largest-ever set of human genomes and the results of various analyses of these data were published. The population of Iceland was chosen for this study because the Icelandic population was founded only 1100 years ago from a relatively small number of people, and full genealogies (family histories) of a very high proportion of the people are known. This question asks you to apply knowledge to an unfamiliar situation, to make some calculations, and to think about ethical issues relating to genome sequencing.

CONTINUED

    iii Many of the loss-of-function mutations were very rare, with an allele frequency of 0.002.

    The Hardy–Weinberg equations are:

    $p + q = 1$

    $p^2 + 2pq + q^2 = 1$

    where p is the frequency of one allele of the gene and q is the frequency of the other.

    Use the Hardy–Weinberg equations to calculate how many individuals in a population of 250 000 you would expect to be homozygous for such a mutation. Show your working. **[2]**

c The Icelandic population from which these data have been collected gave their permission for the information about their genomes to be used for research. The analysis of the genomes has identified many people who carry mutations that may increase the risk of the development of harmful conditions such as breast cancer or Alzheimer's disease.

Discuss the ethical issues involved in making a decision about whether or not these people should be contacted through their doctors and given information about the results of their genome analysis. **[5]**

[Total: 11]

Glossary

Command words

Below are the Cambridge International definitions for commands words which may be used in exams. The italic text provides some additional explanation on the meaning of these words.

The information in this section is taken from the Cambridge International syllabus (9700) for examination from 2022. You should always refer to the appropriate syllabus document for the year of your examination to confirm the details and for more information. The syllabus document is available on the Cambridge International website at www.cambridgeinternational.org.

Assess: make an informed judgement

Calculate: work out from given facts, figures or information – *it is usually a good idea to show all of the steps in your working*

Comment: give an informed opinion – *you will often need to use your biological knowledge and understanding to make a range of statements about the topic*

Compare: identify / comment on similarities and/or differences – *you can often use a table for this; if writing sentences, then use comparative word.*

Contrast: identify / comment on differences – *you can often use a table for this; if writing sentences, then use comparative words*

Define: give the precise meaning – *that is, provide a short but complete description of what the term means*

Describe: state the points of a topic / give characteristics and main features – *for example, use words to say clearly what is shown by a graph, or give a step-by-step account of something*

Discuss: write about issue(s) or topic(s) in a structured way; *it is often a good idea to state points on both sides of an argument, for example, reasons for and against a particular viewpoint, or how a set of results could be interpreted to support or reject a hypothesis*

Explain: set out purposes or reasons / make the relationships between things evident / provide why and/or how and support with relevant evidence – *note that you will often need to use your knowledge of biology to say why or how something happens*

Give: produce an answer from a given source or recall/memory – *for example, using information provided in the question, or from knowledge you have learnt during your course*

Identify: name /select /recognise – *for example, labelling a structure on a diagram or micrograph*

Outline: set out the main points – *that is, give a brief account, picking out the most important points and omitting detail*

Predict: suggest what may happen based on available information

Sketch: make a simple drawing showing the key features.

State: express in clear terms – *that is, give a short, precise answer*

Suggest: apply knowledge and understanding to situations where there is a range of valid responses, in order to make proposals – *that is, use information provided, and your biological knowledge, to put forward possible answers; there is often more than one possible correct answer*

Key words

accessory pigment: a pigment which absorbs wavelengths of light not absorbed by chlorophyll

acetylcholine (ACh): a molecule made up of coenzyme A and a 2C acetyl group, important in the link reaction; a type of neurotransmitter released by cholinergic synapses.

acetylcholinesterase: an enzyme in the synaptic cleft and on the post-synaptic membrane that hydrolyses acetylcholine to acetate and choline actin: the protein that makes up the thin filaments in striated muscle

action potential: a brief change in the potential difference from -70 mV to +30 mV across the cell surface membranes of neurones and muscle cells caused by the inward movement of sodium ions

Glossary

amylopectin: a polymer of α-glucose monomers linked by both 1,4 linkages and 1,6 linkages, forming a branched chain; amylopectin is a constituent of starch

amylose: a polymer of α-glucose monomers linked by 1,4 linkages, forming a curving chain; amylose is a constituent of starch

anaerobic respiration: the enzymatic release of energy from organic compounds in living cells in the absence of oxygen from ethanol or lactate fermentation

anomalous: a result or value that lies well outside the range of other results or values

antibiotic: a substance produced by a living organism that is capable of killing or inhibiting the growth of a bacterium

antibody (plural antibodies): a glycoprotein (immunoglobulin) made by plasma cells derived from B-lymphocytes, secreted in response to an antigen; the variable region of the antibody molecule is complementary in shape to its specific antigen.

bar chart: a graph drawn when the x-axis variable is discontinuous and the y-axis variable is continuous; bars do not touch

Benedict's test: a test for the presence of reducing sugars; the unknown substance is heated with Benedict's reagent, and a change from a clear blue solution to the production of yellow, red or brown precipitate indicates the presence of reducing sugars such as glucose

b-lymphocyte (B cell): a type of lymphocyte that gives rise to plasma cells and secretes antibodies

causal link (relationship): where one factor directly brings about an effect in another

cellulose: a carbohydrate that is a polymer composed of β-glucose units; it is the main component of the cell walls of most plants

chlorophyll: a green pigment that absorbs energy from light, used in photosynthesis

colorimeter: an instrument that measures the intensity of colour of a solution, by passing light of a particular wavelength through it and measuring how much light is transmitted

community: all of the different species living in an area at the same time

condensation: a chemical reaction involving the joining together of two molecules by removal of a water molecule

control group: the group in an experiment or study that does not receive treatment by the researchers and is then used as a benchmark to measure how the other tested subjects do; often referred to as the placebo group

correlation: a relationship or connection between two things based on co-occurrence or pattern of change may be a positive correlation or a negative correlation indicates the extent to which one variable increases as the other decreases

dependent variable: in an experiment, the variable that changes as a result of changing the independent variable

depolarisation: the reversal of the resting potential across the cell surface membrane of a neurone or muscle cell, so that the inside becomes positively charged compared with the outside

digit: a single whole number, for example, 2

disaccharide: a sugar molecule consisting of two monosaccharides joined together by a glycosidic bond

disease prevalence: the number of people who have disease at any one time

electron transport chain: a chain of adjacently arranged carrier molecules in the inner mitochondrial membrane, along which electrons pass by redox reactions

error bar: a line drawn through a point or the top of bar on a graph, extending two standard errors above and below the mean value indicated by the point or bar; you can be 95% certain that the true value lies within the range indicated by the error bar

eyepiece graticule: small scale that is placed in a microscope eyepiece

glucagon: a small peptide hormone secreted by the α cells in the islets of Langerhans in the pancreas to bring about an increase in the concentration of glucose in the blood

glycogen: a polysaccharide made of many glucose molecules linked together, that acts as a glucose store in liver and muscle cells

glycosidic bond: a C–O–C link between two monosaccharide molecules formed by a condensation reaction

grana: (singular: **granum**): stacks of membranes inside a chloroplast

histogram (frequency diagram): a graph drawn when the variable on the x-axis is a series of categories that are continuous, and the y-axis shows the frequency of each category.

homeothermic: maintaining a constant body temperature

hydrogen bond: a relatively weak bond formed by the attraction between a group with a small positive charge on a hydrogen atom and another group carrying a small negative charge, for example, between two $-O^{\delta+}H^{\delta+}$ groups

hydrolysis: a reaction in which a complex molecule is broken down to simpler ones, involving the addition of water

hypoglycaemia: a condition caused by a very low level of blood sugar (glucose)

independent variable: the variable (factor) that is deliberately changed in an experiment

insulin: a small peptide hormone secreted by the β cells in the islets of Langerhans in the pancreas to bring about a decrease in the concentration of glucose in the blood

interval: the 'gap' between the values chosen within the range

Krebs cycle: a cycle of reactions in aerobic respiration in the matrix of a mitochondrion in which hydrogens pass to hydrogen carriers for subsequent ATP synthesis and some ATP is synthesised directly; also known as the citric acid cycle

line graph: a graph drawn when both the x-axis variable and the y-axis variable are continuous

linear correlation: a relationship between two variables such that a straight best-fit line can be drawn when one is plotted against the other

magnification: the number of times greater that an image is than the actual object; magnification = image size ÷ actual (real) size of the object

mark–release–recapture: a method of estimating the numbers of individuals in a population of mobile animals

median: the middle value of all the values in the data set

Michaelis–Menten constant, K_m: the substrate concentration at which an enzyme works at half its maximum rate ($½V_{max}$), used as a measure of the efficiency of an enzyme; the lower the value of K_m, the more efficient the enzyme

mitochondrion: the organelle in eukaryotes in which aerobic respiration takes place

mode (modal class): the most common value, or class, in the set of results

mole: the amount of an element or substance that has a mass in grams numerically equal to the atomic or molecular weight of the substance

monomer: a relatively simple molecule which is used as a basic building block for the synthesis of a polymer; many monomers are joined together to make the polymer, usually by condensation reactions; common examples of molecules used as monomers are monosaccharides, amino acids and nucleotides

monosaccharide: a molecule consisting of a single sugar unit with the general formula $(CH_2O)n$

negative correlation: a relationship between two variables such as when one increases, the other decreases

neurone: a nerve cell; a cell which is specialised for the conduction of nerve impulses

normal distribution: a set of data in which a graph shows a symmetrical distribution around the central value

null hypothesis: a hypothesis that assumes there is no relationship between two variables, or that there is no significant difference between two samples

optical microscope: a microscope that uses light to view a specimen; maximum resolution down to 200 nm

oxidative phosphorylation: the synthesis of ATP from ADP and P_i using energy from oxidation reactions in aerobic respiration (compare photophosphorylation)

pathogen: an organism that causes infectious disease

peptide bond: a C–N link between two amino acid molecule, formed by a condensation reaction

percentage error: the actual error in a measurement calculated as a percentage of the measured value

phagocyte: a type of cell that ingests (eats) and destroys pathogens or damaged body cells by the process of phagocytosis; some phagocytes are white blood cells

phosphorylase kinase: an enzyme that is part of the enzyme cascade that acts in response to glucagon; the enzyme activates glycogen phosphorylase by adding a phosphate group

photosynthometer: apparatus to measure the rate of photosynthesis by collecting and measuring the volume of oxygen produced in a certain time

plasmolysed: a cell where the cytoplasm and vacuole have shrunk so much that the cell membrane has pulled away from the cell wall

polymer: a giant molecule made from many similar repeating subunits joined together in a chain; the subunits are much smaller and simpler molecules known as monomers; polymers may also be referred to as macromolecules; examples of biological polymers are polysaccharides, proteins and nucleic acids; see monomer

polypeptide: a long chain of amino acids formed by condensation reactions between the individual amino acids; proteins are made of one or more polypeptide chains; see peptide bond

polysaccharide: a polymer whose subunits are monosaccharides joined together by glycosidic bonds

protoctist: a member of the Protoctista kingdom; eukaryotic organisms which are single-celled or made up of groups of similar cells

Glossary

quadrat: a square frame which is used to mark out an area for sampling of organisms

random error: uncertainties in experimental results that are caused by variations in variables that should be standardised, or by difficulties in measuring the independent or dependent variables; they can act in different directions for different parts of the experiment

range: the spread between the lowest and highest values of the independent variable

redox indicator: a substance that changes colour when it is oxidised or reduced

replicates: several experiments done in the same situation with the same apparatus and materials

repolarisation: returning the potential difference across the cell surface membrane of a neurone or muscle cell to normal following the depolarisation of an action potential

respiratory quotient (RQ): the ratio of the volume of carbon dioxide produced to the volume of oxygen used

respirometer: a piece of apparatus that can be used to measure the rate of oxygen uptake by respiring organisms

resting potential: the difference in electrical potential that is maintained across the cell surface membrane of a neurone when it is not transmitting an action potential; it is normally about -70 mV inside and is partly maintained by sodium–potassium pumps

R_f value: a number that indicates how far a substance travels during chromatography, calculated by dividing the distance travelled by the substance by the distance travelled by the solvent; R_f values can be used to identify the substance

scanning electron microscope: an electron microscope that provides a three-dimensional view of the surface of a specimen; maximum resolution down to 2 nm

significant figures: the digits that carry meaningful information about the size of the number

spearman's rank correlation: a statistical test to determine if there is a correlation between two variables when one or both of them are not normally distributed

species diversity: a measure of the number and evenness of species living in an area. It indicates which species are present and how abundant each one is

species richness: the number of different species living in an area. It does not measure abundance

stage micrometer: very small, accurately drawn scale of known dimensions, engraved on a microscope slide

standard deviation: a measure of how widely a set of data is spread out on either side of the mean

standard error: a calculation that indicates how close the calculated mean value is likely to be to the true mean value

standard form: a way of writing a number as a value that is always between 1 and 10, and using the power of ten to show how large or small the number is

standardised variable: variables (factors) that are kept constant in an experiment; only the independent variable should be changed

stroma: the chloroplast's equivalent of cytoplasm, and is the site of enzyme reactions relating to photosynthesis

synapse: the junctions between neurones or between neurones and their target tissues such as skeletal muscle

systematic error: uncertainties in experimental results that are caused by lack of precision or accuracy in a measuring instrument; they always act in the same direction

tally count: recording numbers by making a mark for each item counted, using a diagonal mark for each fifth item

tangent: a straight line that just touches a curve at a particular point

thylakoid: a flattened, membrane-bound, fluid-felled sac which is the site of the light-dependent reactions of photosynthesis in a chloroplast

T-lymphocyte: a lymphocyte that does not secrete antibodies; T-helper lymphocytes stimulate the immune system to respond during an infection, and killer T lymphocytes destroy human cells that are infected with pathogens such as bacteria and viruses

transect: a line along which samples are taken, either by noting the species at equal distances (line transect) or placing quadrats at regular intervals (belt transect)

transmission electron microscope: a microscope that uses a beam of electrons to view a very thin section; maximum resolution down to 0.5 nm

triglyceride: a lipid whose molecules are made up of a glycerol molecule and three fatty acids

ultracentrifugation: spinning a suspension at very high speed, so that more dense components settle to the bottom and can be separated

virus: very small (20–300 nm) infectious particle which can replicate only inside living cells and consists essentially of a simple basic structure of a genetic code of DNA or RNA surrounded by a protein coat

V_{max}: the theoretical maximum rate of an enzyme-controlled reaction, obtained when all the active sites are occupied

Skills Grid

AS Level

AO2 Handling, applying and evaluating information

☐ Exercises
▨ Exam-style questions

Skill	1	2	3	4	5	6	7	8	9	10	11
locate, select, organise and present information from a variety of sources		1,4,5,7	1,2	4	1		3	1,2	3,4	1,4	
		6		2,3	3	3		3,5,6	3,4	1,2,3,4,5	4
translate information from one form to another	6	1,3,4	2	4	4		4	1,3	3	2,4	
		5	1,2	1,2,3	3		6	1,2,3,6	2,3	1	4
manipulate numerical and other data		2,3,4	1,3,5		3		4	1,2,3	1,2,3,4	2,4	3
				2			3	2,3	3	5	4
use information to identify patterns, report trends and draw conclusions	6	4	2,3	3,4	1	1,3	4	1,2,3	1,2,3,4	2,4	3
	3	2,3,5,	2	1,2	2,3	1,2	2,3,4,5,6	1,2,3,6	2,3	1,2,3,4,5	4
give reasoned explanations for phenomena, patterns and relationships	6	4	1,2	3	4	3	4	1,2,3	2,3	2,4	3,4
	4	2,3,4,5	1	1,3	3	1,2	3,3,4	1,2,3,6	1,2,3	1,2,3,4,5	1,3
make predictions and construct arguments to support hypotheses			4						1	2	
apply knowledge, including principles, to new situations	6	1,2,3,6	1,3	3,4	1,4	1,3	4	1,2	1,3	2,3,4	3,4
	2,3		1,2,3	1,2,3,4	2,3	1,3	2,3,4	1,2,3,5,6	1,2,3,4	1,2,3,4,5	1,3
evaluate information and hypotheses		4,6	2,4,5	1,3	3,4	3		2	1	2	4
			2,3						3	1	
demonstrate an awareness of the limitations of biological theories and models							6		4	4	
								2			
solve problems		4	2				4	2,3	3,4	2,3,4	
		3,5		2				1,2,3,6	2,3		

AO3 Experimental skills and investigations

Skill	Chapter										
	1	2	3	4	5	6	7	8	9	10	11
Experimental design											
identify independent and dependent variables		6,	4	2	3		4	4			
choose range, number of values and interval for independent variable, and number of replicates		6,	4	2, 3	3		4	4			
describe an appropriate control experiment						3		4		3, 4	
use simple or serial dilution		2, 3	4	2						3, 4	
identify important control variables and how to standardise them		3, 6,	3, 4 3	2	3		4	4		3	
identify methods of measuring dependent variable, including when and how often		6	4 1	2 1			4	4		4	
assess risk and make decisions about safety			4					2		3, 4	
suggest how to modify or extend an investigation to answer a new question			1	3							
Collecting, recording and processing data											
be able to measure an area using a grid								2			
measure using counting, e.g. using tally charts					3			2			
clearly describe qualitative results											
decide how to deal with anomalous results				3				2		4	
measure and record all results to an appropriate number of significant figures											

Skill	1	2	3	4	5	6	7	8	9	10	11
describe how to identify biological molecules, and use a standardised test to estimate quantity		3, 3,5		2							
design, construct and complete results charts	2, 5,		4	1, 2, 3	3			1			
show calculations clearly; record calculated values to correct number of significant figures		2,	1	1, 3			4	1,2	1,2,3,4		3
draw graphs; determine the best type of graph or chart to display results		3,	3	1, 2, 3	3		4	1,3	2,		2, 3, 4
calculate percentage change; calculate rate as 1/time and as gradient of line graph			1								2, 3, 4
find an unknown value using a graph, including extrapolation		3,	1, 3	3			3	3 2,3			
describe patterns and trends shown in tables and graphs			1, 2 / 2, 3	1, 3 / 2, 3	4 / 3		4 / 6	1, 3 / 1,2,3.6	1,2,3,4 / 2,3,4	2,4 / 1,2,3,4,5	4
Evaluating results											
identify significant sources of error, and classify them as systematic or random; calculate percentage error			4	1				2		4	
suggest improvements to reduce error, e.g. by improving standardisation of variables, methods of measurement, quantity of data							4	2		4	
evaluate uncertainty in quantitative results			4					2	4	4	
determine the confidence with which conclusions can be made					4		4	1, 2	1, 2	2, 4	4
make conclusions from data			2, 3 / 3	1	4		4 / 3, 4	2	1, 2 / 3, 4	2, 4 / 1,2,3,4,5	4

Skills Grid

Skill	1	2	3	4	5	6	7	8	9	10	11
Microscopy and observation of specimens											
use a microscope to identify and draw tissues from prepared slides	3						1,2 1,4	1			
calibrate an eyepiece graticule using a stage micrometer, and use it to measure cells and tissues	5										
draw and label plan diagrams							1,2 4,	1,			
draw and label details of cells using high power	3						1	4			
use scale bars and magnifications	2,3 4			3	1		1 4,6	1, 4			
compare observable features of specimens							1,2,3 1,2,4	4	1		

293

A Level

AO2 Handling, information

☐ Exercises
▨ Exam-style questions

Skill	Chapter							
	12	13	14	15	16	17	18	19
locate, select, organise and present information from a variety of sources	1,2,4	1,4	1	1,2	3,4,5,6,7,8			
	1,2,3,4	1,2,3,4	1,2,5	1,2,3,7,8,9	2,3	1,2,3	1,2	1,4
translate information from one form to another	1,3,4	1,2,3,4	1,2,4	1,4	1,9	1,3,4		
	2		4,6,7	1	4	1,2,3	1,2,3	4
manipulate numerical and other data	1,2,4	1,2,3	1,3,4	3	3	1,3,4	1,2	
		1	3,7	1		1,2,3	1,3	4
use information to identify patterns, report trends and draw conclusions	1,2,3,4	1,2,3	1,2,3	1	3,4,5,6,7,8	1,3,4	1,2,3	
	1,2,4,5	1,3,4,5	1,2,3,4,6	1,4,5,6	1,3,4	1,2,3,4,5	1,2	1,3,4
give reasoned explanations for phenomena, patterns and relationships	2,3,4	1,2,3	1,2,3	4	3,4,5,6,7,8			
	2,3,4,6	1,3,4,5	1,2,3,4,6	1,2,3,4,5,6,7	2,3,4,5	1,2,3	2	2,4
make predictions and construct arguments to support hypotheses		1,3			3,7			
		1		1		3	2,3	2
apply knowledge, including principles, to new situations	2,3,4	1,2,3	1,2,3	1,4	3,4,5,6,7,8	3,4	1,2	
	1,2,3,4,5	1,3,4,5	1,2,4,6	12,3,4,5,6	1,2,3,4,5	1,2,3,4,5	1,2,3	1,2,3,4
evaluate information and hypotheses		1,2,3	1,3		7	1	1,2	
		1	2,7	1,5	3	1	1,3	2
demonstrate an awareness of the limitations of biological theories and models		2				4		
solve problems	1,2,3,4	1,2,3	1		3,4,5,6,7,8	3,4		
	1,2,3,4,5	1,3,4,5	1,2,6	1,4,5,6	2,3	1,2,3,5	2,3	1,4

AO3 Experimental skills and investigations

Skills Grid

Skill	Chapter							
	12	13	14	15	16	17	18	19
Planning an investigation and describing method								
construct a testable hypothesis		1,3						
		1	7	1			4	
identify independent and dependent variables	1	1,3	1				3	
		1	2,7	1			4	
choose range, number of values and interval for independent variable, and number of replicates, and describe how to vary the independent variable		3		3				
		1	7				4	
describe any appropriate control experiments	1	1,3	1					
		1	7				4	
make up solutions in %(w/v) or mol dm^{-3} and use serial or proportional dilution			2,3	3			2	
		1						
identify key control variables and describe how to standardise them	1,4	1,3	1				2	
		1	2,7	1			1,4	
identify methods of measuring dependent variable, including when and how often	1,4	1,3	1					
		1	7				4	
describe steps involved in the procedure in a logical sequence		3						
		1	7				4	
assess risk and describe precautions that would be taken to minimise risk		2					4	
Processing and evaluating data								
decide which calculations are necessary in order to draw conclusions	1,4	1	1,3		7		2	
			7	1	2		4	
use tables and graphs to identify key points in quantitative data, including variability	1,4	1,3	1,4			1	2	
	4	1,4,5	2,3	1	2			
draw graphs, including confidence limit error bars	4	1	1			1	2	
		1		1		1	1	

Skill	\multicolumn{8}{c}{Chapter}							
	12	13	14	15	16	17	18	19
choose and carry out appropriate calculations to simplify or compare data	1,4	1,3 1	4 3,7		2,3	1 1	2 44	
use standard deviation or standard error to determine statistical significance of differences between means	1	1 1	1			1	1,3	
choose appropriate statistical tests		1,3			7 3	1	2 4	1
apply statistical methods, including stating a null hypothesis, including t-test, chi-squared test, Pearson's linear correlation and Spearman's rank correlation		3	4 7	1	7 3	1	2 3	
recognise the different types of variable and different types of data presented	1	1	4	1				
identify anomalous results, suggest possible causes and suggest how to deal with them			1					
assess whether replicates were sufficient, and whether range and interval were appropriate		2						
assess whether the method of measuring was appropriate for the dependent variable	4,	1						
assess the extent to which variables have been effectively controlled	4,	1,3						
assess the spread of results by inspection and by using standard deviation and standard error or 95% confidence interval	1	1 1						
make conclusions from data	2,4,5 2,3,4	1,3	3,4 7	6			1,3	
discuss how much confidence can be put in any conclusions, including commenting on the validity of the investigation and how much trust can be placed in the data	1,2,4	1,2	3 7	6			1 1,3	

› Acknowledgements

The authors and publishers acknowledge the following sources of copyright material and are grateful for the permissions granted. While every effort has been made, it has not always been possible to identify the sources of all the material used, or to trace all copyright holders. If any omissions are brought to our notice, we will be happy to include the appropriate acknowledgements on reprinting.

Thanks to the following for permission to reproduce images:

Cover Coldimages/Getty Images

Inside Fig. 1.2, 1.3 STEVE GSCHMEISSNER/SPL; Fig. 1.5 EYE OF SCIENCE/SPL; Fig. 1.6 Art Directors & TRIP/Alamy Stock Photo; Fig 1.8 Cultura RM/Alamy Stock Photo; Fig 1.12 MICROSCAPE/SPL; Fig. 2.9, 7.2, 7.4 DR. KEITH WHEELER/SPL; Fig. 4.2 CLAUDE NURIDSANY & MARIE PERENNOU/SPL; Fig. 5.1 Spike Walker/Wellcome Images; Fig 7.1, 7.9, 8.12 BIOPHOTO ASSOCIATES/SPL; Fig. 8.2, 12.13 CNRI/SPL; Fig 8.3 Alvin Telser/SPL; Fig. 15.7 DR. FRED HOSSLER, VISUALS UNLIMITED /SPL; Fig. 16.3 PROFESSOR T. NAGURO/SPL; Fig 17.2 a, b Bob Gibbons/Alamy Stock Photo; Fig. 18.2 Geoff Jones; Fig. 18.3 Sarah_Cheriton/Getty Images

SPL = Science Photo Library